Materials and Processes for Nuclear Energy Today and in the Future

SCIENCES

Energy, Field Directors – Alain Dollet and Pascal Brault

Raw Materials and Materials for Energy,
Subject Head – Frédéric Schuster

Materials and Processes for Nuclear Energy Today and in the Future

Coordinated by
Fanny Balbaud-Célérier
Céline Cabet

iSTE

WILEY

First published 2024 in Great Britain and the United States by ISTE Ltd and John Wiley & Sons, Inc.

ISTE Ltd
27-37 St George's Road
London SW19 4EU
UK

www.iste.co.uk

John Wiley & Sons, Inc.
111 River Street
Hoboken, NJ 07030
USA

www.wiley.com

Library of Congress Control Number: 2024932616

British Library Cataloguing-in-Publication Data
A CIP record for this book is available from the British Library
ISBN 978-1-78945-186-3

ERC code:
PE8 Products and Processes Engineering
PE8_6 Energy processes engineering
PE4 Physical and Analytical Chemical Sciences
PE4_14 Radiation and Nuclear chemistry

Contents

Preface

Fanny BALBAUD-CÉLÉRIER[1] **and Céline CABET**[2]
[1] *CEA, Service de recherche en Corrosion et Comportement des Matériaux, Université Paris-Saclay, Gif-sur-Yvette, France*
[2] *CEA, Service de Recherche en Matériaux et Procédés Avancés, Université Paris-Saclay, Gif-sur-Yvette, France*

In the actual context of climate change, with the urgent need for low carbon energy technologies to fulfill the global clean energy demand, nuclear energy is an important contributor to the energy technology mix. Indeed, the continuous growth of the earth's population, improvement of living standards and longer life expectancy require the urgent deployment of safe and sustainable low carbon energy technologies. Nuclear power has been part of the global energy mix for more than 50 years; currently it represents 10% of electricity generation, with more than 440 reactors worldwide, and is the world's second largest source of low carbon electricity. In certain countries, it may thus be an important low emission source of electricity, which can complement and back renewable energies in decarbonizing the power sector.

The mean age of the world nuclear fleet is 31 years old (sources: WNISR, with IAEA-PRIS, 2022), and even though the design life of these power plants is typically 40 years, many nuclear power plants will be able to operate longer. This safe and economic operation is essentially due to the initial choices made about the materials used for their construction. Life extension of functioning nuclear power plants is a reality in many countries and new builds are planned. Moreover, many countries have interest in advanced systems, such as small modular reactors and next generation systems. These small modular systems could allow for the production of low carbon heat and hydrogen.

Materials and Processes for Nuclear Energy Today and in the Future,
coordinated by Fanny BALBAUD-CÉLÉRIER and Céline CABET.

In this framework, material science and engineering, including manufacturing processes, play a crucial role in safe, reliable, efficient, economical and overall sustainable nuclear energy. The first priority when performing materials selection for nuclear reactors is to ensure continuous safe operation of the nuclear system, considering the extreme conditions these materials encounter, such as high temperature, high irradiation flux and dose, high stress and chemically aggressive environments. Extending service life requires a thorough understanding of material aging mechanisms to ensure safe operation over time. In many cases, the possibility of component inspection is limited and the lifetime has to be certified over decade-long durations. To this end, modeling and simulation of the behavior of non-replaceable components will be essential, together with surveillance programs.

Moreover, new materials solutions are also under study, including advanced materials processing, innovative materials, and functionalized and architectured systems. The miniaturization of samples and testing systems is also under development to facilitate nuclear testing. Searching for new materials and processes, tailoring them to the desired multifunctional properties and processing them into complex-shape components is central to the nuclear sector. Workability, weldability, cost, etc., are other important aspects that need to be looked into during the materials selection process. These material solutions need to be scalable from lab to pilot to demo industrial size, and must also be cost effective. They must also be developed or designed, considering their overall environmental sustainability. Accelerating the development of materials may be a game changer in reducing the time to market of material innovations. Approaches integrating digital materials design using artificial intelligence tools, high-throughput manufacturing processes and accelerated qualification technologies offer real opportunities in the nuclear field over the next few years.

Access to robust and qualified materials, together with their processing, is critical for many major technological challenges the nuclear sector is facing: (i) the long-term operation of existing Generation II+/III nuclear power plants; (ii) the reprocessing, intermediate storage and final disposal of nuclear wastes; (iii) the construction of the next generation of reactors: generation IV power reactors, small modular reactors, advanced modular reactors; and (iv) the design of fusion power plants of tomorrow.

This book describes the structural materials and processes considered for different existing and prospective nuclear systems. The systems are globally introduced and illustrated. The operating conditions and technological constraints set a series of requirements on materials, which are developed and justified. The current choices of materials and their main properties are explained, together with the manufacturing routes and joining techniques. The main limitations of the materials are examined and the chapters conclude with the current R&D trends that are being used to try to overcome these restrictions. The last chapters more specifically contemplate the innovation pathways in terms of novel materials and the accelerated discovery of materials.

Chapter 1 provides an extended review of materials and processes for light water reactors. Chapter 2 includes materials and processes for nuclear fuel reprocessing. Chapter 3 discusses materials and processes for geological disposal of high-level waste and spent fuel. Chapter 4 develops materials and processes for sodium cooled fast reactors. Chapter 5 is dedicated to materials and processes for heavy liquid metal cooled reactors. Chapter 6 deals with materials and processes for molten salt reactors. Chapter 7 addresses materials and processes for fusion reactors. Chapters 8 and 9, respectively, cover advanced materials in terms of ceramics and composite materials, and oxide-dispersed strengthened steels. Chapter 10 concludes with materials discovery, with three parts on materials discovery/numerical design, advanced manufacturing processes and high throughput characterization.

We would like to extend our warmest thanks to all the authors and contributors to this book. Without their participation, expertise and enthusiasm, this work would not have been possible.

April 2024

1

Materials and Processes for Light Water Reactors

Jean-Paul MASSOUD[1] **and Eric MOLINIÉ**[2]
[1] *EDF/DIPNN/DT, France*
[2] *EDF/DIRES/R&D, France*

1.1. Acronyms

BCC:	body centered cubic (cristallography)
BOP:	balance of plant
BWR:	boiling water reactor
CCS:	components cooling system
CRDM:	control rod drive mechanism
CVCS:	chemical and volume control system
CW:	cold worked
DBTT:	ductile brittle transition temperature
DPT:	dye penetrant testing
EAF:	environmentally assisted fatigue

ECT:	eddy current testing
EDF:	Electricité de France
ESW:	electroslag welding
ESWS:	Essential Service Water System
FAC:	flow-assisted corrosion
FCC:	face centered cubic (cristallography)
GTAW:	gas tungsten arc welding
HAZ:	heat-affected zone
HIP:	hot isostatic pressing
IASCC:	irradiated-assisted stress corrosion cracking
IGSCC:	intergranular stress corrosion cracking
LAS:	low alloy steel
LPSIS:	low-pressure safety injection system
LWR:	light water reactor
MPT:	magnetic particle testing
MSR:	moisture separator-reheater
NPP:	nuclear power plant
NSSS:	Nuclear Steam Supply System
PHT:	preliminary heat treatment
PWR:	pressure water reactor
PWSCC:	primary water stress corrosion cracking
PZR:	pressurizer

RBWMS: Reactor Boron and Makeup Water System

RCS: reactor coolant system

RHR: reactor heat removal

RPV: reactor pressure vessel

RT: radiographic testing

RT_{NDT}: reference temperature Nihil Ductility Temperature

SAW: submerged arc welding

SCC: stress corrosion cracking

SG: steam generator

SMAW: shielded metal arc welding

TGSCC: transgranular stress corrosion cracking

UT: ultrasonic testing

1.2. Introduction with general description of the systems

1.2.1. *Pressurized water reactor (Kaercher 2002; Cattant 2014; Hutin 2017a)*

From the end of the 1960s, light water reactors (LWRs) have been used extensively in many countries around the world for electricity production.

The predominant reactor design worldwide is the pressurized water reactor (PWR), accounting for about two-thirds of the installed capacity, followed by boiling water reactors (BWRs) at about 20%.

Figure 1.1 shows an overview of a typical modern PWR. The condenser is generally cooled by sea, brackish or river water. However, in the latter situation, when the river flow is too low, cooling towers are used, as shown in the picture.

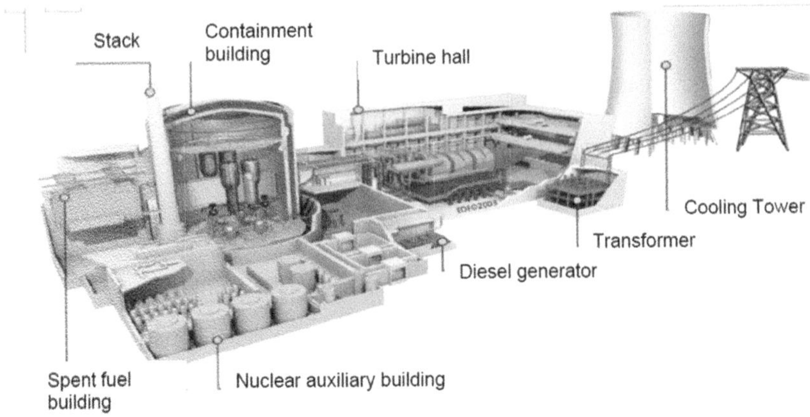

Figure 1.1. *Sketch of a PWR presenting the main buildings and facilities (courtesy of EDF). For a color version of this figure, see www.iste.co.uk/balbaud/nuclearenergytoday.zip*

The flow of primary water is maintained by the pressurizer and flows through the reactor coolant system (RCS) by reactor coolant pumps. The typical RCS water conditions are as follows:

– high temperature: between 285°C (cold leg, reactor pressure vessel inlet) and 325°C (hot leg, reactor pressure vessel outlet);

– high pressure: around 155 bars;

– reducing environment for water radiolysis mitigation by hydrogen injection (typically: 30 cc/kg);

– presence of boric acid for the neutronic reaction control. The primary water $pH_{300°C}$ is typically adjusted to 7.2 by way of lithium hydroxide injection (a few ppm maximum).

The RCS heat energy is exchanged with the secondary circuit through the steam generator tubes. The secondary water is heated to steam that powers the high- and low-pressure turbines, which turn the main generator for electricity production.

The steam is then condensed to water and flows back to the steam generators. For thermodynamic considerations (cycle efficiency), the water is steam-heated before entering the steam generator and the steam is dried and reheated between the high- and low-pressure turbines in the moisture separator-reheaters (MSRs). At the steam generator outlet, the typical steam conditions are as follows:

– temperature: 286°C;

– pressure: 70 bars.

A schematic of the primary and secondary circuits and materials of construction is presented in Figures 1.2 and 1.3 for PWR and BWR, respectively.

There are hundreds of various systems in a PWR; some related to the primary circuit are indicated.

Although optimized, the water and steam environments are still harsh enough to pose a threat to the systems materials, at times leading to corrosion failures. Mechanical loads can also result in mechanical failures, sometimes in conjunction with corrosion, like in the cases of stress corrosion cracking (SCC) and corrosion fatigue.

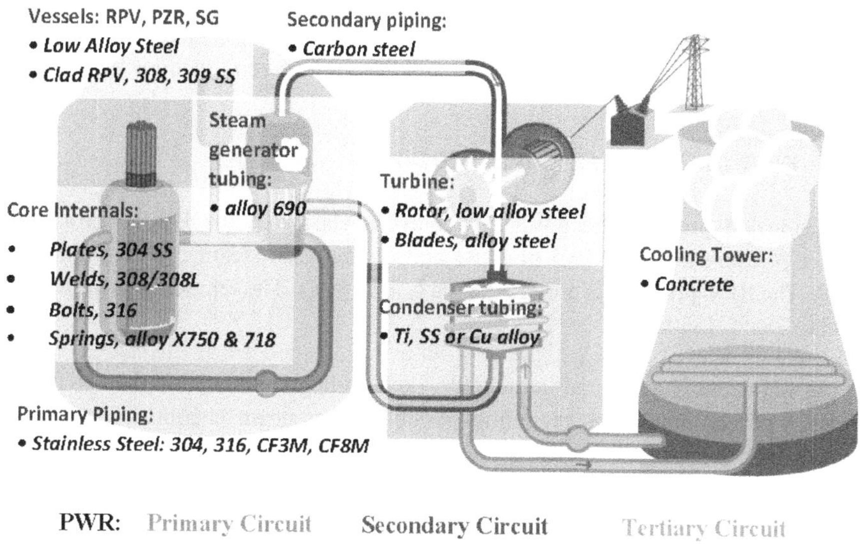

Figure 1.2. *Principle of PWR: main components and materials. In PWR, the water heated in the RPV is used to heat a secondary circuit. For a color version of this figure, see www.iste.co.uk/balbaud/nuclearenergytoday.zip*

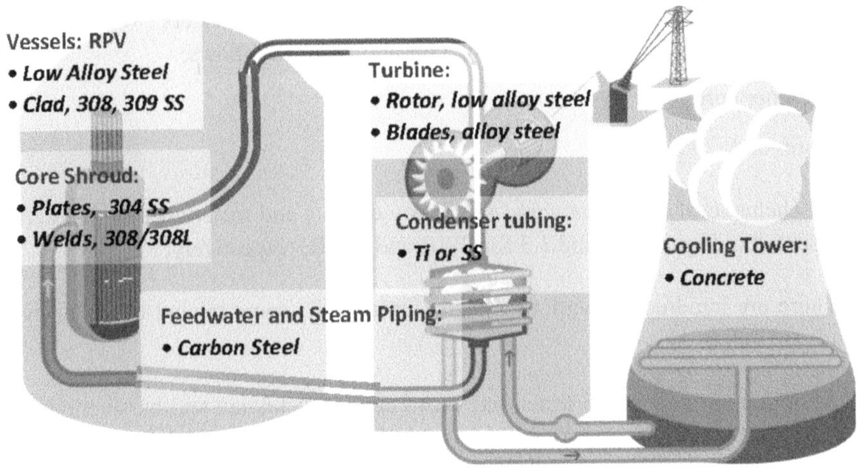

BWR: **Primary Circuit** **Heat Dump Circuit**

Figure 1.3. *Principle of BWR: main components and materials. In BWR, the water heated in the RPV directly feeds the turbine. For a color version of this figure, see www.iste.co.uk/balbaud/nuclearenergytoday.zip*

1.2.2. *BWR (Boiron 2011; Cattant 2014)*

Figure 1.4 presents the BWR power cycle. Basically, the BWR has a lower reactor pressure vessel (RPV) pressure and a simplified steam cycle compared to PWR. The RPV pressure is around 7 MPa (1,020 psig). The temperature in the RPV reaches 288°C (550°F).

As the steam is generated in the RPV, bulk boiling is allowed in the BWR core, which involves limitations regarding possible water/steam chemical conditioning (i.e. no boric acid or lithium hydroxide injection).

One particular component found in BWRs is jet pumps, which provide flow to control reactor power, which enables a higher power level without increasing the RPV size. Jet pumps also provide part of the boundary required to maintain two-thirds of the core height following a recirculation line break event.

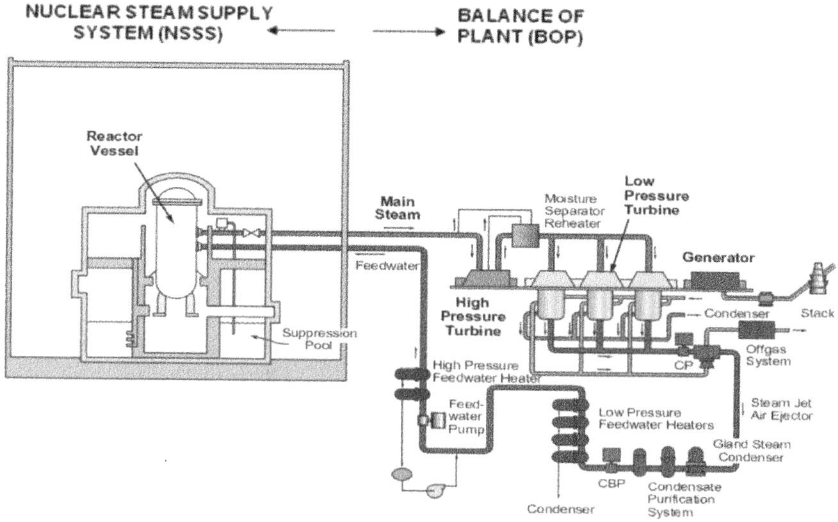

Figure 1.4. *Advanced BWR power cycle (Cattant (2014), courtesy of GE). For a color version of this figure, see www.iste.co.uk/balbaud/nuclearenergytoday.zip*

Figure 1.5 presents a typical BWR lower plenum, with the presence of:

– control rod drive guide tubes;

– control rod blades;

– control rod drive housings;

– stub tubes;

– in-core housings;

– guide tubes;

– flux monitor dry tubes.

Another major component of BWR core internals is the core shroud, which is a large austenitic stainless steel cylinder surrounding the core. It separates upward flow through the core from downward flow in the downcomer annulus. It also provides a two-thirds core height floodable volume.

Another big piece of equipment sitting at the top of BWR RPV internals is the steam dryer. The steam dryer provides a 99.9% steam flow to the main turbine. In the dryer, wet steam is forced through the dryer panels horizontally:

– the steam is forced to make a series of rapid changes in direction;

– the moisture is thrown to the outside.

Note that initial power uprate plants experienced flow induced vibrations, which have been minimized by design improvements.

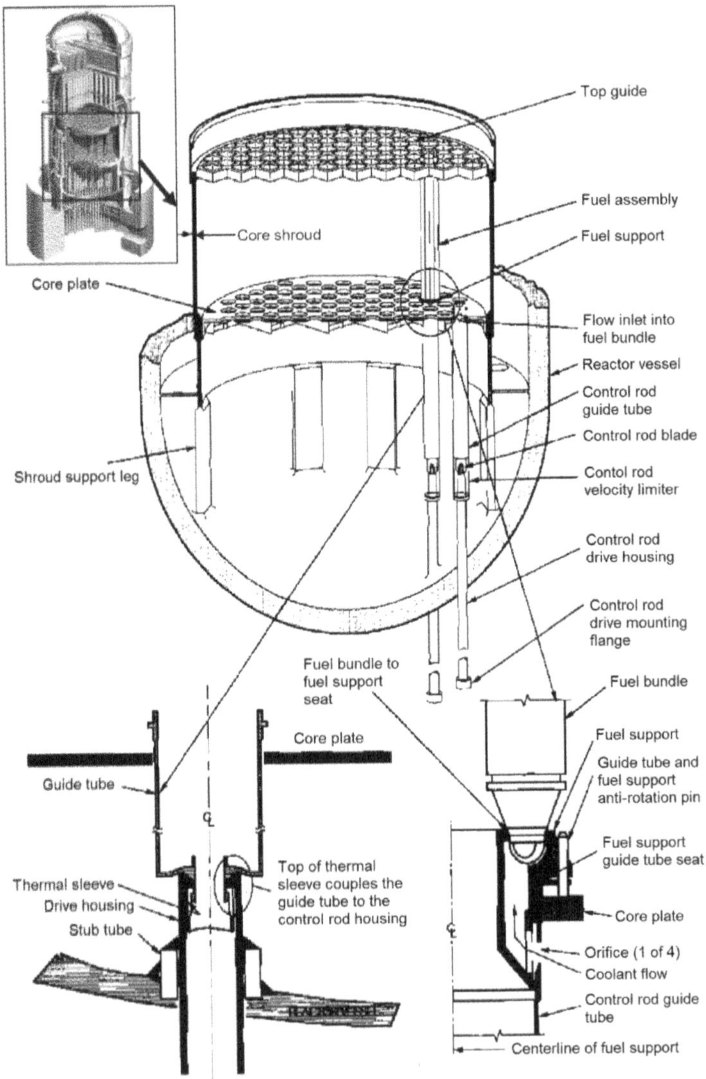

Figure 1.5. *BWR typical lower plenum (Cattant (2014), courtesy of GE). For a color version of this figure, see www.iste.co.uk/balbaud/nuclearenergytoday.zip*

1.3. Requirements for materials (Meyzaud and Vieillard-Baron 1998; Champigny 2005; Lemaignan 2010; Zinkle and Was 2013)

The selection of materials must take into account some drivers that are specific to LWR, as summarized below:

– large size components (with homogeneous properties), such as the reactor vessel, the primary pump, the pressurizer or the steam generators;

– good corrosion resistance (primary and secondary circuits);

– knowledge of materials in-service behavior, in accordance with regulator requirements concerning fatigue, irradiation, thermal aging, corrosion, etc.;

– prevention of the risk of brittle fracture;

– limitation of activation and contamination of materials (avoiding Co, etc.).

In order to reach the required quality level, nuclear construction codes (ASME, RCC-M, etc.) require:

– reduced chemical composition ranges for main components and strict control of residual elements;

– non-destructive examinations at various stages of manufacturing;

– technical qualification for the first manufactured component (additional testing, technical qualification);

– identification of main parameters controlling material properties, in technical manufacturing programs of suppliers.

1.3.1. *Brittle fracture*

One of the most significant regulatory requirements is the prevention of the risk of brittle fracture, defined as a sudden fracture, without any preliminary significant overall deformation. As far as the material is concerned, brittle fracture prevention is controlled through dedicated requirements for notch impact toughness (Charpy tests) at several temperatures, at least down to 0°C, and ductility (i.e. elongation at rupture and reduction of area through tensile tests).

To prevent any risk of brittle fracture, strict requirements are imposed on materials at the procurement stage. As an example, for heavy components, additional margins have been introduced in the equipment specification by imposing a maximum reference Nihil Ductility Temperature (RT_{NDT}) value equal to –20°C.

Nevertheless, this is not sufficient in cases where aging may reduce the resistance to brittle fracture through either an increase of the brittle-ductile transition temperature or an intrinsic reduction of fracture toughness. Further precautions have been taken to limit the effects of aging, as described below.

1.3.2. *Corrosion*

Materials corrosion is also an important issue not only from the viewpoint of the damage to the component itself, but also due to the generation of corrosion products. The materials must provide a high degree of protection against general corrosion, as well as the different forms of localized corrosion such as stress corrosion cracking, pitting or crevice corrosion.

The undesirable consequences of the migration of corrosion products may also be significant. Deposition of such products on fuel elements and heat exchanger surfaces may lead to ineffective heat transfers and the formation of radioactive isotopes after activation inside the reactor core.

Strong efforts have been applied to limit the cobalt contents of materials, with the aim to reduce activation, especially in steam generator tubes, for which an upper limit of 0.015% Co has been imposed. Besides, for main primary component surfaces directly in contact with the primary coolant, an upper limit of 0.060% Co has been set in stainless steels. The prevention of potential cobalt release has also lead to efforts to develop alternative materials to replace cobalt base alloys for hard facing. When available, these alternative materials can be used, particularly in valves.

The selection of materials for new reactors has resulted from choices made in the past for French and German reactors (N4 and Konvoi, respectively). The specific requirements for these materials take field experience and improvement in technical knowledge into account. They are de facto more severe now than they used to be for original units.

The materials used in nuclear power plants (NPPs) were naturally chosen from the experience of fossil power plants; it was indeed essential to rely on a set of reliable and very long-lasting data in order to focus innovation on a few specificities related to the operating conditions of nuclear reactors:

– effect of neutron bombardment on reactor pressure vessel materials or vessel internal structures;

– new nature of certain chemical environment (boric water of the primary water or steam that is very charged with moisture in turbines, etc.);

– difficulty of intervention for repair in radioactive environment.

The materials used are mostly steels. The logic is to consider the most usual steels (with the lowest cost and whose conditions of use are the largest) in the first place and to only choose more sophisticated steels or alloys to meet particular constraints of use.

Non-alloy ferritic steels are the most conventional boiler construction steels. These steels are suitable for most pipes of water-steam circuits, outside the creep range. However, the use of other steels or alloys is necessary when higher mechanical characteristics, improved corrosion resistance or special physical properties are required.

Low alloy steels are used when higher mechanical characteristics are needed in the field of temperatures below 350°C, for example, for pressure capacities of a large diameter (concerning limit thicknesses): case of the reactor pressure vessel. However, these materials have several disadvantages: the increase in tensile characteristics is generally to the detriment of brittle fracture resistance (resilience, toughness); they are sensitive to hydrogen embrittlement, which can be connected to corrosion or welding operations; finally, the quench rate varies greatly between the skin and the core during the manufacture of thick parts, which could involve detrimental heterogeneities of structure and properties.

Austenitic stainless steel grades selected for PWRs are usually "very low carbon" grades or stabilized (i.e. containing titanium or niobium) grades to prevent sensitization during manufacturing operations. Most often, considering corrosion susceptibility leads to the choice of stainless steels (addition of chromium with contents greater than 12–13%). This choice may be necessary in primary circuits where the pH cannot be set in the range minimizing general corrosion of carbon steels. The selection of stainless steels helps to avoid a high potential corrosion rate and limit the harmful consequences of an excessive release of corrosion products in the circuit, which could be activated in the irradiated areas. Although highly resistant to general corrosion, stainless steels can be susceptible to various localized corrosion phenomena such as intergranular corrosion, stress corrosion cracking or pitting corrosion.

Among stainless steels, martensitic stainless steels are used for their high mechanical characteristics, which can be adjustable by heat treatment (bolting, valve shaft, etc.). However, those steels, in particular the high chromium grades, may undergo thermal aging and embrittlement at reactor operating temperatures (300–350°C), which may strongly limit their practical use.

Austenitic stainless steels have, for the basic grades, low tensile characteristics, but good weldability and easy formability. They are widely used for pipes subject to severe corrosion conditions (primary circuit piping, etc.). For castings and welding filler metals, austenitic-ferritic grades are selected, as a minimum amount of ferrite prevents hot cracking during solidification. However, ferrite is sensitive to thermal aging at working temperatures and its content must be limited.

Ferritic stainless steels are sometimes selected in very singular cases.

In the case of the specific corrosion conditions encountered by steam generators, tubes and other components (stress corrosion expected), nickel-based alloys are selected.

All materials must provide a high degree of protection against general corrosion and the various forms of localized corrosion, such as stress corrosion cracking, pitting or crevice corrosion. Indeed, the migration and deposition of corrosion products may lead to ineffective heat transfers and the formation of radioactive isotopes after activation inside the reactor core.

Zirconium alloys were selected for the fuel cladding based on their neutron transparency and good resistance to corrosion in water, and when in contact with the fuel rods. Discussing materials and processes for the fuel assembly is beyond the scope of this chapter.

The main difference between PWRs and BWRs is that the latter consist of single water circuits, which are designed to allow boiling to occur in the core, with steam flowing directly to the turbine, which eliminates the steam generator and pressurizer found in PWRs. Operating temperatures remain close for reactors (lower for BWRs), with similar stress and radiation environments (lower for BWRs), and with the specificity of a more oxygenated environment for BWRs. Consequently, most of the structural alloys are very similar in both reactor types.

1.4. Material choices and basic properties

All of the considerations on materials requirements lead to the choice of commercial materials grades, which are well known by manufacturers and easy to produce, such as carbon steels, low alloy steels (LASs), austenitic stainless steels and nickel base alloys. Concrete (e.g. for containment building and cooling towers), polymers (e.g. for cables, composite materials and HDPE) and miscellaneous materials (e.g. hard-facing materials (stellite), copper and titanium base (condenser tubes), high strength materials (bolting)) are briefly considered at the end of this chapter.

	Carbon steels	Low-alloy steel	Austenitic stainless steel	Ni alloys
Cristal structure	Body centered cubic (cristallography) (BCC)	BCC	Face centered cubic (cristallography) (FCC)	FCC
Microstructure	Ferritic, bainitic	Ferritic, bainitic	Austenitic	Austenitic
Main alloying elements	0.05–0.35% C, 0.5–1.5% Mn, Si, etc. Total (w/o Mn) < 1%	C < 0.27% Mn, Ni, Mo, Cr, etc. Total < 5%	~18% Cr, ~10% Ni Mn < 2%, Si < 1% C < 0.08%	~15% Cr, ~10% Fe ~30% Cr, ~10% Fe 0.04–0.08% C
Main grades	TU, A-48; TU, A-42, P355 NH A106, A333, A515, etc.	16MND5, 20MND5 A533 Cl.1, A508 Cl.2/3, 15Kh2MFA, 15Kh2NMFA	304 L, 316 (L), 321, 347 Cast: CF3M, CF8M, etc. A800	Alloy 600 Alloy 690, others; X750, 718, etc.
Filler products	A42, A48, SA105, SA216, SA350, etc.	SA508 Gr 2/3, SA533 Type A/B	308L, 309L	182/82, 152/52
Fabrication	Forged, rolled or cast	Forged or rolled, cladded	Forged, rolled or cast	Forged, rolled
Heat treatment	Austenitizing, air cooling	Quenched + tempered + post-weld HT	Solution annealed (SA) (+ cold work)	Mill annealed Thermally treated
Weldability	Cold cracking sensitivity	Cold cracking sensitivity	Hot cracking sensitivity	Hot cracking sensitivity
Yield stress	250–450 MPa	250–450 MPa	~200 MPa (SA) if cold worked (CW), YS increases	~300 MPa (SA) if CW, YS increase

Toughness/ ductility	High ductile brittle transition temperature (DBTT)	High, DBTT	Very high, no DBTT	Very high, no DBTT
Irradiation resistance	N/A	Moderate (% Cu, P low)	High	High
Thermal aging	Moderate (% P)	Moderate (% P)	CF3M, CF8M	N/A
Corrosion resistance	Risk for flow-assisted corrosion (FAC) High resistance SCC	Moderate	High	High
Main components	Secondary piping, tubes and plates for exchangers	Vessels (RPV, pressurizer, steam generator)	Primary piping, surge line, RPV internals, pump casing, etc.	Steam generator (SG) tubing, divider plate, BMI, Head penetration, etc.

Table 1.1. *Main basic properties and application of the different classes of used materials in an LWR*

1.4.1. *Carbon and LASs (SFEN 2008; Cattant 2014; Gras 2017)*

Carbon and LASs have been selected for LWR applications, mainly because of the following properties:

– rather low cost;

– good mechanical properties;

– good weldability;

– high SCC resistance;

– rather low neutron embrittlement;

– good thermal conductivity (tubing applications).

Carbon steels and LASs can be supplied as:

– black and hot-dipped galvanized steel pipe;

– carbon steel forgings for piping applications;

– seamless carbon steel pipe for high-temperature service;

– pipe, steel, electric-fusion (arc)-welded (sizes NPS 16 and over);

– steel castings, carbon steels, suitable for fusion welding for high-temperature service;

– pipe fittings of wrought carbon steel and LAS for moderate and elevated temperature service;

– low and intermediate tensile strength carbon steel plates;

– seamless and welded steel pipe for low-temperature service;

– pipe fittings of wrought carbon steel and LAS for low-temperature service;

– tubes for heat exchangers;

– pressure vessel plates, carbon steel, for intermediate-temperature and higher temperature service;

– pressure vessel plates, carbon steel, for moderate-temperature and lower temperature service.

1.4.2. *Austenitic stainless steels (SFEN 2008; Cattant 2014; Hutin 2017b)*

Austenitic stainless steels have been selected in LWRs because of the following properties:

– corrosion resistant to the surfaces directly in contact with the primary coolant;

– good weldability;

– high mechanical properties when cold worked;

– rather easy casting with some amount of ferrite.

Austenitic stainless steel grades are now "very low carbon" grades or stabilized (i.e. containing titanium or niobium) grades to prevent intergranular corrosion sensitization.

For primary components with surfaces directly in contact with the primary coolant, an upper limit of 0.060% Co has been set in stainless steels.

Austenitic stainless steels can be supplied as:

– forged: valves, fittings, pipe flanges;

– rolled: plates, sheets, tubes;

– castings: elbows, pump casings, pipes;

– tubes (also: ferritic stainless steels).

Alloy 800 is also considered as an austenitic stainless steel, mainly used for steam generator tubing.

1.4.3. *Nickel base alloys (SFEN 2008; Cattant 2014)*

Nickel-base alloys have been selected for LWR applications, mainly because of the following properties:

– corrosion resistance (Cl^-, general, primary water stress corrosion cracking (PWSCC));

– high mechanical properties;

– expansion coefficient close to those of LASs (interesting for heat exchanger tubing).

To reduce activation, the cobalt content was limited to 0.015% Co, for instance, in steam generator tubes.

After the first evidence of the SCC susceptibility of former alloy 600 exposed to primary PWR environment as long as secondary corrosion and SCC, improvements for steam generator tubes have been put in place: first with a dedicated heat treatment at 715°C, and finally with the choice of 690 TT alloy with higher chromium content.

Nickel-base alloys can be supplied as:

– rods, bars and wires;

– seamless pipes and tubes;

– plates, sheets and strips;

– forgings.

1.4.4. *Other materials*

1.4.4.1. *Hardfacing materials*

Hard-facing alloys have been selected for LWR applications, mainly because of the following properties:

– high wear resistance;

– high mechanical properties;

– good corrosion resistance;

– good thermal shock resistance.

Potential cobalt release also triggers efforts to develop alternative materials to replace cobalt base alloys for hard facing (Norem).

1.4.4.2. *Condenser tubes (Gras 2017)*

In the 1970's, the construction of the tube bundles of PWR condensers was based, for power plants cooled by river water, on conventional technology, with brass tubes (copper-zinc alloy), inherited from the experience gained from fossil power plants. These condensers quickly presented generic problems that past experience did not predict:

– erosion by wet steam of impact tubes and peripherals;

– vibration of the tubes, resulting in fatigue ruptures or contact wear;

– ammoniacal corrosion of the brass tubes of the central part of the beams where the non-condensables are concentrated;

– stress corrosion in the rolling zone that can be initiated from the inside of the tubes (refrigeration water side) or from the outside (ammonia-conditioned steam side);

– flow-assisted corrosion (FAC) on the raw water side that results in a regular thinning of the inner skin tube at a speed of about 50 µm/year, affecting the entire bundle of tubes and determining the life of the condenser (about 10 years).

These problems led to the adoption of AISI 316L austenitic stainless steel for the rehabilitation of damaged condensers and the construction of new condensers. It should be noted that the gradual replacement of copper alloys, a bacteriostatic material (i.e. preventing the creation of a biofilm), by austenitic stainless steel, has given rise to a new concern: the appearance of pathogenic amoebae in the environment in summer.

For power plants cooled by sea water, the construction was based on the choice of titanium. The results of the first years of operation showed that the technological choices made for condensers cooled by sea water were overall satisfactory. These devices do not pose major problems for the operator and are not expected to undergo major changes in the future, as long as the cathodic protection is properly adjusted and monitored over time.

1.4.4.3. *High-strength alloys*

The typical high-strength alloys used in NPPs are as follows: 17-4 PH; X750; X718; A286; cold worked (CW) austenitic stainless steels; martensitic stainless steels; Ni-base alloys.

These alloys have been selected for LWR applications, mainly because of the following properties:

– high mechanical properties;

– quite often: good corrosion resistance.

High-strength alloys can be supplied as (not all products for all alloys) bars, plates, sheets, strips, wires, forgings and forging billets, seamless tubes, rings, turbine blades and castings.

The attachment of the closer head to the reactor pressure vessel is carried out using studs screwed into the main flange on one side and through the lid flange on the other, so as to ensure proper tightening by nuts placed above the lid. These studs are machined in forged bars made of low-alloy steel, containing nickel, chromium, molybdenum and vanadium to give the metal the necessary high tensile characteristics, especially in temperature, maintaining good resistance to fast fracture.

The bolting of reactor coolant pumps, whose diameter is much smaller, is carried out using forged steel bars containing nickel, chromium, molybdenum, and those of steam generators and pressurizers are made using forged steel or rolled with chromium and molybdenum with a possible addition of vanadium. In each case, the codes (RCC-M, etc.) provide precise recipe acceptance conditions regarding the mechanical characteristics and non-destructive tests to be carried out.

1.4.4.4. *Concrete (Bouniol 2001)*

Concrete and, more generally, cementitious materials are used in the civil engineering works of NPPs (reactor building, cooling tower, etc.). The aging mechanisms of cementitious materials are numerous due to their heterogeneous nature and porous structure. Thus, because of this porosity, the species present in the external environment are likely to penetrate the material, which can cause a depassivation of steel in reinforced concrete structures. In particular, carbon dioxide from the atmosphere causes a degradation of concrete, called carbonation, which modifies its physical and chemical properties, and, in particular, leads to the pH of the interstitial solution falling to values of about 8. Another common cause of concrete degradation comes from the diffusion of chlorides. The stages of corrosion and the different stages of the evolution of the degradation of a reinforced concrete structure can be summarized as follows:

– a transport step of aggressive agents (mainly carbon dioxide and chlorides; it could also be sulfates) from water and oxygen;

– a corrosion initiation step (depassivation of reinforcements). The two main phenomena at the origin of the depassivation of steel in cementitious medium are the decrease in the pH of the interstitial solution due to carbonation and the penetration of chloride ions;

– a stage of rust growth (in an airy environment), which gradually extends over the steel. The resulting volume expansion leads to damage to the concrete (chips, splinters and cracks).

1.4.4.5. *Polymers*

Polymers, in the broad sense of the term (elastomers, thermoplastics, composite matrix, ion exchange resins, etc.), are widely used in NPPs. The mechanisms of aging depend on the material and the conditions of use or exposure. Physical aging is often distinguished from chemical aging, although in reality both phenomena interfere. Irradiation aging combined with thermal aging is also possible for cables.

1.5. Fabrication and joining

Nuclear power imposes particular developments and manufacturing requirements:

– The large sizes of some of the forged pieces (pressure vessel shells, closure head, etc.) reach the practical limits of feasibility. It is then necessary to control the composition of very large ingots and their homogeneity. Macro-segregation must be minimized during the solidification of ingots (hence the use of hollow ingots). Finally, it is necessary to have presses adapted to the size of these components.

– The composition requirements/specifications impose full control of the metallurgy, with many preliminary tests and validation steps in the reactor, which require examinations in “hot laboratories”.

1.5.1. *Forging of Ferritic components: from ingot to component (Vieillard-Baron and Meyzaud 1998; Destre and Izard 2015)*

Pressure vessels of a large diameter (with concern to limit thicknesses) – reactor pressure vessels, steam generator (SG) parts, tube sheets and presssurizer shells – are generally forged from LAS ingots.

Since the 1960s, the various changes in the grade of steel in PWRs reflect the improvement in the steelmakers' processes (reduction in the maximum values of impurities as S and P, narrowing of the variation ranges of certain elements, etc.), and a better knowledge of the effect of additives elements on the properties of steels.

Advances in steelmaking techniques have been very beneficial for these steels. In particular, from the 1980s, the techniques of vacuum degassing and ladle refining have made it possible to obtain:

– better analytical precision;

– very low contents of elements such as hydrogen (harmful for cold cracking), sulfur (responsible for the anisotropy of the characteristics and a reduction in the level of the ductile bearing) and phosphorus (influence on the fragility tempering, resistance to irradiation and thermal aging);

– better control of the aluminum content for the trapping of free nitrogen by formation of AlN;

– minimization of metal inclusions, endogenous or exogenous.

Whatever their geometry, components intended for the nuclear industry must satisfy increasingly demanding design specifications. Among all of the criteria required, we can cite those relating to the inevitable phenomena of segregation that particularly concern forged parts from large ingots:

– limitation of the rate of macro-segregation remaining;

– absence or limitation of area with "ghost lines" (segregation lines) on faces to be coated by welding.

The heterogeneity of large forge ingots, induced by solutal and thermal convection currents, and exacerbated by long solidification times, is well known (Figure 1.6).

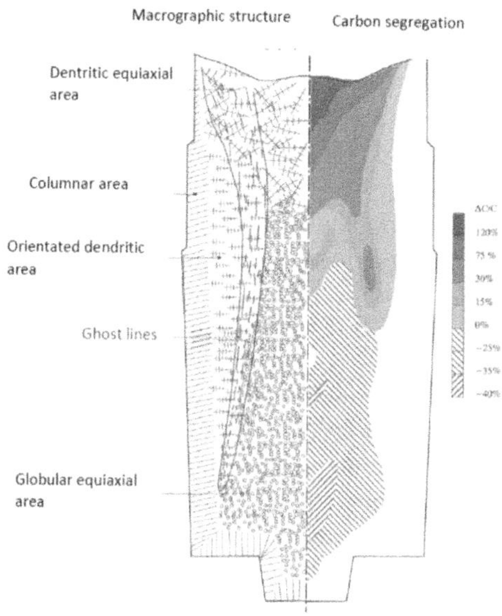

Figure 1.6. *Phenomena of segregation in large ingots: improvement of numerical simulation. For a color version of this figure, see www.iste.co.uk/balbaud/nuclearenergytoday.zip*

Although for most hollow parts, most of the macro-segregation is removed by hot drilling during forging, it is impossible to remove the entire area of "ghost lines". Gradually, over time, the approach was therefore directed toward adapting the types of ingots according to the parts to be delivered, in order to reduce the level of macro-segregation and minimize the disadvantages linked to segregation and "ghost lines". This was done by locating them in parts where they are less detrimental, that is, within the thickness. For this, different types of ingots were developed at the end of the 1970s, such as the directional solidification ingot and the hollow ingot. Nevertheless, for very thick components such as SG tube sheets, it is still not possible to eliminate all of the heterogeneities.

The forging of LAS is carried out between 1,300 and 1,000°C in several operations, such as blooming, crushing, hot drilling, drawing on a mandrel and horning, in order to obtain a deformation rate, allowing the desired properties to be obtained. The forging ranges are optimized regularly to obtain parts as close as possible to finished parts, to eliminate welds, or to move them to more accessible areas. For example, we can cite the manufacture of steam generator heads with pipes extruded by hot punching, and more recently, the development of forging the vessel head of 900 MW monobloc reactors, which makes it possible to eliminate the flange-cap weld.

1.5.2. *Casting... to forging (Vieillard-Baron and Meyzaud 1998)*

A major part of the components of the main coolant lines are made from forged austenitic stainless steels tubes.

But, historically, for primary pump casing, elbows, fittings, valves, etc., cast duplex austenitic-ferritic stainless steels were mostly used. As the production progressed, the sizing of the coolers and risers made it possible to limit the appearance of shrinkage cavities during the solidification of these cast components. These steels have a ferrite content ranging from about 10 to 25% to avoid solidification cracking and obtain higher tensile properties. At operating temperature however, due to thermal aging, hardening of the ferrite phase takes place, inducing embrittlement.

A molybdenum-free grade (CF3 type), less susceptible to aging, has been selected; more recently, cast duplex steels in primary pipe branches have been replaced by forged austenitic grades that are not susceptible to thermal aging. Then, for the recent fabrications, mostly the pump casings are made from cast duplex steel.

1.5.3. *Forging of PWR SG tube bundle*

The U-tube bundle transfers heat from the primary system to the secondary system. The ends of the U-tubes are inserted in a tube sheet 500–600 mm thick. Their entire inserted lengths are roll-expanded and the tube-ends are then welded to the tube-sheet. Straight leg portions of the tube bundle are held in place by tube support plates and bent portions by anti-vibration bars.

Tube blanks are made from billets obtained by hot forging of ingots. The billets are cut, machined and extruded, then converted by drawing or rolling into blanks. The blanks are then cold worked by drawing or rolling in several passes, with intermediate annealing. They undergo final annealing at a specified temperature, then are subjected to a supplementary 715 ± 15°C heat treatment, followed by bending. U-bends with radii less than 10 times tube diameter are subjected to an additional stress-relieving heat treatment for 2 h at 715°C.

1.5.4. *Welding and cladding (Buisine and Perrat 201)*

1.5.4.1. *Welding*

The main welding techniques used are arc welding with filler material for all materials (submerged arc welding (SAW), shielded metal arc welding (SMAW) and gas tungsten arc welding (GTAW)). For the record, electroslag welding was formerly used for pump casing welding.

Component/materials	Welding technique	Source	Protection
RPV, SG, PZR, etc. (*LAS*) MCL/*austenitic stainless steel*	Shielded metal arc welding (SMAW)	Arc	Slag
RPV, SG, PZR, etc. (*LAS*) MCL/*austenitic stainless steel*	Submerged arc welding (SAW)	Arc	Slag
MCL/*austenitic stainless steel*	Gas tungsten arc welding (GTAW)	Arc	Gas
Pump casing parts *Duplex stainless steel*	Electroslag welding (ESW)	Joule effect	Slag

Table 1.2. *Main welding techniques used*

1.5.4.2. *Cladding*

The inner wall of the vessel is coated with one or two layers of austenitic stainless steel, which protects it from corrosion by the cooling water. To avoid intergranular corrosion, the number of layers and the mode of deposition are defined so that the carbon content of the inner layer is at most equal to 0.035%, hence the need to produce a two-layer coating.

Two deposition processes were used: the conventional flux arc welding process with strip electrode (strip width 60 or 90 mm), then the electroconductive slag welding process with strip electrode (so-called Maglay process, strip width 75 or 150 mm).

1.5.5. *Heat treatment (Vieillard-Baron and Meyzaud 1998; Destre and Yzard 2015)*

After forging, LAS components undergo normalization heat treatment and tempering at around 650°C to facilitate subsequent ultrasonic inspection operations in the semi-finished state.

After roughing and inspection, the last quality heat treatment includes (i) austenitization at 875°C, which, among other things, makes it possible to refine the grain, (ii) quenching by immersion in water, which provides the required bainitic microstructure and (iii) tempering around 650°C, followed by slow cooling.

After welding, relaxation heat treatment is required for unalloyed or low alloyed ferritic steels around 600°C, but is not required (because it is potentially unfavorable) for austenitic stainless steels (due to potential risk of sensitization or formation of embrittling phases).

Austenitic stainless steels are generally only solution annealed at 1,050°C and water cooled.

1.5.6. *Manufacturing defects (links with fabrication)*

Among the different types of welding defects that can affect the components (geometrical defects, technological defects and metallurgical defects), a few of them are more specific to LWR components: for example, underclad defects.

1.5.6.1. *Underclad defects*

Some weld procedures used for stainless steel cladding of pressure vessels (reactor, steam generator and pressurizer) resulted in manufacturing defects in the heat affected zone of the base metal under the cladding.

Two types of underclad cracking have been identified: reheat cracking and cold cracking.

Reheat cracking occurred as a result of post weld heat treatment of austenitic stainless steel cladding applied using the high-heat-input welding process. Reheat cracking was attributed to high residual stresses near the yield strength in the weld metal and base metal interface, combined with a sensitive microstructure of the base metal: coarse grains and segregated lines. The cracks are in the base metal region directly beneath the cladding and are confined to a region approximately 0.125 inches deep and 0.4 inches long. This problem was eliminated by modifying the welding sequence of the second coating layer in the affected areas.

Cold cracking occurred after deposition of the second and third layers of cladding, where no pre-heating or post-heating was applied during the cladding procedure. The cold cracking was attributed to residual stresses near the yield strength in the weld metal and base metal interface after cladding deposition, combined with a crack-sensitive microstructure in the heat-affected zone (HAZ) and a high level of diffusible hydrogen in the austenitic stainless steel. The hydrogen diffused into the HAZ and caused cold (hydrogen-induced) cracking as the HAZ cooled. Corrective measures consisted of heating the parts to more than 150°C in the first and second layers of coating. Another measure consisted of ensuring that the segregations were not present on the wall to be coated using a hollow ingot.

Other manufacturing defects such as hydrogen flakes sometimes affect LAS components due to insufficient degassing treatments. Welding defects as hot cracking can also affect austenitic stainless steels. Cast components are affected by solidification shrinkage defects.

1.5.7. *Non-destructive testing (links with expected defects) (Champigny 2005)*

Non-destructive techniques are specifically developed and used to detect the most current defects that can affect the reactor components.

According to the type and location of the defect to be detected, the techniques presented in Table 1.3 are used.

	ND technique	Main components
Global	Visual testing (VT) Acoustic emission (AE) Thermography	RPV cladding, vessel head cladding, PZR cladding, SG channel head cladding, etc.
Local (surface)	Dye penetrant testing (DPT) Magnetic particle testing (MPT)	Instrumentation nozzles, RPV lifting lugs PZR, reactor coolant pump closure studs, etc.
Local (sub-surface)	Eddy current testing (ECT)	SG tubes, RPV bottom head instrumentations, RPV closure studs, etc.
Local (volume)	Ultrasonic testing (UT) Radiography testing (RT)	RPV welds, RPV closure studs, main primary circuit dissimilar steel welds, PZR welds, SG channel head welds, etc. Main primary circuit dissimilar steel welds, cast elbows, PZR welds, etc.

Table 1.3. *Main non-destructive techniques used to detect the most current defects affecting reactor components*

As an example, Table 1.4 lists the non-destructive examinations carried out during the various stages of manufacture of large forged components of low-alloy ferritic steels.

Manufacturing step	Characteristic concerned	Quality risk	Parameters influencing heterogeneity	Non-destructive examinations associated with the risk of heterogeneity
Elaboration: casting	Internal health	Exogenous inclusions and hydrogen flakes	Elaboration: casting	Ultrasonic testing (UT) exploration 100%
Forging: cropping	Internal health	Exogenous inclusions and compactness defects	Minimum cropping rate head/food and/or drilling diameter (axis of the ingot)	UT exploration 100%
Forging: wrought	Internal health	Compactness defects	Wrought rate	UT exploration 100%
Forging range	Surface health	Forging folds	Forging range	Surface examination
Preliminary heat treatment (PHT)	Internal health	Hydrogen flakes	Minimum temperature before PHT and degassing cycle	UT exploration 100%
Welds	Internal health	Weld defects (cold cracking, etc.)	Welding parameters	UT exploration 100%
Cladding	Internal health	Underclad cracking, intergranular reheat cracking	Welding parameters	UT exploration 100%

Table 1.4. *List of non-destructive examinations carried out during the different stages of manufacture of large forged components of low-alloy ferritic steels*

1.6. Main material limitations

Like any industrial installation, pressurized water nuclear boilers are exposed to mechanical, thermal (operating temperature below 350°C) or chemical (presence of boricated water, lithine and dissolved hydrogen in the primary water; ammonia or morpholine and hydrazine in the secondary water) limitations when in service for very long operating times, 40 years, or even 60 years. More specifically, materials located in the vicinity of nuclear fuel are exposed to intense neutron bombardment.

The damages that are likely to result from these operating conditions are as follows: the evolution of the properties of materials by thermal aging or embrittlement by irradiation, thinning by wear or generalized corrosion, initiation of cracks by fatigue or corrosion under stress, propagation of manufacturing defects, etc. This damage is taken into account, when possible, from the design stage of the equipment through materials selection materials and assessment of mechanical performance required by regulations and codes. Identified damage is monitored in service. If other damages appear when they are not expected, corrective measures make it possible to stop them or at least to monitor their evolution.

This results in two types of problems to be dealt with: the mechanical and physicochemical resistance of the materials used, as well as the evolution of their properties during operation in sometimes aggressive conditions, in particular because of irradiation phenomena. We will focus in particular on this aspect in the rest of the text.

1.6.1. *BWRs*

From a material point of view, this type of reactor mainly presents problems of SCC or generalized corrosion related to the slightly oxygenated environment. In advanced BWRs, many methods have been used to prevent this form of stress corrosion cracking; in particular, the compression of the walls in contact with the environment and the introduction of hydrogen into the feed or demineralized water proved sufficient to stop certain cases of corrosion in the recirculation loops.

Even though they receive much lower neutron doses than PWRs, the internal structures of austenitic stainless steel BWRs are highly affected by irradiated-assisted stress corrosion cracking (IASCC) (from very low neutron doses). This low threshold of sensitization to IASCC appears to be related to chromium depletion of grain boundaries; it has been limited by adding hydrogen to the primary water and reducing in-service stresses.

Finally, to reduce the radiological activity of the circuits (especially due to cobalt) in the new reactors, more stringent specifications of very low cobalt were adopted, and some modifications were done regarding water chemistry specifications: starting from the purest oxidizing water possible, injection of oxygen, hydrogen, iron, zinc and surface treatments with palladium were proposed.

1.6.2. *PWRs*

The aging of the components in service of PWR nuclear boilers results from a wide variety of degradations which can occur, isolated or in combination. These forms of aging are considered from the design stage of the equipment. Some damage is inevitable, such as thermal aging or neutron irradiation. In this case, the most resistant materials are selected, evolution of the properties in service is monitored in representative samples and the components are inspected to detect or monitor the evolution of possible defects. Some damage not expected at the design stage may however occur in service, for example, under unusual or transitory operating conditions. The same inspection techniques are used to determine the causes and solve the problem: in-service monitoring of the damage, repairs, or even replacing the component.

This paragraph presents the main aging mechanisms based on the knowledge available to date obtained through R&D studies, field experience and laboratory examinations.

1.6.3. *Irradiation embrittlement (Champigny 2005; Massoud 2008; Lemaignan 2010; Barbu and Cattant 2014)*

The irradiation of steel by neutrons has the effect of driving many atoms from their initial site into the crystal lattice. These atoms are then inserted into the network (they are called interstitials) and thus leave vacant sites called vacancies.

If an interstitial and a vacancy are very close to each other, by simple elastic interaction due to network distorsion, they recombine and disappear. Vacancies and interstitials are point defects that can also move by jumps in the network if the temperature is high enough. These defects are known to diffuse, with the rate of diffusion increasing rapidly with temperature. Again, when an interstitial and a vacancy meet, they can recombine. Vacancies and interstitials can also disappear on wells of defects: step on the surface of the crystal, dislocation, grain joint, etc. Defects of the same type can also be gathered into clusters:

– clusters of vacancies or interstitials that grow to form loops of dislocations (additional disks of atoms from the interstitials or on the contrary disks devoid of atoms in the case of agglomerations of vacancies) or;

– three-dimensional vacancies called cavities. Finally, irradiation creates point defects, clusters of defects, dislocation loops and cavities.

Point defects or clusters of defects, by introducing disorders into the crystal lattice, block the dislocations that allow the steels to deform. The yield point, that is, the stress threshold to enable the mobility of dislocations, increases; there is then a hardening and embrittlement under irradiation.

If the defects created by irradiation can directly be the cause of hardening and embrittlement, other effects of irradiation can also be observed:

– If there is formation of cavities, everything happens as if the atoms that were initially contained in the cavity had passed to the surface of the crystal: its volume increases. Swelling of the material can be observed.

– Irradiation results in the creation of vacancies in steels. This phenomenon leads to an acceleration of atomic diffusion, which occurs in particular in the low-temperature domain where it does not exist in the absence of irradiation. We can then observe, under irradiation apparition of precipitates, demixing, which should normally only occur at higher temperatures. We can also identify the creation/emergence of phases that do not exist in the original equilibrium diagram. This is particularly the case when an impurity is attracted by the vacancies: gathering of the vacancies results in the accumulation of the atoms of the impurity. It is by this type of mechanism that we explain the formation of microscopic copper clusters in the low alloy ferritic steels of PWR pressure vessels exposed to irradiation. All of these phases, formed thanks to an accelerated diffusion by irradiation, are obstacles to the mobility of dislocations and produce, as do irradiation defects, hardening and embrittlement.

– Neutron irradiation causes an increase in the concentration of point defects that can also cause segregation effects on the different sinks. Segregation will manifest itself mainly on grain boundaries, but also on dislocation loops, helium bubbles, cavities, etc. The effects of segregation under irradiation will be maximum for the temperature and flux range, for which the concentration of point defects and their mobility are high: at low temperatures, the high density of the wells and the low mobility of the point defects reduce the segregation under irradiation; at high temperatures, the recombination vacancies-interstitials prevail. The entire field of operating temperatures of PWRs is then included in the segregation area. Segregation under irradiation is associated with incomplete or interstitial mechanisms. On the one hand, some elements with a high diffusion rate will become poorer in the vicinity of the wells in order to balance the flow of vacancies (lacunar mechanism). This mechanism, called the "reverse Kirkendall effect", would explain the depletion of chromium and iron and the enrichment in nickel: chromium diffuses faster than iron, which itself diffuses faster than nickel. On the other hand, some elements with a negative size effect (silicon and phosphorus) will preferentially associate with the interstitials and segregate with the wells (interstitial mechanism). This segregation at grain boundaries can lead to a weakening of the grain boundaries both mechanically and from the point of view of corrosion sensitivity.

After recalling the different types of defects created by irradiation in metals, we will focus more particularly on the specific mechanisms of embrittlement under irradiation of low alloy ferritic steels and then austenitic stainless steels.

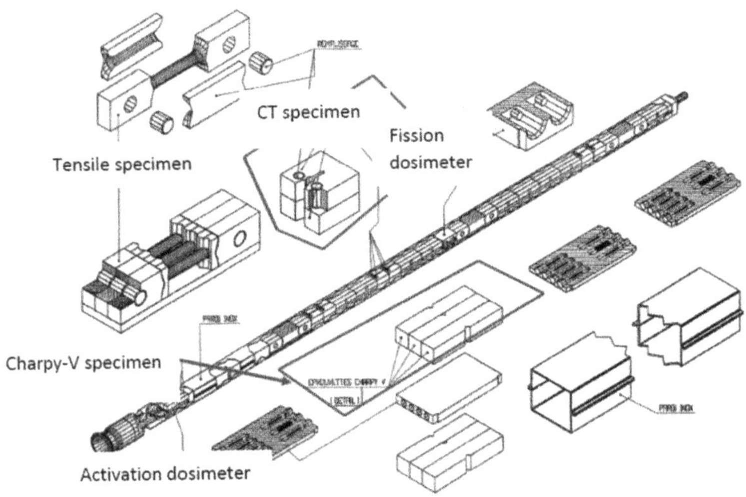

Figure 1.7. *Monitoring program irradiation capsule: different kinds of irradiated specimens. For a color version of this figure, see www.iste.co.uk/balbaud/nuclearenergytoday.zip*

For ferritic steels (such as reactor pressure vessel steels), which can be exposed to a maximum neutron irradiation of the order of 6.5×10^{19} neutrons/cm^2 (energy > 1 MeV), residual elements like copper and phosphorus that interact with point defects are the cause of most of the embrittlement. In addition to the chemical composition of the material and its structure, other significant parameters are the number of neutrons received per unit area (fluence) and temperature. The flux and energy spectrum of neutrons do not appear to have a significant effect. The consequences of irradiation on low alloy ferritic steels are an increase in yield strength, which leads to an increase in the ductile-brittle transition temperature (DBTT), and therefore to an increase in the risk of brittle fracture by cleavage. The factors determining the magnitude of irradiation embrittlement are well known. These are related to the contents of residual elements copper and phosphorus. Some alloying elements (Ni, etc.) can enhance the effect of these elements. From there, forecast formulas (for base metals and welded joints) could be proposed. These formulas are of the type DBTT = f(% Cu,% P,% Ni, etc.).

In addition, in order to monitor the evolution under irradiation of the reactor pressure vessel, a monitoring program has been set up. It consists of placing, in capsules, in appropriate places in the reactor pressure vessel, specimens representative of the materials used. These specimens are extracted and examined at regular time intervals according to a program to monitor and anticipate the evolution of the main properties of these materials during the life of the RPV (Figure 1.7).

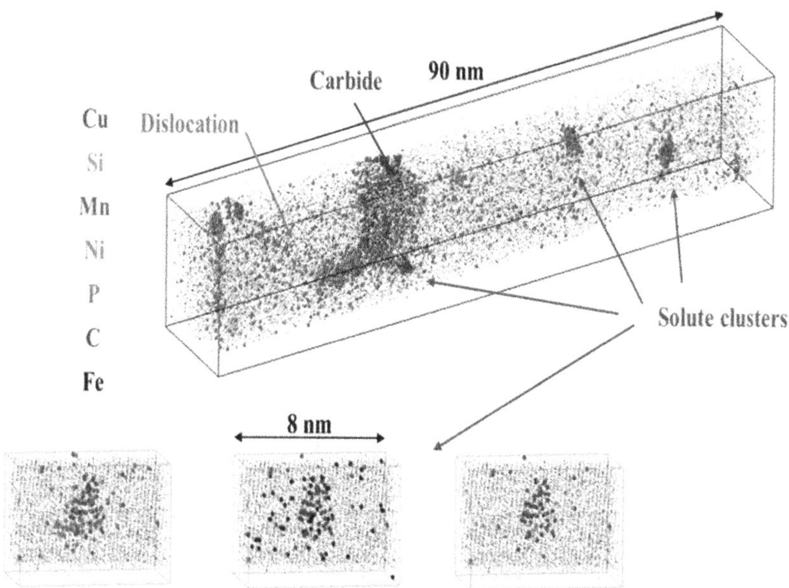

Figure 1.8. *3D-atom probe analysis of n-irradiated RPV steels (courtesy of EDF/U-Rouen). For a color version of this figure, see www.iste.co.uk/balbaud/nuclearenergytoday.zip*

The analysis by atomic probe of the microstructure of irradiated low alloy ferritic steels makes it possible to analyze the spatial distribution of the solutes present in the reactor pressure vessel steel. It shows that the neutron irradiation of a PWR reactor pressure vessel steel leads to the formation of clusters of solutes (Cu, Ni, Mn, Si, P, etc.) whose size (3–4 nm), density and chemical composition (with the probable exception of copper content) seem to evolve with fluence. The ferritic solid solution after neutron irradiation presents a decrease in the concentration of copper up to 0.03%. The copper concentration of the clusters formed increases with its nominal content in the steel. These clusters contribute, like point defects and segregation at grain boundaries, to the embrittlement of steel exposed to neutron irradiation; their respective contributions are not fully known at this time (Figure 1.8).

Numerical simulation has led to significant progress in recent years in understanding the mechanisms of damage formation. Atomic scale simulations of the transfer of energy from the initial neutron kinetic energy to the atoms of the material confirmed the formation of displacement cascades. These cascades lead to the production of a few dozen pairs of point defects in a volume of a few nanometers in about 10 ps. These studies have shown that the number of defects created as a function of the transferred kinetic energy is lower than that commonly predicted.

The evolution of these defects over longer times is simulated by the so-called "kinetic Monte Carlo" method. To achieve significant durations, the so-called "cluster dynamics" method is used (a code explicitly dealing with the solute has been developed with the CEA). Each of these techniques using information obtained at the lower scale make it possible to simulate irradiation damage in a model reactor pressure vessel steel. The input data are the neutron spectrum, the irradiation time, the chemical composition of the material, and in output, the results obtained are the evolution of the microstructure and the increase in yield strength. These developments are carried out through an international framework in Europe, the United States and Japan.

For austenitic stainless steels (such as the internal structures of PWR pressure vessels), at temperatures between 300°C and 370°C, neutron irradiation causes significant changes in the microstructure:

– Irradiation restores the network of work hardening dislocations in favor of a high density of small faulted "Frank" dislocation loops and "black dots" (small clusters of point defects). It seems that the evolution is of the asymptotic type, with a saturation that is reached quite quickly. As the temperature rises, the density of loops decreases while their size increases.

– Irradiation also causes a strong segregation of certain elements at sinks, especially grain boundaries. There is a significant depletion of chromium, molybdenum and iron, and enrichment in nickel and silicon. There seems to be no segregation of impurities such as sulfur or phosphorus.

– Finally, neutron irradiation can cause small gas bubbles and cavities to appear, with dimensions close to a few nanometers. Cavities are voids of matter resulting from the accumulation of vacancies created under irradiation. Gas bubbles contain helium mainly generated by the transmutation of nickel following its interaction with thermal neutrons during irradiation. This element is indeed poorly soluble in steel and will accumulate on preferential sites to form gas bubbles under very high pressure. These bubbles and cavities can lead to swelling of the material. But this mechanism is not proven in western PWR conditions.

As a result of the formation of the Frank loop dislocation substructure under irradiation, these materials undergo a substantial increase in yield stress (hardening) and reduction in ductility. The yield stress and mechanical strength become very close to 900–1,000 MPa. At the same time, the material undergoes a very strong degradation of its ductility, possibly due to a localization of the deformation in some bands. The tensile properties saturate at doses in the range of 5–10 dpa, as does microstructural evolution.

Figure 1.9. *PWR core lower internals, baffle to former bolt (CW 316). IASCC circumferential crack propagating toward the head top (courtesy of EDF)*

In addition, neutron irradiation beyond a threshold makes austenitic stainless steels sensitive to stress corrosion in a PWR environment. This is a typical case of damage coupling both aging under irradiation and corrosion under stress, called

IASCC. As an example, Figure 1.9 shows baffle to former bolt (CW 316) cracking due to IASCC (circumferential crack propagating toward the head top).

1.6.4. *Thermal aging (Barbu and Massoud 2008; Cattant 2014)*

Most of the components used in NPPs undergo thermomechanical forming treatments (quenching, solution annealing, rolling, tempering heat treatments, recrystallization, etc.), which give them the required properties. However, the materials then stay in unstable or metastable thermodynamic equilibrium states. Under operating temperature (300–350°C) and for operating times as long as 40 or 60 years, they will tend to evolve to recover their state of thermodynamic equilibrium more or less quickly.

The term "thermal aging" therefore covers a whole series of mechanisms based on the diffusion of chemical elements and which allow metallurgical "systems" to return to a more stable thermodynamic state of equilibrium. This return to equilibrium results in a microstructural evolution and an evolution of the properties of materials. Often, thermal aging leads to hardening and embrittlement (loss of ductility and/or shift of the fragile-ductile transition temperature).

The main mechanisms of thermal aging that may appear on the components of NPPs are as follows:

– Precipitation: it is the germination of a second phase (or several phases), then the growth of this phase until the achievement of thermodynamic equilibrium and finally, coalescence. If the precipitation is fine and uniformly distributed inside the grains, this mechanism causes a significant hardening. Precipitation rarely occurs for the low operating temperatures of PWRs (around 300°C). However, we can mention the precipitation of a chromium-rich hardening phase in martensitic stainless steels (13-4 PH, 17-4 PH) of valve rods or the coarser precipitation of carbides at the grain boundaries of austenitic stainless steels, which can cause their sensitization to intergranular corrosion (usually at temperatures higher than operating temperatures for short periods of time).

– The "spinodal" demixion: it is a mechanism quite close to precipitation, but which starts instantly without involving the energy necessary for the formation of germs. The initial phase is gradually and very slowly broken down into two new phases. This mechanism is observed in the chromium-rich ferritic phase of cast austenitic-ferritic stainless steels constituting the primary circuit (elbows, piping, pump scrolls, valves, nozzles, etc.). This evolution has a very hardening and weakening effect (Figures 1.10–1.12).

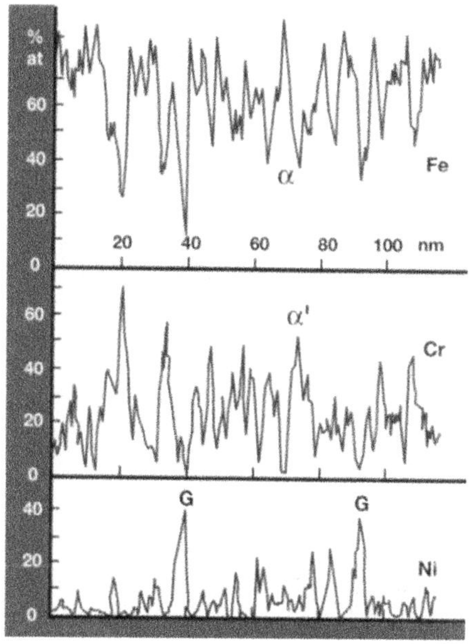

Figure 1.10. *Atomic probe concentration profiles of ferrite aged for 10 h at 400°C; Fe-rich α zones and Cr-rich α' zones (courtesy of EDF). For a color version of this figure, see www.iste.co.uk/balbaud/nuclearenergytoday.zip*

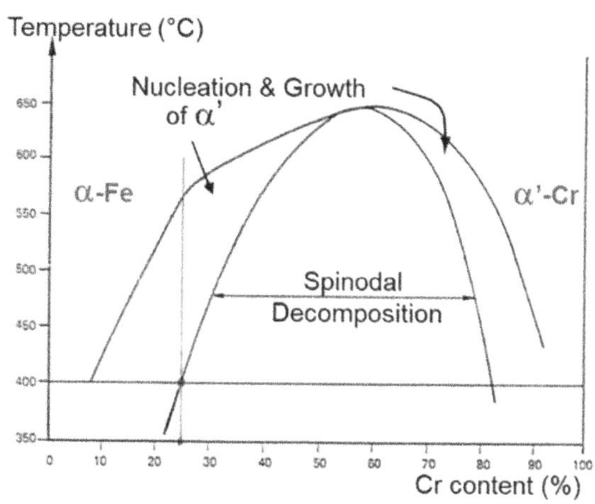

Figure 1.11. *Microstructural evolution: unmixing of the Fe-Cr solid solution. For a color version of this figure, see www.iste.co.uk/balbaud/nuclearenergytoday.zip*

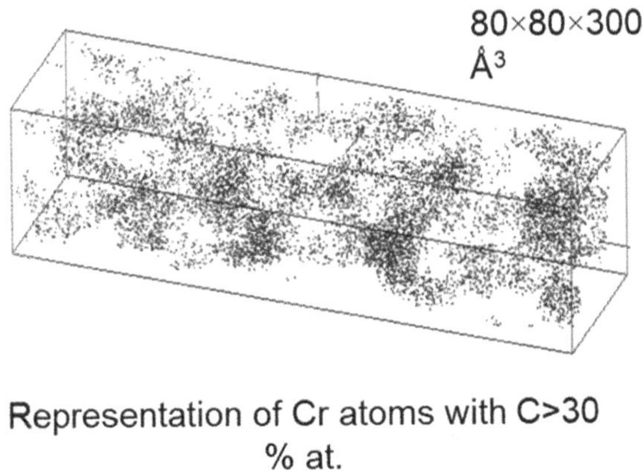

Figure 1.12. *Precipitation of α'-Cr (very fine < 10 nm) (courtesy of EDF/University of Rouen). For a color version of this figure, see www.iste.co.uk/balbaud/nuclearenergytoday.zip*

– Segregation: the segregation of certain chemical species, if it takes place at grain boundaries, can lead to a decrease in the cohesion of the interfaces (favoring intergranular rupture) or to a sensitization of the grain boundaries to corrosion (in the case of dechromization of the grain boundaries). For example, the segregation of phosphorus at grain boundaries is a potential weakening problem for the PWR pressurizer, whose operating temperature is about 345°C.

– Dynamic aging: it is a coupling between the deformation (mobility of dislocations) and the diffusion of solute atoms such as nitrogen or carbon that "anchor" the dislocations and, consequently, harden the material. This type of event is found in carbon-manganese steels of secondary pipings.

Special attention should be paid to the areas affected by welding, as well as the dilution zones of dissimilar welds assembling heteregeneous materials. The kinetics and effects of the aging mechanisms mentioned above can indeed be modified because of the structural specificities characterizing these areas (coarse grain structures or particular compositions).

All the mechanisms being based on diffusion, the important parameters are obviously the time and temperature of operation, but also the chemical composition of the material, and in general, the manufacturing parameters (presence of heterogeneities).

The main problem for the evaluation of the consequences of thermal aging is the accuracy of the microstructure evolution and mechanical characteristics over time. Generic models are available to predict phase change kinetics, but very often the systems to be studied are so complex that available models must be applied with caution.

In addition, neutron irradiation creates, among other damages, vacancies in the material. It will therefore tend to accelerate the diffusion of chemical elements and, consequently, promote thermal aging. When neutron irradiation creates specific damage, the mechanisms involved interfere with those of thermal aging. It is then difficult to isolate the thermal aging effects from those assisted by irradiation.

1.6.5. *Stress corrosion cracking (Lemaignan 2010; Cattant 2014; Combrade and Vaillant 2014)*

When a material is exposed to a chemical environment and, at the same time, to mechanical tensile stresses, cracking can occur: the material is said to have undergone "stress corrosion cracking" (Figure 1.13). Most materials (metals and alloys, etc.) are likely to be affected by stress corrosion cracking, and for each material, several different chemical environments could generate the phenomenon (aqueous solutions, gases, molten metals, etc.).

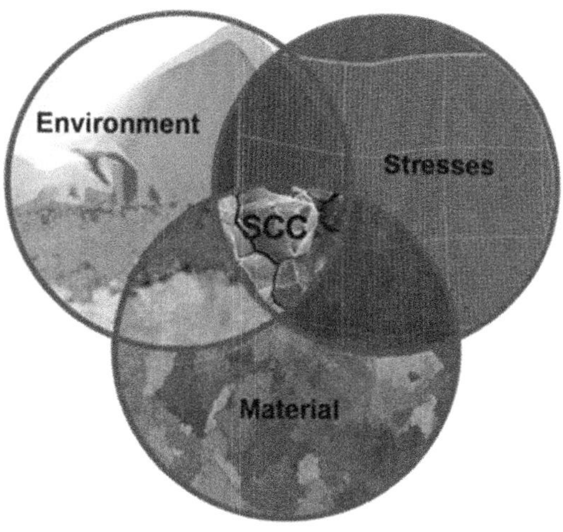

Figure 1.13. *SCC phenomenon. For a color version of this figure, see www.iste.co.uk/balbaud/nuclearenergytoday.zip*

Stress corrosion mechanisms are very numerous, but two main families can be distinguished. The first family includes the mechanisms involved in the dissolution of the material in the environment: the oxide that protects the material is locally damaged and, at the location of this damage, the environment produces a microscopic dissolution. The oxide progressively reforms and stops this dissolution, but another step of damage occurs, and the process repeats itself to produce cracking. The second family mixes together mechanisms that attribute the origin of cracking to a succession of small successive mechanical ruptures. The latter become possible because mechanical properties are degraded by the environment or by corrosion products entering the material.

Two types of degradation can then take place. Sometimes, there is a strong local reduction in the plasticity of the metal, which undergoes cracking under the effect of deformation (example of oxidation by oxygen from water). Sometimes, on the contrary, the environment or corrosion products locally increase the plasticity of the metal and the deformations that concentrate in the areas thus altered are at the origin of cracking.

For both cases of reduction and increase in plasticity, the mechanical cracks that initiate are microscopic, but the processes described are iterative and can lead to macroscopic cracks. This phenomenon particularly affects the nickel alloy tubes constituting the steam generators that operate in both the primary and secondary waters, and also stainless steels.

Austenitic stainless steels are highly resistant to stress corrosion cracking, but SCC can occur in different cases:

– in slightly polluted primary water: due to the deleterious effect of reduced species of sulfur in non-hydrogenated primary water at 200°C, or due to the effect of chlorides in oxidant primary environments, a decrease in the chloride threshold when the temperature increases;

– in non-polluted primary water: significant strain hardening is necessary for the initiation and propagation of SCC for $7.2 < pH_{320°C} < 9.1$.

Illustrations of different types of SCC in different materials are proposed in Figures 1.14–1.19.

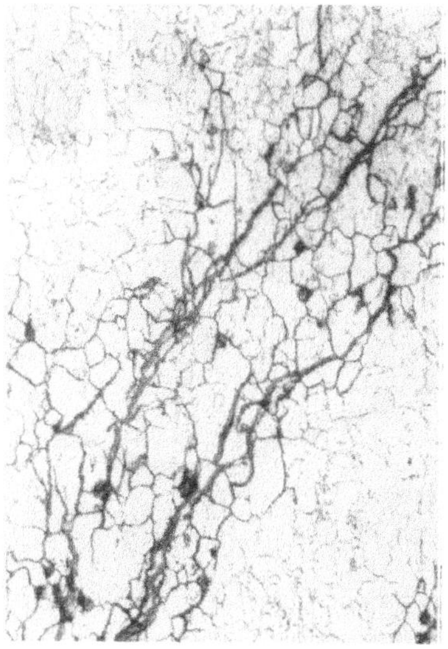

Figure 1.14. *Mix of intergranular stress corrosion cracking (IGSCC) and transgranular stress corrosion cracking (TGSCC) cracks observed on the outside surface of a SG tube (MA 600, flattened specimen, secondary water) (Cattant (2014), courtesy of EDF)*

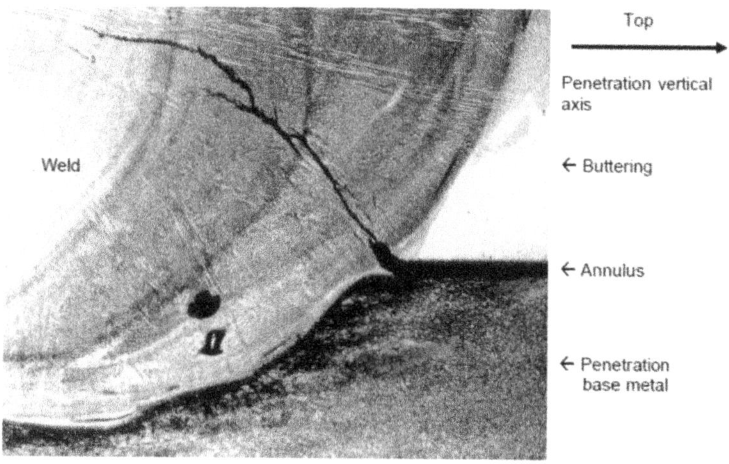

Figure 1.15. *RPV head leaking penetration: illustration of PWSCC in A182 weld (Cattant (2014), courtesy of EDF)*

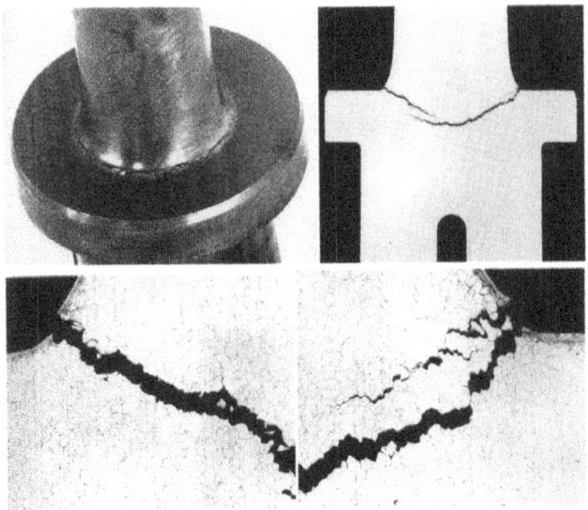

Figure 1.16. *Reactor upper internals. X750 split pins: IGSCC 360° in the shank/flange radius (Cattant (2014), courtesy of EDF)*

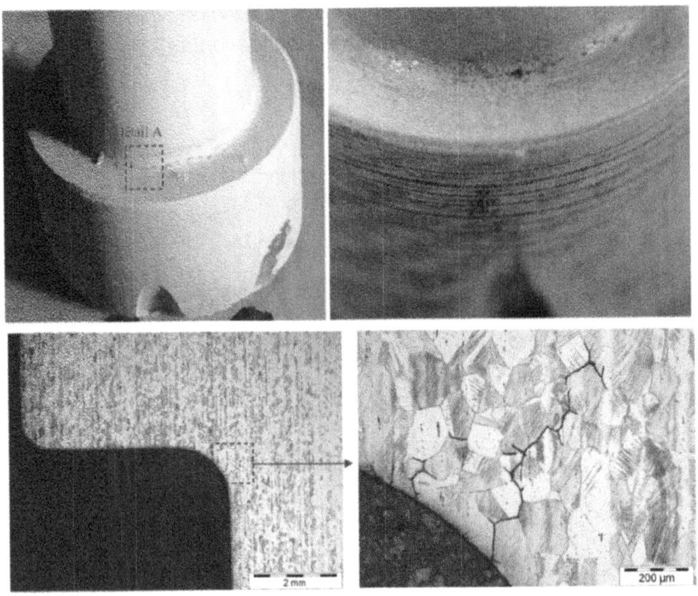

Figure 1.17. *RCP hydrostatic bearing bolt (CW 316). IGSCC in the head/shank radius (Cattant (2014), courtesy of EDF). For a color version of this figure, see www.iste.co.uk/balbaud/nuclearenergytoday.zip*

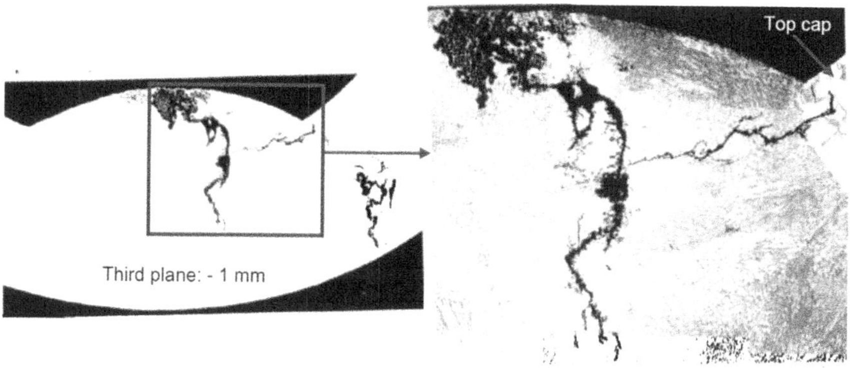

Figure 1.18. *Upper omega seal (304L). TGSCC in polluted environment (chlorides + oxygen) (Cattant (2014), courtesy of EDF). For a color version of this figure, see www.iste.co.uk/balbaud/nuclearenergytoday.zip*

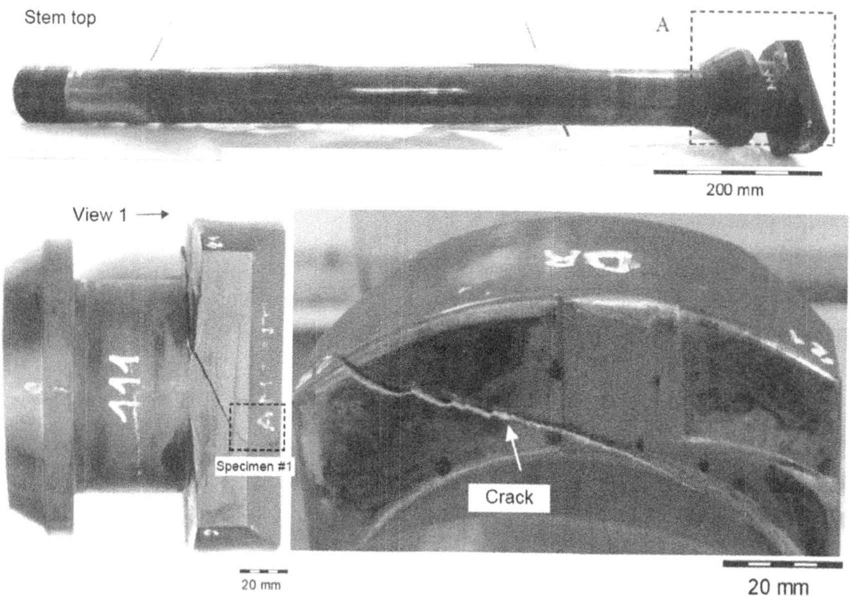

Figure 1.19. *IGSCC of a 17-4PH mean steam valve stem (Cattant (2014), courtesy of EDF). For a color version of this figure, see www.iste.co.uk/balbaud/nuclearenergytoday.zip*

1.6.6. *Flow-assisted corrosion (Mansour and Trévin 2014)*

FAC is another type of general corrosion which is accelerated by a water flow. FAC occurs when the following conditions are met:

– The component is made of carbon steel or very LAS.

– The water flow is single phase or de-aerated dual phase (oxygen content < 5 ppb), with a neutral or slightly alkaline pH, the flow rate being high (flow rate > 1.5 m/s).

– The temperature is in the following range: 100°C < temperature < 300°C.

FAC is a chemical-based phenomenon; it is not a mechanical process such as erosion, cavitation or abrasion, which all end up with a loss of thickness as well.

FAC can be roughly described by Fe dissolution in flowing water. FAC is a four-step process:

– oxidation of Fe into either ferrous ions (Fe^{2+}) or magnetite (Fe_3O_4) at the metal-oxide internal interface;

– diffusion of soluble species (ferrous iron) through the porous oxide layer;

– dissolution–reduction of the magnetite at the oxide–water external interface;

– transport of iron soluble species to the water flow.

FAC occurs only if the metal surface has a magnetite layer. The release of soluble Fe into the water is controlled by the Fe concentration gradient in the diffusion-limiting layer at the oxide/water interface (Figure 1.20).

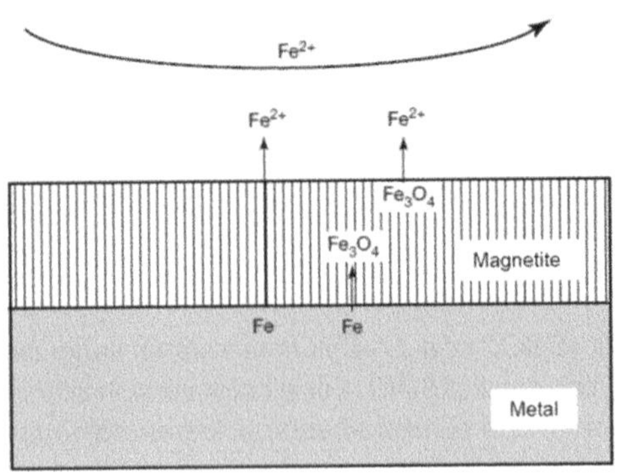

Figure 1.20. *Mechanism of FAC*

FAC is a steady state for which the oxide thickness does not change over time. It corresponds to equilibrium between the formation of magnetite at the metal-oxide interface and an oxide reduction at the oxide–water interface.

FAC has no real incubation or initiation period. However, FAC occurs after a short period of time, a period during which the required chemical conditions and thermal hydraulics are established. For example, after a startup following an outage, hematite may be present at the surface. Once the reducing conditions are back again, FAC occurs only when all the hematite has turned back to magnetite.

1.6.7. *Other main possible mechanisms (Cattant 2014)*

1.6.7.1. *Wear*

When two metallic pieces are in contact, the surface roughness, as the result of machining, interacts. When these two pieces move, the two surfaces scratch each other. If the relevant mechanical conditions are met, small debris is torn away from one surface or the other (sometimes from both, but this is not frequent). The result is a loss of material, also called wear.

Wear is surface state dependent. But the environment can also play a role by forming surface layers and modifying the surface state.

The contact conditions also play a role: sliding objects will wear differently than impacting objects. Various wear modes are possible: the type of contact determines the mode of wear. The two modes of wear are known as two-body and three-body wear. Two-body wear occurs when grit or hard particles remove material from the opposite surface. Three-body wear occurs when the particles are not constrained and are free to roll and slide down a surface.

– Adhesive wear: if the contact interface between two surfaces has enough adhesive bonding strength to resist relative sliding, large plastic deformation is introduced in the near surface reciprocal contact surfaces under compression and shearing. As a result of such large deformation in the contact region, a crack is initiated and propagates thanks to the combined fracture mode of tensile and shearing. When the crack reaches the contact interface, a wear particle is formed and adhesive transfer is completed. This type of wear damage, which occurs when there is enough adhesive bonding at the contact interface, is called adhesive wear.

– Abrasive wear: if the contact interface between two surfaces has interlocking of an inclined or curved contact, ploughing takes place in sliding. As a result of ploughing, a certain volume of surface material is removed and an abrasive groove is formed on the softer surface. This type of wear is called abrasive wear. Here, we assume a single contact point model where a hard, sharp abrasive is indented against the flat surface and forms a groove on it by ploughing. When the worn material has a ductile behavior, a ribbonlike, long wear particle is generated by the mechanism of micro-cutting. In the case of brittle material, however, a wear particle is generated by a crack propagation. In abrasive wear, the respective values of hardness between the two materials in contact is paramount.

– Fatigue wear (see also fretting corrosion below): repeated cycles of contact are not necessary in adhesive and abrasive wear for the generation of wear particles. There are other cases of wear where a certain number of repeated contacts are essential for the generation of wear particles. Wear generated after such contact cycles is called fatigue wear. When the number of contact cycles is high, the high-cycle fatigue mechanism is expected to be the wear mechanism. When it is low, the low-cycle fatigue mechanism is expected.

– Corrosion wear: when sliding takes place, especially in corrosive liquids or gases, reaction products are formed on the surface mainly by chemical or electrochemical interactions. If these reaction products adhere strongly to the surface and behave like the bulk material, the wear mechanism should be almost the same as that of the bulk material. In many cases, however, such reaction products behave very differently from the bulk material. Therefore, wear is quite different from that of the bulk material and is dominated by the reaction products formed by the interaction of solid materials with the corrosive environment. Thribo-chemical wear accelerated by corrosive media is called corrosive wear. In corrosive wear, tribo-chemical reaction produces a reaction layer on the surface. At the same time, such a layer is removed by friction. Therefore, relative growth rate and removal rate determine the wear rate of the reaction layers and, as a result, of the bulk material.

1.6.7.2. *Fatigue*

Fatigue is a well-known degradation mechanism. Cyclic loading may affect the lifetime of components. Cumulative straining induces slip bands on the surface and crack initiation. Fatigue is a consequence of mechanical and thermal loading, but environmental effects can also play an important part.

Under mechanical or thermal cyclic loads, material microstructure evolves and exhibits cracks after a certain length of time called initiation (characterized by well known and codified Wohler curves). These cracks then propagate through the material powered by the next cyclic loads.

Depending on the level of loading, fatigue has two domains:

– high-cycle fatigue, where the strain is typically elastic at a macro scale;

– low-cycle fatigue, where macroscopic plastic strain occurs at every cycle.

Rupture typically occurs after a few dozen to a few thousand cycles for low-cycle fatigue, but can last several million or even several hundred million cycles for high-cycle fatigue.

Apart from the material origin and its environment, the main parameters which control fatigue life are as follows:

– the maximum and minimum loads applied every cycle;

– the existence of stress raisers, because of the component particular geometry or due to the presence of defects.

From a practical standpoint, loading cycles are usually characterized by their amplitude, their mean value, the ratio between the minimum and the maximum loads and sometimes by their frequency.

Fatigue life is generally split into two periods: an initiation phase and a cracks propagation phase. The duration of each phase depends on the fatigue type: low or high cycle.

Regarding initiation, dimensioning methods are based on experience, which means some rules of thumb are applied. Regarding propagation, up to date fracture mechanics is used for NPP design and fatigue life assessment.

Fatigue can be sorted into four groups. However, mechanical fatigue is the common driver for all of these various types of fatigue:

– mechanical fatigue;

– thermal fatigue;

– environmentally assisted fatigue (EAF);

– fretting corrosion.

Fatigue is an important cause of incidents in PWRs, including mechanical loadings due to vibrations (most frequent type of occurrence ~70%); for instance, high-cycle vibration fatigue of socket welds (Figure 1.21).

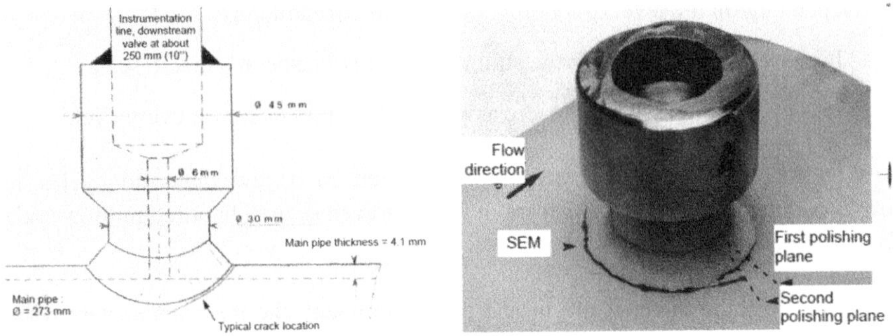

Figure 1.21. *Low-pressure safety injection system (LPSIS) pump: instrumentation nozzle: illustration of high-cycle fatigue vibration of socket weld (Cattant (2014), courtesy of EDF). For a color version of this figure, see www.iste.co.uk/balbaud/nuclearenergytoday.zip*

Thermal fatigue is considered a potential problem in the following cases: when thermal expansion is impeded, it can get stresses in materials; for instance, a thermal barrier or flow nozzle labyrinth from the reactor coolant pump (Figure 1.22). Biaxial surface stress solicitations often give very specific bidimensional multiple cracking observed for thermal fatigue.

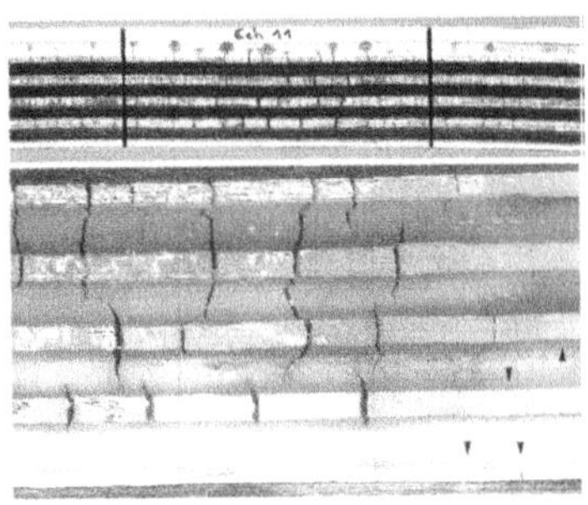

Figure 1.22. *RCP flow nozzle labyrinth; PT thermal fatigue indication (Cattant (2014), courtesy of EDF). For a color version of this figure, see www.iste.co.uk/balbaud/nuclearenergytoday.zip*

Thermal fatigue is generally due to one or several of three phenomena: mixing zones, vortex penetration and thermal stratification (this can appear for any pipe size, depending on the thickness of the pipe (Figure 1.23):

– thermal fatigue in hot and cold water mixing areas. For instance, fatigue cracks occurred in the mixing areas of high (180°C) and low (30°C) temperature fluids of the reactor heat removal (RHR) system;

– thermal striping in isolable or unisolable piping sections;

– stratification (inside surge line and steam generator feedwater lines);

– EAF: when frequency or crack propagation is low, environment can play an important role in fatigue initiation and propagation, for instance, in reactor auxiliary systems (Figure 1.24). Fatigue corrosion can occur for materials susceptible to SCC for the same environmental conditions, but fatigue solicitations can extend the susceptibility domain of cracking;

– fretting corrosion (see also fatigue wear above): repeated small displacements (>10^6) under PWR conditions can lead to wear; for instance, control rod drive mechanism (CRDM) gripper latch arms (Figure 1.25).

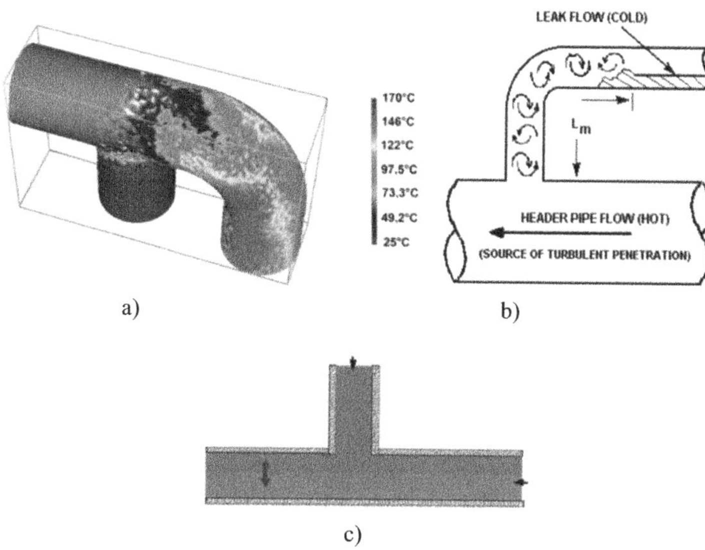

Figure 1.23. *Illustrations of different thermal fatigue phenomena: (a) mixing zones, (b) vortex penetration and (c) thermal stratification (courtesy of EDF). For a color version of this figure, see www.iste.co.uk/balbaud/nuclearenergytoday.zip*

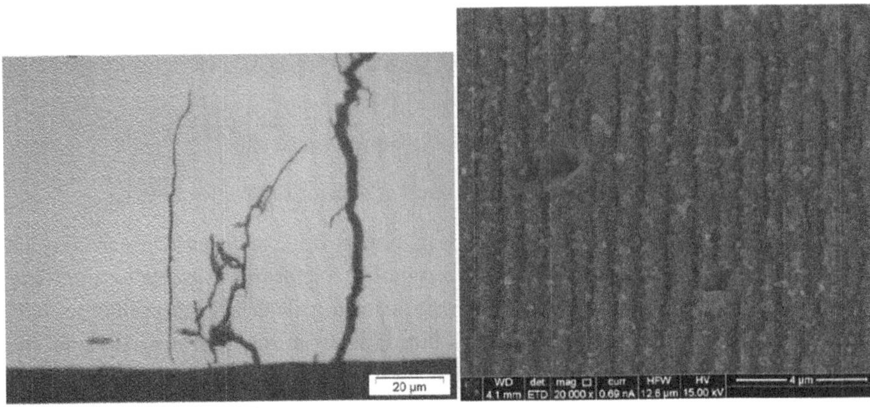

Figure 1.24. *Bifurcation and striation of environmentally assisted fatigue in AISI 304L ($d\varepsilon/dt = 4 \times 10^{-3}\ s^{-1}$, $d\varepsilon/2 = 2.5 \times 10^{-3}\ s^{-1}$) (courtesy of EDF/ENSMA Poitiers). For a color version of this figure, see www.iste.co.uk/balbaud/nuclearenergytoday.zip*

Figure 1.25. *CRDM gripper latch arms. Left: profile of a non-bearing face of the tooth of the latch arm (stationary gripper). Right: profile of the bearing face of the tooth of the latch arm (moveable gripper) (courtesy of EDF)*

1.7. Main material challenges and solutions under evaluation (Lefrançois 2010; Zinkle and Was 2013)

1.7.1. *Precautions to take during manufacturing to ensure the homogeneity of the through wall property of huge components*

Advances in steelmaking techniques have been very beneficial for the development of low alloy ferritic steels:

– They decrease the level of harmful elements, such as hydrogen responsible for cold tapping and cracking; sulfur responsible for an anisotropy of characteristics and a lowering of resilience, in particular at the ductile level; phosphorus, tin and antimony acting on resistance to irradiation and reversible tempering embrittlement.

– They reduce the level of inclusions harmful to ductility and isotropy, for example, in the direction of thickness on sheets.

– They improve analytical accuracy and cast-to-cast dispersion.

Normal processing is conducted in an electric furnace with highly selected scrap metals, with drastic precautions on the copper content, which cannot be eliminated during processing. Ladle metallurgy (refining in heating ladle), vacuum degassing, aluminum calming and vacuum casting, used alone or successively, allow the chemical composition and quality level to be regulated. The casting and cooling conditions of the ingots were also carefully examined. The heterogeneity of large ingots, the phenomena of macro- and micro-segregation, and the distribution of inclusions were studied on ingots exceeding 100 t, so as to limit their harmfulness. The hydrogen content will be reduced to less than 0.0001% (1 ppm), oxygen to 0.004% (40 ppm), sulfur to 0.001% and phosphorus to at least 0.01%.

The improvement of the methods of making LASs by refining in vacuum heated ladles has made it possible to take a step further in lowering the sulfur and phosphorus content in low alloy ferritic steels used for nuclear reactor vessels. Contents as low as 0.001% become available in industrial production. It is then necessary to pay attention to the harmful side effects that could result from these low levels, and we realize that it is the uniformity of sulfur and phosphorus to the nearest 0.001%, more than the minimum value reached, that makes the quality of the elaboration.

An alternative method of manufacturing industrial ingots was used to make the shells of reactor pressure vessels and improve the inner surface quality of these walls, which were then coated with austenitic stainless steel, and thus reduce the risk of cracks under coating. In addition, a quasi-isotropy of the mechanical properties is obtained on the finished part. This method consists of casting cylindrical hollow ingots that are simultaneously inner/outercooled, so that both faces have a quality similar to what is intended on both sides of a thick sheet. This hollow ingot technique has been used in particular for reactor pressure vessels of the latest generations.

1.7.2. *Precautions to take during manufacturing to prevent or mitigate in-service thermal aging, irradiation embrittlement, fatigue and stress corrosion cracking*

This paragraph presents the precautions taken during manufacturing to prevent or mitigate in-service irradiation embrittlement, thermal aging, fatigue, stress corrosion cracking and FAC.

In fine, the design of the new LWR reactors takes advantage of the significant advances derived from R&D performed on mechanisms of materials aging, field experience and improvements gained during the repair or replacement of components of units in operation. This brings confidence to ultimately face the in-service aging of materials for the full lifetime of 60 years.

1.7.2.1. *Irradiation embrittlement*

RPV

Embrittling elements (copper, phosphorus and possibly nickel, which may have a synergistic effect) have been strictly limited in equipment specification for a long time.

For the EPR reactor, because of the large pressure vessel diameter, the neutron fluence for the RPV is reduced to low values. The choice of a heavy reflector for RPV internals improves the situation.

Chemical limitations on copper and phosphorus contents in the steels have been made tighter. Thus, the ductile-brittle transition temperature RT_{NDT} expected after 60 years of operation remains below 30°C.

Internals

For the EPR, a heavy reflector made of stainless steel was selected for the lower internals that reduces neutron "leaks" and the neutron fluence for the reactor pressure vessel. A design based on the use of forged blocks, kept in place by tie rods kept away from the core, limits the risks of exposing bolts or welds to high neutron fluxes.

1.7.2.2. *Thermal aging*

More severe specifications on residual elements contents, replacing cast components by forged ones and avoiding steel grades susceptible to thermal aging.

For some materials, diffusion and precipitation of some elements occur at operation temperatures. The "thermal aging" induces a loss of ductility and a decrease in toughness. The two most significant instances are phosphorus precipitation at grain boundaries in ferritic steels and chromium demixtion in the ferritic phase of duplex austenitic-ferritic stainless steels.

Low alloy steels

Low alloy steels selected for the pressure boundary of heavy components (RPV, steam generators and pressurizers) may be affected by thermal aging due to phosphorus segregation at grain boundaries that takes places under some circumstances. Aggravating factors are embrittling residual element content (P, Sn, Sb, As), a higher temperature and coarse grains in HAZs. To minimize this last effect, improvements in manufacturing (welding) processes have been introduced now for quite a few years.

For the EPR reactor, specifications on residual elements content have been made even more severe, which was made possible by the significant progresses made in the steelmaking processes.

Cast austenitic-ferritic stainless steels

Cast austenitic-ferritic stainless steels were mostly used for primary piping and the primary pump casing. These steels have a ferrite content ranging from about 10% to 25% to avoid solidification cracking and obtain higher tensile properties.

Over time at service temperature, hardening of the ferrite phase takes place, inducing embrilttlement. Cleavage can then be observed, even at elevated temperatures (300°C).

Substantial research programs were undertaken in the 1980s to get a better understanding of the phenomenon and develop formulae for predicting fracture toughness of aged materials.

Significant changes took place in the past few decades. First, internal cleanliness of steels (inclusion content) has greatly improved, and most of all, molybdenum containing grades are now forbidden for operation temperatures higher than 250°C.

For the EPR reactor, the molybdenum-free grade (X3CrMo20-09M), that is less susceptible to aging, has been selected as the only material for the pump casing. The fracture toughness values that can now be guaranteed for this material, after extensive experimental programs, are now deemed satisfactory.

The most significant change is that cast austenitic-ferritic steels are no longer used for primary pipe branches, now replaced by forged austenitic grades, not susceptible to thermal aging.

Martensitic stainless steels

Martensitic stainless steels are mostly used for internals, bolting and operating stems in valves. The consequences of thermal aging have been observed in service, mostly valve stems ruptures. In some instances, this phenomenon may occur very soon, sometimes after as little as two years in service, for higher temperatures (350°C).

Therefore, martensitic stainless steels containing more than 13% Cr have been forbidden in the construction of some components exposed to loads at low temperature, but generally operating over 250°C in practice.

For applications where higher mechanical properties are needed, a new grade was developed, which also benefits the EPR reactor.

1.7.2.3. *Fatigue*

The prevention of fatigue crack initiation is obtained by using good design practices (avoidance of sharp transition geometries, etc.), fatigue analyses performed for stress concentration zones, considering all possible loading situations, and proper use of in-service inspection.

Thus, fatigue cases met in service are, as a general rule, not related to transient loading of the reactor, but are mostly due to unexpected phenomena, such as pipe vibrations, thermal oscillation in mixing zones when two fluids at different temperatures mix, stratification and changing fluid levels.

Lessons learnt from field experience have been taken into account for the design of the EPR reactor. Among other things, further requirements for the surface condition (roughness) of specific zones have been introduced to improve fatigue behavior. New methods have also been developed for a better understanding of phenomena occurring in mixing zones. They are now fully understood and controlled. Fatigue monitoring tools can also be used to identify thermal cycling and its cause.

1.7.2.4. *Stress corrosion cracking*

Stress corrosion cracking may occur in service, with various features, when applied loads or residual stresses are high:

– strain induced corrosion of C-Mn steels or LASs;

– chloride corrosion, with oxygen, for austenitic stainless steels;

– intergranular corrosion of nickel base alloys and austenitic stainless steels in caustic media;

– intergranular stress corrosion of nickel base alloys (alloy 600) in pure water at elevated temperature.

This type of damage was first observed for steam generator tubes in transition zones, then for pressurizer instrumentation nozzles and reactor pressure vessel control rod drive mechanism nozzles.

After 15 years of accelerated tests in several laboratories worldwide, including those of CEA, Electricité de France (EDF) and AREVA, it was shown that new alloys developed, mainly alloy 690, as well as alloys 52 and 152 for welds, are not susceptible to stress corrosion cracking during operation in the primary water.

Sometimes, zinc is also added to the RCS either for PWSCC mitigation or dose reduction, or both.

For the EPR reactor, alloy 690, and alloys 52 and 152 for weld, have been selected for the steam generator tube bundle, pressure vessel penetrations and internal supports.

Strict requirements for chemical compositions and heat treatments ensure a good reproducibility of properties and behavior, close to the possible optimum.

1.7.2.5. *Flow-assisted corrosion*

For corrosion and FAC control, the secondary water $pH_{25°C}$ is adjusted between 9.1 and 9.8 (depending on the condenser tubes material) using ammonia or various amines such as morpholine or ethanolamine.

Increasing Cr content (beyond 0.5 wt%) improves FAC resistance.

1.7.2.6. *Future prospects*

Furthermore, substantial research programs are still in progress worldwide to get a better understanding of these phenomena and develop formulae for predicting in-service behavior.

1.7.3. *New possible fabrications under evaluation: hot isostatic pressing and additive manufacturing*

Recent developments in metallurgy could radically change the way metal components are designed, manufactured and repaired. On the one hand, hot isostatic pressing (HIP) could be a good alternative to the manufacture of large historically forged or molded components; on the other hand, additive manufacturing is developing rapidly.

The application of these manufacturing processes is currently underway in the aeronautical, automotive and medical fields. In the energy domain, additive manufacturing would make it possible to innovate in terms of maintenance of fleets in operation, and would allow an optimization of the availability of means of production, the reduction of the storage cost of spare parts and enlarged repair solutions.

The nuclear fleet needs a wide range of spare parts to ensure the maintenance of facilities. Due to that and in order to optimize maintenance costs, but also in a context of obsolescence of the original equipment, it is necessary to explore alternatives and, if necessary, develop and qualify new manufacturing processes. The operator must justify the maintenance of the mechanical components and the qualification of the equipment for longer periods ("Grand Carénage") in a more stringent regulatory context (ESPN order, reduction of the cobalt inventory, etc.). The field of nuclear materials is often subject to many safety constraints that result in some evolutions for codification, qualification and non-destructive examinations. The performance of the components must be more robust to face these more stringent requirements, but in some cases conventional manufacturing methods no longer provide answers to these constraints. In this context, a recently mature industrial technology such as HIP could make it possible to implement more reliable and less risky manufacturing of large components, for which previous methods (forging or molding) would appear less and less susceptible to comply with new regulations.

As additive manufacturing is in the development phase, its use in the nuclear field can now be considered initially for components that do not have a preponderant role in safety. In addition, some additive manufacturing processes could provide innovative ways to repair damaged surfaces under optimized conditions compared to more conventional hard-facing solutions.

1.8. References

Barbu, A. and Massoud, J.P. (2008). Comportement des matériaux dans le cœur des REP. *Techniques de l'Ingénieur*, BN3760.

Boiron, P. (2011). Réacteurs à eau ordinaire bouillante. *Techniques de l'Ingénieur*, BN3130.

Bouniol, P. (2001). Bétons spéciaux de protection. *Techniques de l'Ingénieur*, BN3740.

Buisine, D. and Perrat, G. (2016). Soudage des matériaux nucléaires à eau pressurisée. *Techniques de l'Ingénieur*, BN3732.

Cattant, F. (2014). *Materials Aging in Light-Water Reactors: Handbook of Destructive Assays*. Editions Lavoisier, Cachan.

Champigny, F. (2005). Inspection en service des REP. *Techniques de l'Ingénieur*, BN3305.

Combrade, P. and Vaillant, F. (2014). Corrosion des alliages de nickel et des aciers inoxydables en milieux REP pollués ou confinés. *Techniques de l'Ingénieur*, BN3756.

Destre, D. and Izard, J.P. (2015). Construction des centrales REP. Équipements primaires. *Techniques de l'Ingénieur*, BN3270.

Gras, J.M. (2017). Matériaux du nucléaire. Revue et introduction. *Techniques de l'Ingénieur*, BN3679.

Hutin, J.P. (2017a). Maintenance des centrales nucléaires. Dispositions générales, les hommes, les organisations. *Techniques de l'Ingénieur*, BN3295.

Hutin, J.P. (2017b). Maintenance des centrales nucléaires. Composants de la chaudière. *Techniques de l'Ingénieur*, BN3296.

Kaercher, M. (2002). Installation générale des réacteurs à eau sous pression. *Techniques de l'Ingénieur,* BN3260.

Lefrançois, A. (2010). Aging of materials during plant operation: Preventive measures in the design of the EPR™ reactor. In *Conference Fontevraud 7*, O01-A163T00.

Lemaignan, C. (2004). *Science des matériaux pour le nucléaire*. EDP Sciences, Les Ulis.

Lemaignan, C. (2010). Matériaux pour le nucléaire. *Techniques de l'Ingénieur*, BN1280.

Mansour, C. and Trévin, S. (2014). Corrosion-érosion des aciers ferritiques dans les centrales nucléaires. *Techniques de l'Ingénieur*, BN3751.

Société Française d'Energie Nucléaire (2008). Evolution du choix des matériaux dans les REP : petites et grandes histoires. In *Section Technique no. 2: "Science et technologie des matériaux, CND, chimie", Journée Technique*, November 25.

Vieillard-Baron, B. and Meyzaud, Y. (1998). Structures des réacteurs nucléaires. Aciers spéciaux. *Techniques de l'Ingénieur*, BN3730.

Zinkle, S.J. and Was, G.S. (2013). Materials challenges in nuclear energy. *Acta Materialia*, 61, 735–758.

2

Materials and Processes for Nuclear Fuel Reprocessing

Nathalie GRUET[1], Benoit GWINNER[1], Pierre LAGHOUTARIS[1], Beatriz PUGA[1], Pierre FAUVET[1], Hervé ANTONY[2], Valentin ROHR[2], Eric CAVALETTI[3], Pierre CHAMBRETTE[3] and Laurent JUNOD[4]

[1] CEA, Service de recherche en Corrosion et Comportement des Matériaux, Université Paris-Saclay Gif-sur-Yvette, France

[2] Orano Recyclage, Etablissement de la Hague, La Hague, France

[3] Orano Recyclage, Chatillon, France

[4] Orano Projets, Cherbourg-en-Cotentin, France

2.1. Introduction and general description of systems

2.1.1. *Description of the process and the main functions*

The PUREX process used in the two spent fuel reprocessing-recycling plants located in La Hague (UP2-800 and UP3 facilities) involves a number of different steps, which are summarized in Bourgeois (2000) and (Monographie CEA 2008) and are illustrated in Figure 2.1. The first step involved in the spent fuel reprocessing-recycling process is to shear the fuel assemblies, dissolve the uranium and plutonium oxide pellets in nitric acid, and thus separate the structural components of the fuel assemblies from the latter. The dissolution solution treatment is based on the chemical process of liquid–liquid extraction, which is used to separate fission products (FP) and transuranic products from uranium and plutonium, before purifying the recyclable components (uranium and plutonium) so they can be used to manufacture new recycled fuels. Before the vitrification step, the FP in aqueous nitric acid are concentrated in a fission product

Materials and Processes for Nuclear Energy Today and in the Future, coordinated by Fanny BALBAUD-CÉLÉRIER and Céline CABET.

evaporator-concentrator to reduce their volume for intermediate storage before vitrification, and also to recover the nitric acid so it can be recycled in the process.

The construction of the UP3 (line A) and UP2 800 plants was authorized via the French Ministerial Decree of May 12, 1981. The main fuel treatment and waste conditioning workshops (Figure 2.1) were commissioned between 1986 for the first fuel storage pool, and 2001 for the last R4 plutonium reprocessing workshop in the UP2-800 facility.

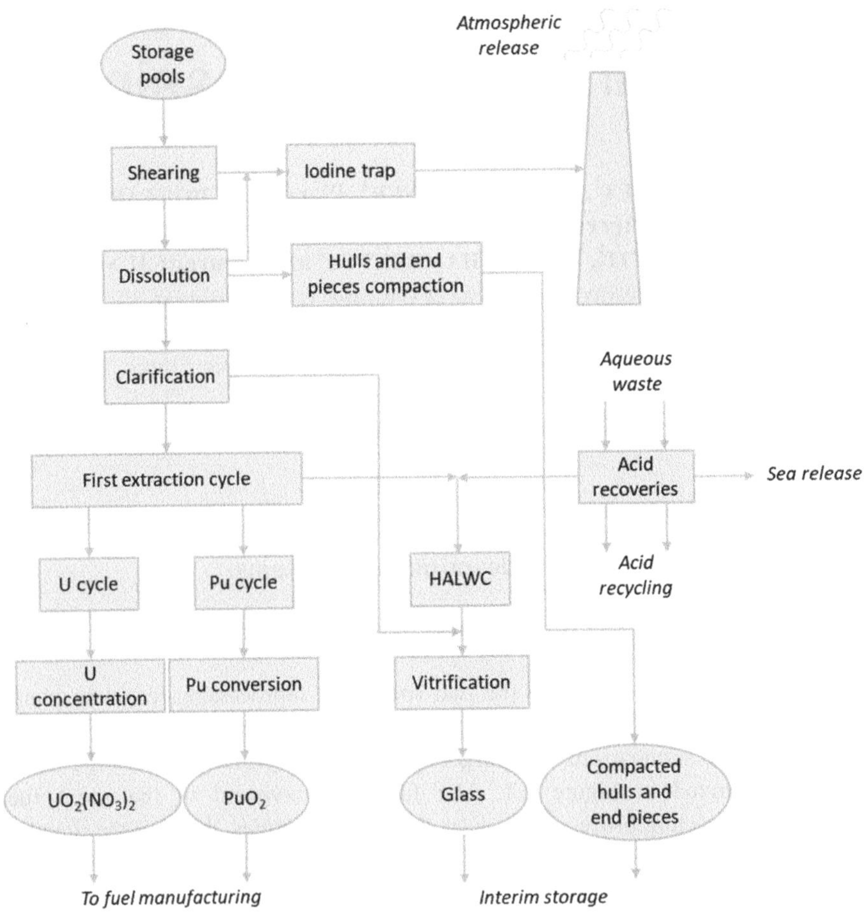

Figure 2.1. *Representation of the main steps of nuclear fuel reprocessing (high activity level waste concentration [HALWC]). For a color version of this figure, see www.iste.co.uk/balbaud/nuclearenergytoday.zip*

These two facilities have unquestionably met their objectives linked to expected lifetime and availability after 30 years of operation. They processed more than 34,000 tons of spent fuel from light water reactors (PWR and BWR) and conditioned about 24,300 standard containers for vitrified high activity level liquid waste and 19,480 standard containers for end pieces and hulls compacted waste, all by the end of 2021.

Numerous equipment (mechanically constructed or made by metal work) are required to carry out the operations mentioned in Figure 2.1, thus ensuring a recycling capacity of 3–4 tons of heavy metals per day. The lifetime control of such equipment permits the fulfillment of the requirements of nuclear safety, facility availability and financial risk management. Although the design of mechanical equipment allows for regular maintenance that can be performed remotely, metal work welded equipment, chemical contactors and vessels, which contribute to the first containment barrier, must be designed for the same lifetime as the facility itself. Their design must also be appropriate for a suitable aging monitoring.

2.1.2. *Description of aging conditions (temperature, nitric acid concentration, etc.)*

A broad range of corrosive environments can be found in the equipment and pipes at the La Hague plants. Other than the Pu conversion and vitrification operations, practically all the corrosive environments in these plants involve nitric acid solutions. The parameters describing the corrosive nature of the environment and thus governing the choice of materials are as follows:

– type of environment: liquid, condensates;

– nitric acid and nitrate concentrations. These concentrations can reach up to 13.6 $mol.L^{-1}$;

– type of minor elements:

- halides (F^-, Cl^-, I^-),

- catalyst and oxidizing ions (Np^{VI}, Pu^{VI}, Fe^{III}, Cr^{VI}, platinum group metals, etc.),

- other ions (sulfates, phosphates, etc.);

– temperature of the environment and the material. Temperatures up to the environment's boiling point can exist, while material surface temperatures can be significantly higher than the environment's boiling point in areas heated by the heat flux from equipment, for example, double-walled heaters of evaporators or tubes of boilers;

– redox potential;

– flow velocity of the fluid;

– renewal rate;

– any deposits;

– solid particles (dissolution fines, suspensions of solids, etc.).

2.2. Requirements for materials

2.2.1. *Functional requirements associated with the process: mechanical resistance, corrosion resistance and erosion resistance*

The design assumptions for La Hague plants corresponded to a lifetime of 20–30 years, depending on the equipment involved. Some safety justifications depend on the estimated lifetime of equipment considered at the design of the facilities. The following aging phenomena have therefore been taken into consideration when designing metal work welded equipment:

– effect of corrosion, mainly in a nitric acid environment, on the thermomechanical resistance of equipment (pressure, fatigue and/or fatigue-creep resistance of equipment subject to thermal or pressure cycles related to the process, seismic compliance, etc.);

– effect of corrosion on the geometry of sub-critical equipment.

Design operations mainly consist of defining the additional corrosion allowance thicknesses to be included in the requirements that result from the thermomechanical and/or criticality safety design; these additional corrosion allowances will be "consumed" by general and uniform corrosion or by general and intergranular corrosion (IGC) mechanisms. In both cases, the expected corrosion rates must be compatible with the equipment's lifespan. Local corrosion mechanisms due to pitting and crevices must be avoided by managing the chemical composition of the solutions, the geometry of equipment, etc. These demonstrations of the absence of risks of localized corrosion require considerable in-depth R&D knowledge and experience feedback.

2.2.2. *Regulatory context and industrial strategy*

Several factors explain the establishment of the equipment monitoring approach implemented onsite at La Hague:

– the goal of lifetime extension up to 2040 for the facilities in the UP2-800 and UP3 plants now requires the careful management of equipment operability levels;

– the safety reviews of the different licensed nuclear facilities (INB for "Installations Nucléaires de Base" in French) on the site, the first having been held in 2007. This requires establishing the conformity of equipment from a nuclear safety perspective of 10 years lifetime extension;

– nuclear pressure equipment regulations (ESPN for "Equipements Sous Pression Nucléaires" in French) applicable to most of the biggest equipment of the recycling process (evaporator, dissolver, desorber, etc.), equipped with pressurized heating and cooling compartments welded to the body of the vessel containing the radioactive solutions (Articles R.557-12-1 to R. 557-12-9 of the French Environment Code, Ministerial Order of December 30, 2015 amended on nuclear pressure equipment and safety accessories intended for their protection, Article 5.1 from the Ministerial Order of February 7, 2012 amended on pressure equipment specifically designed for licensed nuclear facilities).

Within this context, an approach was defined in the early 2000s based on an analysis of the risks for industrial security, nuclear safety and operations. Taking inspiration from the business risk management concept and considering aging and obsolescence phenomena, this risk analysis was enriched by experts, especially in technology and equipment, who helped define the priorities and methods for integrating the subjects identified (impact, proximity of risk, etc.). As a priority, the subjects were handled according to their level of risk, leading to the development of technological bricks and methods that could sometimes take several years of research and development before being ready (e.g. remote non-destructive tests and welding in inaccessible areas, robotics and steam jet ejector repairs).

The regulatory context surrounding the safety reviews was progressively organized within the Orano Group, in line with the following milestones:

– The French law on nuclear safety and transparency (TSN for "Transparence et à la Sécurité en matière Nucléaire" in French), dated June 13, 2006, stipulates the requirement to carry out a safety review every 10 years in every INB.

– The INB Ministerial Order of February 7, 2012 defines the notion of protection-important components (EIP for "Eléments Importants pour la Protection" in French), together with their defined requirements and qualification principle.

This principle has been defined to last over time: *Provisions for their design, construction, testing, inspection and maintenance enable this qualification to be maintained for as long as necessary.*

– As of August 1, 2018, the Orano Group began reviewing its directive that defines the continuous "facility conformity verification" process. This process is designed to control facility conformity at any period of its lifetime by making sure that any changes to the facility or to the operating conditions, particularly those due to modifications or aging, do not question the facility's initial conformity.

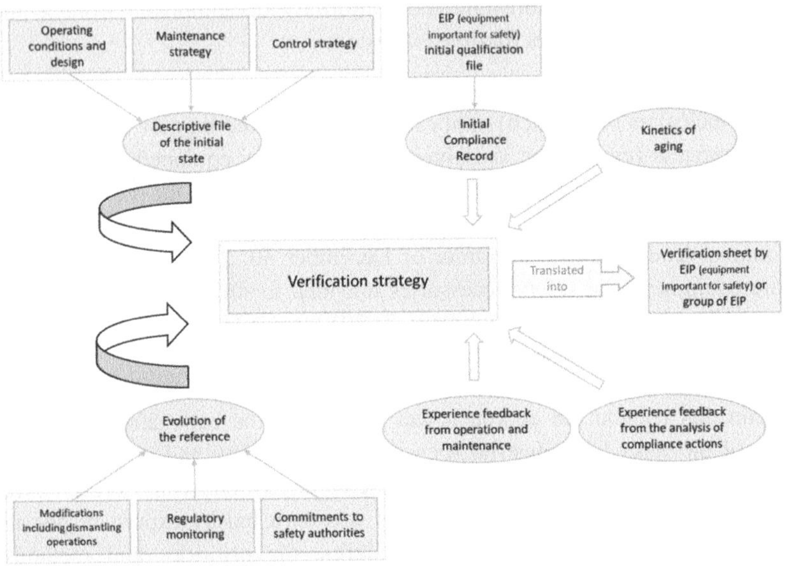

Figure 2.2. *Strategy for checking equipment conformity. For a color version of this figure, see www.iste.co.uk/balbaud/nuclearenergytoday.zip*

The purpose of the conformity and aging studies (ECV for "Etudes de Conformité et Vieillissement" in French) is to demonstrate the conformity of equipment important for nuclear safety (EIP) with respect to the nuclear safety requirements by identifying the operational measures in place, achieving in situ investigations, and studying to control their aging on the basis of design data and identifying degradation mechanisms. The capacity of these equipment to remain operational for the next 10 years must be justified (justification of "fitness for purpose").

The methodology used involves carrying out studies of conformity and of aging on a selection of EIP test specimens per licensed nuclear facility and per subject. These subjects are as follows: containment, instrumentation and control, electricity, radiation protection monitoring, nuclear process monitoring, civil engineering, heat transfers, radiolysis, handling and ventilation. Following these analyses, monitoring and action plans can be recommended for the EIP test specimens or extended to cover all of the EIPs. The first safety reviews conducted in the UP3 and UP2 800 plants led to the definition of actions plans, including more than 4,000 actions, some of which are still underway.

The in situ investigations – needed to support these action plans – were difficult to implement owing to the location of fabricated equipment in hot cells made inaccessible by design (e.g. hot cell providing radiological shielding and contributing to the dynamic containment). These recurring investigations required defining, developing and qualifying non-destructive testing (NDT) techniques. New techniques were therefore recommended in addition to TV inspections and thickness measurements using ultrasounds with longitudinal waves: phased-array ultrasounds, ultrasounds on double-walls, ultrasonic listening devices, guided-wave ultrasonic testing, thermography, etc. New carrier systems were also developed to extend access to the surfaces to be inspected, for example, multi-axis rods/arms, remotely operated robotic systems and drones.

Such developments have allowed us to measure the thickness of equipment wall and internal components (coils, tubes, welds, casing, etc.) to locate and characterize indications on equipment that were completely inaccessible before. These developments also mean that nuclear pressure equipment now have additional inspections that ensure a level of safety equivalent to traditional pressure equipment, despite the important accessibility constraints.

This surveillance has considerably enhanced the knowledge of the corrosion kinetics of the alloys used in the equipment (mainly austenitic stainless steels and zirconium grade 702) over periods of several years, which are inaccessible to laboratory testing.

2.3. Current material choices, fabrication and joining

2.3.1. *Current material choices*

The processing of spent fuel requires the use of nitric acid in varying concentrations and temperatures that can reach boiling point. The environment is composed of a broad range of chemical elements, most of which come from the spent fuel (about 30 elements of significant concentrations, depending on the environments).

The choice of materials varies, depending on the location of the equipment in the reprocessing line and is conditioned by the parameters given in section 2.1.2.

Most of the facilities are made from conventional austenitic stainless steels (ASS) (Table 2.1 and Figure 2.3(a), grades: 1.4306/1.4307, 1.4404) with a corrosion allowance thickness. These ASS generally show a good behavior in a nitric acid environment due to their passivation. In these conditions, corrosion is uniform (called "general") with controlled corrosion rates related to their passive behavior. In the most severe cases (HNO_3 + oxidizing ions), that is, high acidity and temperature levels, these ASS can exit their region of passivity and move into a transpassive behavior region. In this case, they develop IGC with corrosion rates that are sometimes incompatible with the target lifespan of the facility. More resistant materials must therefore be used: ASS with a high chromium content (Table 2.1, 1.4335), ASS enriched with silicon (Table 2.1, 1.4361, Figure 2.3(b)), and zirconium and its alloys (Table 2.2, alloy 702 and Figure 2.3(b)) in the most severe cases (and with no free fluorine).

ASS grade	Max C	Cr	Ni	Si	Max S	Max P	Mo	Max Mn	Max N
1.4306	0.030	18–20	10–12	1.0 max	0.015	0.045		2.0	0.1
1.4307	0.030	17.5–19.5	8.0–10.5	1.0 max	0.015	0.045		2.0	0.1
1.4404	0.030	16.5–18.5	10.0–13.0	1.0 max	0.015	0.045	2.0–2.5	2.0	0.1
1.4335	0.020	24.0–26.0	20.0–22.0	0.25 max	0.010	0.025	0.20 max	2.0	0.10
1.4361	0.015	16.5–18.5	14.0–16.0	3.7–4.5	0.010	0.025	0.20 max	2.0	0.10

Table 2.1. *Mass composition of SS (in wt.%) used in spent fuel reprocessing-recycling plants (Norme NF EN 1088-1 2014)*

Zirconium grade	Zr + Hf	Max Hf	Fe + Cr	Max C	Max H	Max N	Max O
702	99.2	4.5	Max: 0.2	0.05	0.005	0.025	0.16

Table 2.2. *Example of the mass composition (in wt.%) according to standard ASTM B551/B551M-12 (ASTM International 2021) for the zirconium alloy used in spent fuel reprocessing-recycling plants*

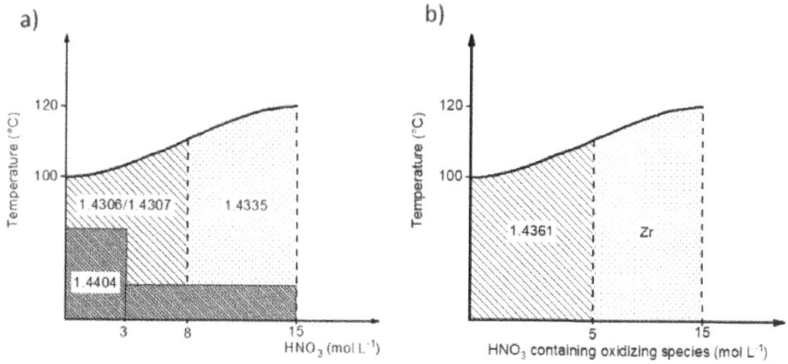

Figure 2.3. *Recommended range of use for metallic materials in nitric acid environments: (a) without oxidizing ions and (b) with oxidizing ions (Fauvet et al. 2008; Fauvet 2012)*

2.3.2. *Fabrication and joining*

2.3.2.1. *Austenitic stainless steels: description of standards*

In addition to the design and thermomechanical calculation rules, the internal fabrication and assembly procedures used to manufacture the ASS equipment and pipes cover all of the following steps:

– procurement of semi-finished products;

– handling and storage;

– traceability and marking;

– grinding and machining;

– forming, bending and straightening;

– heat treatments;

– welding, paying attention to the welding conditions of some ASS grades sensitive to hot cracking and associated controls;

– polishing, surface treatment (cleaning and passivation) and both factory and onsite cleaning.

Stringent requirements are associated with these procedures in terms of the type and scope of destructive and non-destructive tests used, as well as the acceptance criteria. These inspections include systematic tests to check the sensitivity of materials to IGC (on semi-finished products and weldments).

The management and control of these different steps allow us to ensure the respect of the following aspects, up to installation on site:

– conformity of the mechanical properties of semi-finished products with respect to the relevant applicable standard and former calculations;

– good metallurgical quality of semi-finished products (compacity, inclusions, grain size, residual ferrite, etc.);

– cleanliness of surfaces, absence of pollution (e.g. chlorinated or ferritic) and absence of surface defects (e.g. no shortage of material, deep scratches, impacts, etc.);

– acceptable level of cold deformation and quality of hyper-quenched heat treatment where appropriate;

– high level of quality insurance for welded joints;

– no sensitivity to IGC or significant surface oxidation during welding operations.

2.3.2.2. *Zirconium: description of standards*

Zirconium 702 alloys are also governed by a set of design, procurement, fabrication and inspection rules. These rules take into account the specificity of the grade. For instance, the fabrication rules must be adapted for the following operations:

– hot and cold forming (the bend radius and temperature are adapted during deformation operations);

– fusion welding (shielding and quality of the gas cover).

2.4. Basic properties

2.4.1. *Corrosion of austenitic stainless steels*

In a nitric acid environment, the polarization curve of ASS can be illustrated as shown in Figure 2.4 (Pinard-Legry 1989).

Three regions can be observed in the anodic part corresponding to the oxidation of the ASS (mainly Fe, Cr and Ni). Polarization in any one of these regions depends on the more or less oxidizing nature of the environment and thus on the cathodic process.

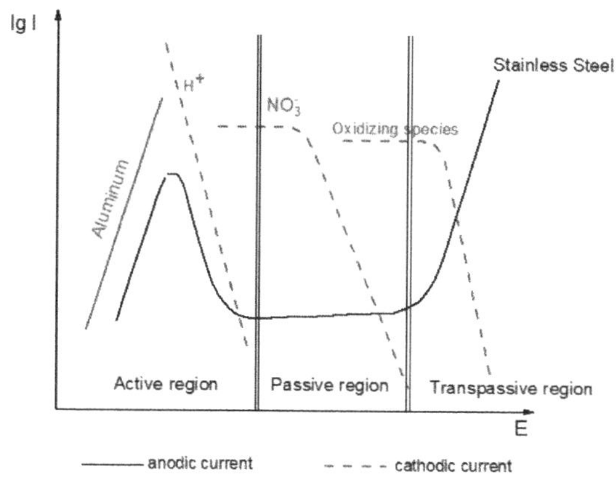

Figure 2.4. *Schematic diagram of the polarization curve for ASS in nitric acid (black). Cathodic current is in red (reduction in protons, nitric acid or oxidizing ions) and anodic current is in green. For a color version of this figure, see www.iste.co.uk/balbaud/nuclearenergytoday.zip*

When the environment is weakly oxidizing or coupled with a less noble material (see section 2.4.1.4), ASS are in the active region where they show a uniform corrosion with a corrosion rate that can be high.

When the conditions are moderately oxidizing (HNO_3 only), the corrosion potential is in the passive region of the ASS characterized by lower and uniform corrosion rate (see section 2.4.1.1). This is the classic field of use for ASS.

When the environment is strongly oxidizing (HNO_3 with other oxidizing species), these ASS are in the transpassive region characterized by corrosion that tends to be intergranular (see section 2.4.1.2).

2.4.1.1. *Uniform corrosion*

All the ASS exhibit uniform corrosion in moderately oxidizing environments (for example, HNO_3 at low concentration and temperature, without oxidizing ions, see Figure 2.3(a)). These are protected by a passive oxide layer that is mainly composed of chromia (Cr_2O_3) for 1.4306/1.4307 ASS grade (Robin et al. 2008; Tcharkhtchi-Gillard et al. 2016). Silicon enrichment in the passive layer is observed on the extreme outer surface of 1.4361 ASS initially enriched with 4% silicon (Robin et al. 2008; Laurent 2017; Balbaud-Célérier et al. 2020), which provides specific behavior in strongly oxidizing medias. Cross-section examination of the ASS surface in Figure 2.5 shows this type of corrosion.

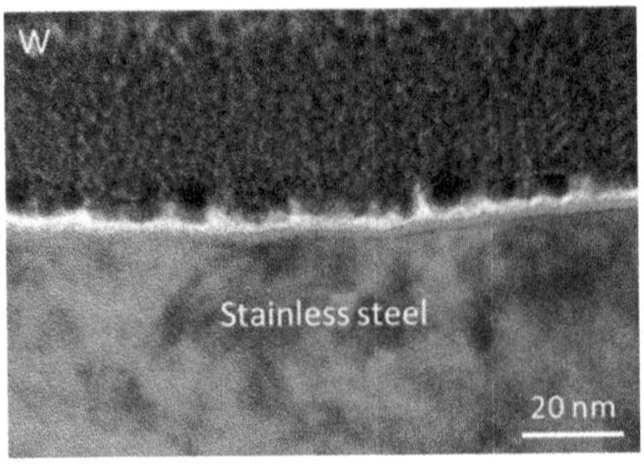

Figure 2.5. *Transmission electron microscopy cross-section micrographic after 24 h of immersion in HNO_3 4 mol.L^{-1} at 100°C (Laurent 2017)*

For this type of corrosion, the mass loss kinetics is linear in stationary conditions. The corrosion rate is then determined by relating the mass loss per unit area to the duration of the test. For instance, this corrosion rate can be expressed in terms of the annual loss in thickness taking into account the material's density.

The uniform corrosion phenomena at the La Hague plants are currently monitored by means of ultrasonic measurements and predictive models. This non-destructive analysis technique requires the shutdown of some facilities and cannot reach some difficult areas in the plant. This is why predictive models are implemented. They are validated by the results of lab-conducted experiments during which the operating conditions were assumed to be stable and predefined, simulating the different physicochemical medias encountered on the plants.

2.4.1.2. *Intergranular corrosion*

This section discusses the IGC of non-sensitized ASS, that is, with no chromium carbides in the grain boundaries, for instance. The solution annealed ASS used for reprocessing equipment contain a maximum carbon content (0.025 wt% for 1.4306/1.4307 and 1.4404 ASS, 0.020 wt% for 1.4335 ASS, and 0.015 wt% for 1.4361 ASS, see section 2.3.1) to prevent the formation of chromium carbides, particularly due to the thermal effect (sensitization) of welding operations (Cunat 2000).

Except for 1.4361 ASS, the IGC of non-sensitized ASS occurs in nitric acid environments when they are sufficiently oxidizing to shift the corrosion potential of the ASS into the transpassive region. The passive film dissolves, due to the oxidation of Cr(III) into Cr(VI), to produce dissolved $Cr_2O_7^{2-}$. This greatly speeds up of the dissolution of ASS. Increased reactivity in the grain boundaries compared with the grains themselves also provokes intergranular grooves. This can be expressed by the existence of two different corrosion rates, that is, V_s for the grain surfaces corrosion rate and V_{gb} for the grain boundary corrosion rate (see Figure 2.6). When these rates are constant over time, a triangular groove will form, with its angle being related to the ratio of the two corrosion rates:

$$\frac{V_{gb}}{V_S} = \frac{1}{sin(\alpha/2)} \quad [2.1]$$

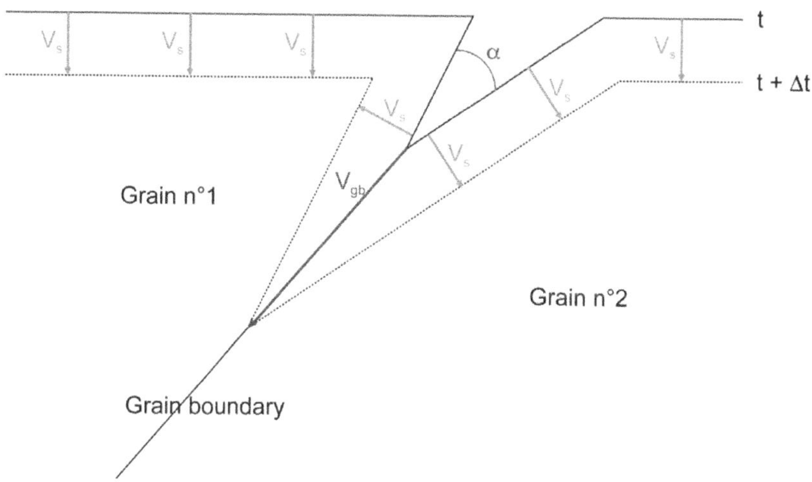

Figure 2.6. *Schematic diagram of a groove between two grains when there are two different corrosion rates (V_s for the grain surfaces corrosion rate and V_{gb} for the grain boundary corrosion rate). For a color version of this figure, see www.iste.co.uk/balbaud/nuclearenergytoday.zip*

This specific reactivity can be associated with a chemical composition (segregation of elements, see Figure 2.7) and/or a specific crystal lattice (coordination numbers, misorientation angle, etc.) in the grain boundary (Beaunier et al. 1994; Emery 2019).

The addition of silicon (>3.5%) to the ASS seems to avoid this specific sensitivity in the grain boundary, as is the case of 1.4361 ASS.

This specific reactivity (i.e. corrosion rate V_{gb}) appears globally homogeneous and continuous for all of the grain boundaries in the material, excluding welds (Emery 2019). For this reason, the growth of these grooves is consistent throughout the material. Once the intergranular grooves completely surround a grain, it breaks away from the material. As a result of this mechanism, there is a variation in the mass loss of the material that exhibits transient behavior before becoming steady state (see Figure 2.8): the mass loss rate first increases from a corrosion rate of V_s (corresponding to the dissolution of material on a plane, groove-free surface) to a mean corrosion rate of V_{gb}/τ, where τ represents the mean tortuosity of the intergranular path. This apparent transient behavior results from the gradual increase of the surface in contact with the solution due to the growth of intergranular grooves (which increases the dissolution of material) and the gradual detachment of the grains. When grain detachment affects the whole surface, the detachment regime reaches a maximum steady-state situation, as does the surface in contact with the solution (grain detachment compensates for the increased surface area related to the growth of grooves). The transition time can be estimated by $T = 3/2\ D/V_{gb}$ (with D representing the mean grain size of the ASS), the time needed to detach the largest grain initially found on the surface of the material.

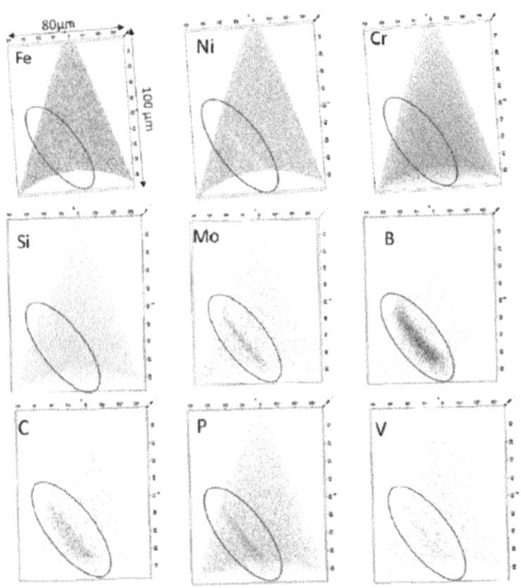

Figure 2.7. *Chemical mapping of a grain boundary (red circle) in a 1.4307 ASS using a tomographic atom probe (TAP) (Emery 2019). For a color version of this figure, see www.iste.co.uk/balbaud/nuclearenergytoday.zip*

More recently, simulation work was carried out to represent this IGC according using a 2D (Gwinner et al. 2016, 2017) or 3D approach (Guiso 2020; Guiso et al. 2020, 2021, 2022) in order to accurately reproduce the behavior observed experimentally.

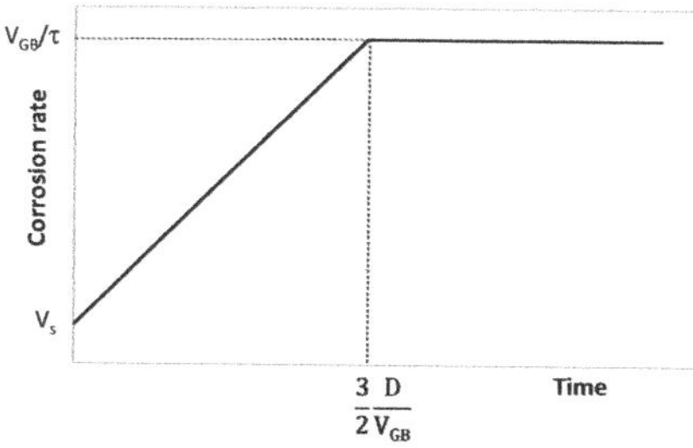

Figure 2.8. *Qualitative change in the corrosion rate measured by mass loss for an ASS subjected to intergranular corrosion*

Even though the model above can be used to predict IGC in the long term (supported by "short-term" tests), equipment monitoring used to measure material thickness with ultrasonic techniques is still effective.

2.4.1.3. *End-grain corrosion*

ASS used in spent nuclear fuel reprocessing-recycling equipment can exhibit different behaviors, depending on the surface exposed to the nitric acid environment. More specifically, end-grain corrosion (EGC) or tunneling corrosion can affect the surface perpendicular to the metal forming direction (rolling, extrusion, forging, etc.) (Kain et al. 1992; Jeng et al. 1993; Kaneko and Abe 1995; Takeuchi and Whillock 2004; Whillock et al. 2005; Chandra et al. 2008; Khan et al. 2012). This can be found in particular in forged rod, wire and tube-like products (Chandra et al. 2008). Consequently, it does not affect the walls of equipment but rather areas where the surface is exposed, such as equipment internals (e.g. machined nozzle on the bubbler pipe/thermocouple thermowell) (Chandra et al. 2008). In nitric acid environments, this type of corrosion appears with IGC (Kaneko and Abe 1995; Khan et al. 2012). It seems to be promoted by the environment itself (presence of oxidizing ions) (Kaneko and Abe 1995; Takeuchi and Whillock 2004; Whillock et al. 2005; Khan et al. 2012).

EGC is characterized by local penetrations in the direction in which the material was thermomechanically formed (Figure 2.9) (Kaneko and Abe 1995; Takeuchi and Whillock 2004; Whillock et al. 2005; Khan et al. 2012). The penetration front in the material seems rather heterogeneous (Takeuchi and Whillock 2004; Whillock et al. 2005) with respect to its depth. The variation kinetics of this EGC have not been determined. It seems to progress at a faster pace than IGC (Takeuchi and Whillock 2004; Whillock et al. 2005).

The origin of this corrosion is not entirely established. The way the ASS was manufactured could play a role. EGC may be related to the existence of chemical heterogeneities aligned in the same direction as the main thermomechanical deformation in the material. These heterogeneities may be (Jeng et al. 1993; Takeuchi and Whillock 2004; Khan et al. 2012):

– inclusions, for example, MnS type (Kain et al. 1992; Chandra et al. 2008);

– segregation of elements such as phosphorus (Kaneko and Abe 1995), chromium (Chandra et al. 2008) or silicon.

The higher reactivity of these heterogeneities with respect to the nitric acid environment could be responsible for this localized corrosion. Then, a lateral attack (such as IGC) causes the tunneling to widen.

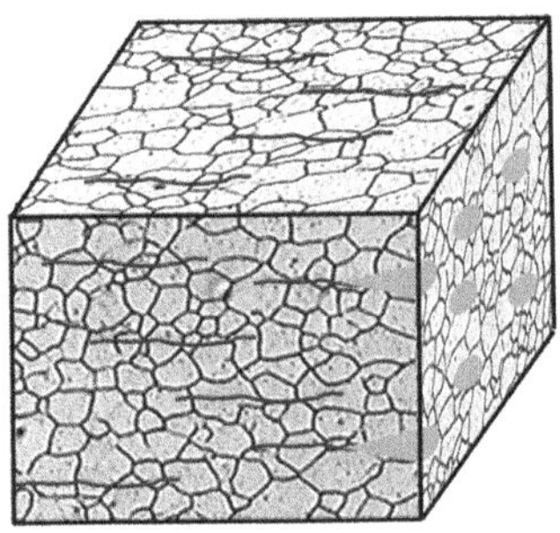

Figure 2.9. *EGC (in orange) and alignment of inclusions (in red). For a color version of this figure, see www.iste.co.uk/balbaud/nuclearenergytoday.zip*

There are technical solutions designed to limit (or even avoid) this type of corrosion:

– The minimized impurity rate, such as P < 0.015% (Kaneko and Abe 1995) makes it possible to limit its segregation in the alloy.

– During the ASS manufacturing process, annealing is used to re-homogenize the ASS from a chemical perspective (Chandra et al. 2008; Khan et al. 2012). The higher the temperature and the longer the material is kept at this temperature, the better the ASS's resistance to tunneling corrosion (Kaneko and Abe 1995). The cooling conditions are equally important (Kaneko and Abe 1995): water cooling is more effective because it is quicker than air cooling.

– For instance, the end product can be treated by re-melting the surface with a laser (Jeng et al. 1993; Chandra et al. 2008), making it possible to re-homogenize the chemical elements, that is, removing any alignment (Jeng et al. 1993). Welding overlay (Chandra et al. 2008) can be done to add a different metallurgical structural layer that is protective.

2.4.1.4. *Corrosion by galvanic coupling*

Within the scope of the spent nuclear fuel reprocessing-recycling process, corrosion by galvanic coupling can occur when there is contact between the equipment's structural materials, such as ASS, and other metallic elements resulting from:

– structural materials of some specific nuclear fuels (graphite, Al);

– fission products from irradiated fuels, such as platinoids (Pd, Rh, Ru, Tc).

The type of galvanic coupling between ASS and another metal depends on the relative position of the corrosion potential for both materials in a nitric acid environment.

In the case of ASS coupling with less noble metal elements, such as unpassivated Al, Ti or Zr and Ta (Brunet et al. 1970; Mori et al. 2003), the ASS potential shifts into its active region. Consequently, the corrosion rates of ASS can increase significantly (around several $mm.year^{-1}$).

In the case of ASS coupling with more noble components than itself, such as graphite, platinoids (Pd, Rh and Ru) and passivated Zircaloy, the corrosion potential shifts into the transpassive region of ASS where IGC can occur (Brunet et al. 1971; Robin et al. 2004; Lange 2012) (see section 2.4.1.2). It has been observed that the effect of coupling platinoids depends on their characteristics. This effect decreases in the following order: Ru > Rh > Pd. This classification can be explained by the cathodic reaction kinetics (reduction of nitric acid), which seems to be slower with Rh and Pd than Ru (Fauvet et al. 2008).

As is the case in other aggressive environments, galvanic corrosion depends on the concentration and temperature of nitric acid, as well as on the ratio between the surface areas of the two coupled materials and the chromium content in the ASS (an ASS with 25% chromium is more resistant than one with 18% chromium).

The risk associated with this type of corrosion by galvanic coupling has been taken into account with respect to the spent fuel reprocessing-recycling facilities. This kind of risk is detected in R&D studies carried out beforehand and is conditioned by the choice of materials.

2.4.2. Corrosion of zirconium

2.4.2.1. Uniform corrosion

The French spent fuel reprocessing-recycling industry uses zirconium grade 702; its specifications are given in Table 2.2. This material exhibits excellent corrosion resistance, regardless of the oxidizing nature of the environment (see section 2.3.1).

From a thermodynamic perspective, zirconium can easily oxidize into species with a valency of 4 (mainly ZrO_2) owing to the low standard potential of the related electrochemical couples (Fauvet 2012). For this reason, zirconium will spontaneously form a nanometric layer of ZrO_2 when it comes into contact with most oxidizing environments. This quasi-insulator layer provides protection over a broad range of potentials as demonstrated by the low passive currents (0.1–10 $\mu A.cm^{-2}$) in the polarization curves in Figure 2.10. The corrosion rate is typically below 1 $\mu m.year^{-1}$, except in the presence of fluorides as zirconium exhibits high sensitivity to these ions (see section 2.5.3). Variations in the passive layer are controlled by a mechanism related to high electric fields in this layer (about 10^6 $V.cm^{-1}$), which causes ions to migrate through the layer. In the case of zirconium, this migration practically only concerns O^{2-} ions from the oxide–solution interface to the metal-oxide interface. Owing to this mechanism, the thickness of the oxide layer can increase slightly as a function of time and potential, which in turn slows down corrosion. A number of models have been developed to take into account this phenomenon, for example, a model by Cabrera, Mott and Verwey (Verwey 1935; Cabrera and Mott 1949; Meisterjahn et al. 1989) or the high-field model (Lohrengel 1993; Gwinner et al. 2015; Benoit 2016).

Good corrosion behavior can end at a certain potential level (known as the depassivation potential, see Figure 2.10). Above this level, the oxidation current densities are several orders of magnitude higher than those measured in the passive region. Contrary to the passive region, the resulting oxide layer is micrometric in size, black and brittle, with a tendency for spalling. The oxidation kinetics are fast and do not decrease over time. The mechanism responsible for the change in the oxidation kinetics is known as dielectric rupture or breakdown (Darwish et al. 1989; Lohrengel 1993; Gwinner et al. 2015; Benoit 2016). It seems to correspond to the transition from an oxidation regime controlled by ionic conduction to a regime controlled by electronic conduction.

With the increased concentration of nitric acid, the corrosion potential of zirconium increases and the depassivation potential tends to decrease. Consequently, the margin with respect to the possibility of depassivation tends to reduce with an increase in the nitric acid concentration. However, this margin remains sufficient to protect zirconium up to the azeotropic concentrations of nitric acid. Additionally, zirconium retains its excellent corrosion resistance in the presence of oxidizing ions and platinoids.

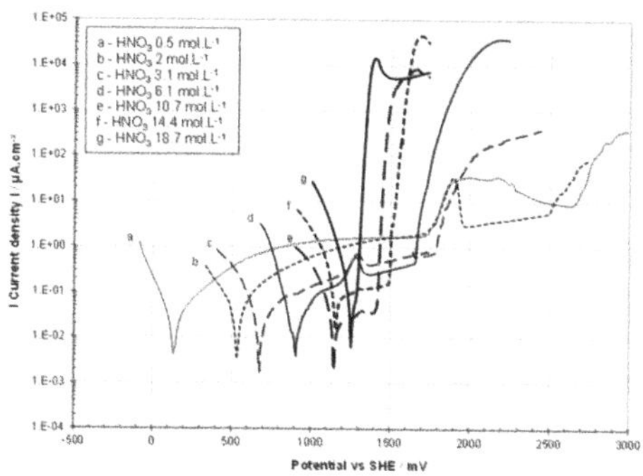

Figure 2.10. *Polarization curves of zirconium 702 in nitric acid at boiling point after 24 h of immersion (v = 0.14 mV.s^{-1})*

2.4.2.2. *Stress corrosion cracking*

The sensitivity of zirconium to stress corrosion cracking (SCC) has been studied in nitric acid environments using constant strain tests, under a constant charge or at a low constant strain rate (Fauvet 2012). Its sensitivity is characterized by reduced maximum elongation (or ductility) and by a brittle break mode due to transgranular fractures.

Its sensitivity seems to be directly linked to the passivity of zirconium (see section 2.4.2.1 and Figure 2.11). The fracture surface is ductile in the passive region; whereas, it becomes brittle with reduced elongation in the transpassive region. Consequently, just like the transpassive potential (see section 2.4.2.1), the sensitivity of zirconium to SCC also depends on the nitric acid concentration: more specifically, the critical threshold potential of SCC reduces when the nitric acid concentration increases.

The sensitivity to SCC also seems to depend on:

– crystal orientation;

– rolling direction, with the cross-direction seemingly more sensitive than the lengthwise direction;

– surface preparation method;

– stress level; in particular, the stress levels for which SCC occurs are higher than those permitted for the initial thermomechanical design of equipment;

– presence of species, such as hydrofluoric acid (HF).

Different assumptions have been proposed to explain the mechanism behind SCC:

– Beavers et al. (1981) suggest that wedging of the resulting oxide plays a role in crack propagation.

– Yau (1988) believes that the mechanism may be local anodic dissolution promoted by stress. He suggests an SCC sensitivity assessment criterion based on the comparison of strain rates, dissolution rates and re-passivation rates.

– Ishijima et al. (2013) suggest a 'tarnish rupture' mechanism based on the initiation and propagation of microcracks under the resulting oxide layer.

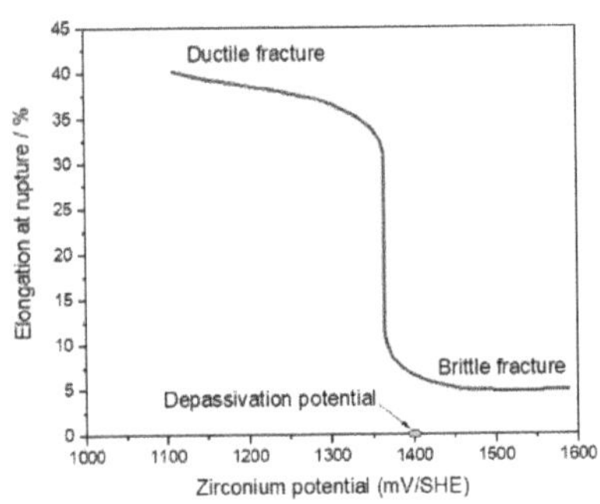

Figure 2.11. *Elongation at rupture of Zr as a function of the potential in constant elongation tests (HNO_3 14.4 mol.L^{-1}, at boiling point). For a color version of this figure, see www.iste.co.uk/balbaud/nuclearenergytoday.zip*

As zirconium remains passive (see section 2.4.2.1) and critical stresses are not reached under spent fuel reprocessing-recycling conditions, zirconium SCC is not at risk under the industrial conditions in which it is currently used.

2.5. Parameters influencing corrosion

The corrosion kinetics of a material is a result of its oxidation kinetics and the reduction kinetics of the environment. This can be affected by different parameters, such as:

– corrosion products (resulting from the material's oxidation or reduction in the environment);

– oxidizing ions (other than nitric acid);

– halides (mainly fluoride and chloride ions);

– temperature and/or heat flux;

– irradiation;

– effect of the metallurgical state, welds in particular.

2.5.1. *Corrosion products*

2.5.1.1. *Impact of products from the reduction of nitric acid*

In a concentrated nitric acid environment, it has been shown that the nitric acid reduction process (Vetter 1950a, 1950b, 1959; Schmid 1959, 1961, 1964, 1969; Schmid and Delfs 1959; Razygraev 1988, 1990; Balbaud-Célérier et al. 2000; Lange et al. 2013; Sicsic et al. 2014; Benoit 2016) can be autocatalytic and involves a charge transfer step for which the electroactive species is not nitric acid (a species with oxidation state III, probably nitrous acid, HNO_2). This is followed by several chemical reactions that allow the electroactive species to regenerate. These reactions may be occurring at the electrode or in solution. Therefore, the main species involved in the reduction of nitric acid on a noble electrode are HNO_2, NO and NO_2, while the charge transfer step is the HNO_2 reduction reaction.

Figure 2.12 illustrates the effect of adding nitrite ions to the 8 mol.L^{-1} nitric acid environment at 100°C on the polarization curves for a 1.4306/1.4307 ASS. A very significant increase in the corrosion potential can be observed. The corrosion potential is therefore located at the bottom of the transpassive region (Schosger 1996).

It follows that the presence of nitric acid reductions products, such as HNO_2, accelerates the reduction kinetics of nitric acid, which in turn would provoke an increase in the corrosion potential (Schosger 1996; Balbaud-Célérier 1998; Whillock et al. 2010) (Figure 2.13). This increase in the corrosion potential can, in certain cases, lead the ASS to shift from its passive region to its transpassive region, resulting in the appearance of IGC (see section 2.4.1.2). For a noble electrode, such as a platinum electrode, these reduction products lead to a decrease in the redox potential.

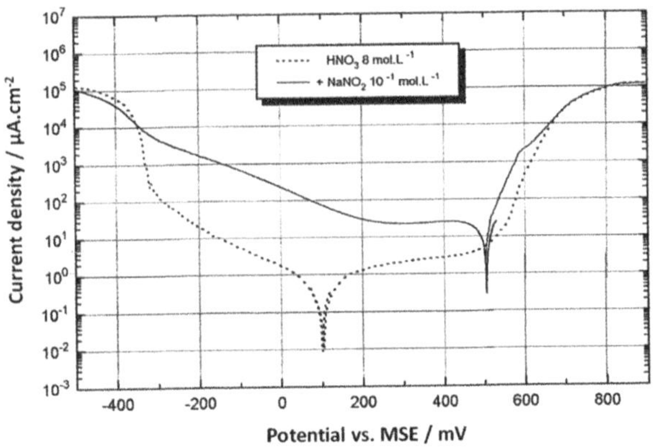

Figure 2.12. *Effect of adding nitrite ions at a 10^{-1} mol.L^{-1} concentration on the polarization curves of a 1.4306/1.4307 ASS in 8 mol.L^{-1} nitric acid at 100°C*

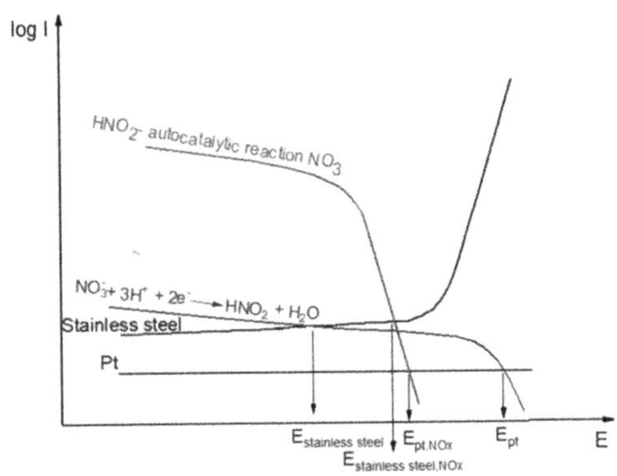

Figure 2.13. *Anodic (black) and cathodic (blue and red) polarization curves. The red cathodic curve represents the effect of NOx bubbling on the corrosion process in nitric acid. Adapted from Whillock et al. (2010). For a color version of this figure, see www.iste.co.uk/balbaud/nuclearenergytoday.zip*

2.5.1.2. *Effect of austenitic stainless steel corrosion products*

The effects on the reduction of nitric acid can be observed in the presence of ASS corrosion products (Fe(III), Cr(III) and Ni(II)). The effect is even more pronounced in the presence of Fe(III) ions. This is explained by their catalytic effect on the reduction of nitric acid on the ASS (Figure 2.14; Schosger 1996). In the same way, the fact that the corrosion potential shifts into the transpassive region following the increase in the reduction rate of nitric acid can provoke the severe corrosion of the material (Whillock et al. 2010).

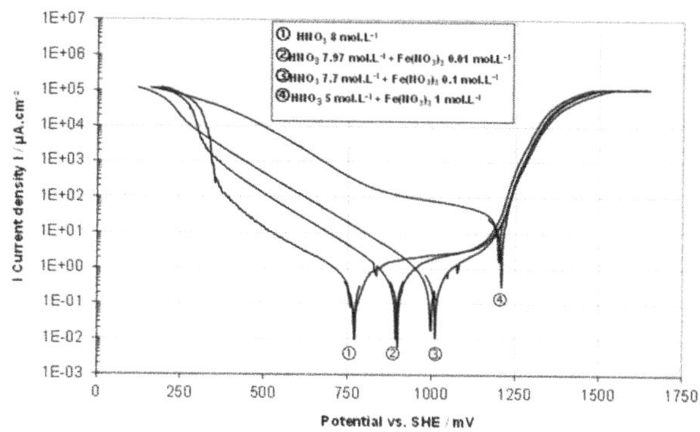

Figure 2.14. *Effect of adding ferric iron on the polarization curves of a 1.4306/1.4307 ASS in 8 mol.L^{-1} nitric acid at 100°C, (v = 0.04 mV.s^{-1}) (Schosger 1996)*

The increase in the reduction rate of nitric acid – which can shift the material from its passive region into its transpassive region in the absence of any oxidizing species, due to the autocatalytic reaction of the reduction – can occur in confined areas where there is little renewal, so the products resulting from the material's oxidation and the solution's reduction can accumulate.

2.5.2. *Oxidizing ions*

The spent nuclear fuel dissolution environment is very complex and contains numerous species capable of being more oxidizing than the nitric acid environment (Table 2.3).

Studies have investigated the behavior of ASS in nitric acid environments containing oxidizing ions relevant to spent fuel reprocessing-recycling processes: Np (VI) (Motooka and Kiuchi 2002; Andreoletti et al. 2006) and Pu (VI) (Ozawa et al. 1985; Okubo et al. 1987; Takeda 1996). They confirm the aggressive nature of these ions depending on the standard potentials given in Table 2.3: IGC (section 2.4.1.2) can occur in 1.4306/1.4307 ASS grades with only several tens of mg.L^{-1} of Np (VI) and several g.L^{-1} of Pu (VI) in nitric acid at boiling point (the orders of magnitude vary according to the temperature and concentration of nitric acid).

Figure 2.15 shows two corrosion mechanisms that can be expected to occur with the presence of oxidizing ions in a nitric acid environment.

Element	Origin	Redox couple	Redox reaction	Standard potential $E^0_{25°C}$ (mV/SHE)
Nitrogen	Nitric acid	N(V)/N(III)	$NO_3^- + 3H^+ + 2e^- = HNO_2 + H_2O$	934 (Weast 1988–1989)
Palladium	Fission product	Pd(II)/Pd(0)	$Pd^{2+} + 2e^- = Pd$	951 (Weast 1988–1989)
Plutonium	Fuel actinide	Pu(IV)/Pu(III) Pu(VI)/Pu(IV)	$Pu^{4+} + e^- = Pu^{3+}$ $PuO_2^{2+} + 4H^+ + 2e^- = Pu^{4+} + 2H_2O$	982 (Pascal 1962) 1043 (Pascal 1962)
Vanadium	Pu and Np simulant for corrosion	V(V)/V(IV)	$VO_2^+ + 2H^+ + e^- = VO^{2+} + H_2O$	1004 (Weast 1988–1989)
Neptunium	Fuel actinide	Np(VI)/Np(V)	$NpO_2^{2+} + e^- = NpO_2^+$	1150 (Pascal 1962)
Chromium	Corrosion	Cr(VI)/Cr(III)	$Cr_2O_7^{2-} + 14H^+ + 6e^- = 2Cr^{3+} + 7H_2O$	1232 (Weast 1988–1989)
Ruthenium	Fission product	Ru(VII)/Ru(IV) Ru(VIII)/Ru(VII)	$RuO_4^- + 4H^+ + 3e^- = RuO_2 + 2H_2O$ $HRuO_5^- + H^+ + e^- = RuO_4^- + H_2O$	1533 (Nagano 1988a) 1660 (Nagano 1988a)
Cerium	Fission product/ Deconta. reagent	Ce(IV)/Ce(III)	$Ce^{4+} + e^- = Ce^{3+}$	1610 (Weast 1988–1989)
Silver	Reagent (electrogene-rated)	Ag(II)/Ag(I)	$Ag^{2+} + e^- = Ag^+$	1980 (Weast 1988–1989)

Table 2.3. *List of oxidizing elements and ions likely to be found in the spent fuel reprocessing process, as well as some chemical analogues used in R&D, including redox couples and standard potentials*

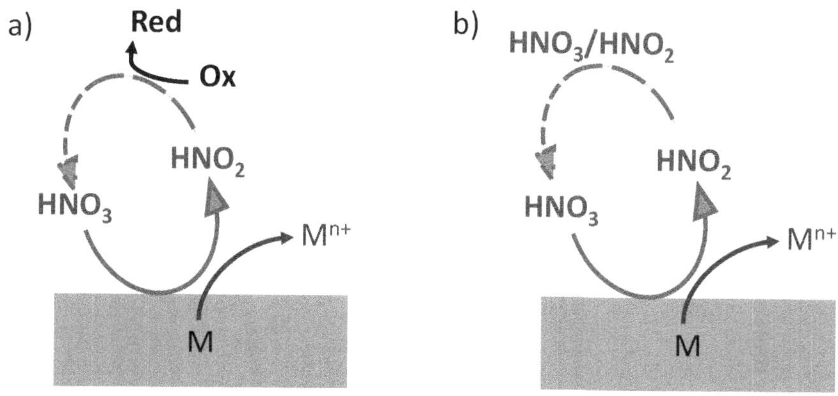

Figure 2.15. *Corrosion process likely to occur with an ASS in a nitric acid (HNO_3) environment containing oxidizing ions (Ox). For a color version of this figure, see www.iste.co.uk/balbaud/nuclearenergytoday.zip*

In the first case (Figure 2.15(a)), nitric acid controls the corrosion mechanism and reduces on the surface of the ASS. The oxidizing species is assumed to be "inert" in solution (with respect to corrosion); however, nitrous acid resulting from the reduction of nitric acid can react with the oxidizing species in the solution to reduce it. In the second case (Figure 2.15(b)), the oxidizing species reduces on the ASS surface and causes the oxidation of the ASS. Nitric acid can then only play the role of electrolyte or re-oxidize the reduced species and maintain the corrosion process.

A thermodynamic approach can be used to determine the oxidizing ability of couples in solution, but kinetics determines the corrosiveness of the environment. This is the case for concentrated nitric acid solutions at boiling point in which nitrous acid desorbs, promoting the re-oxidation of reduced species. In such conditions, all the ions listed in Table 2.3 – except Ag(I) and Ce(III) – can be oxidized by nitric acid despite their standard apparent potentials higher than the standard potential of the N(V)/N(III) couple.

Inversely, numerous oxidizing species can be reduced in the presence of nitrous acid and the corrosion of ASS can decrease in their presence (Coriou et al. 1961; Okubo et al. 1987; Sasahira et al. 1991).

A study carried out on a 1.4306/1.4307 ASS with several oxidizing ions showed that the corrosion processes were only controlled by the cathodic reaction (Coriou et al. 1961), with the anodic curve not being impacted by the presence of these ions (Figure 2.16).

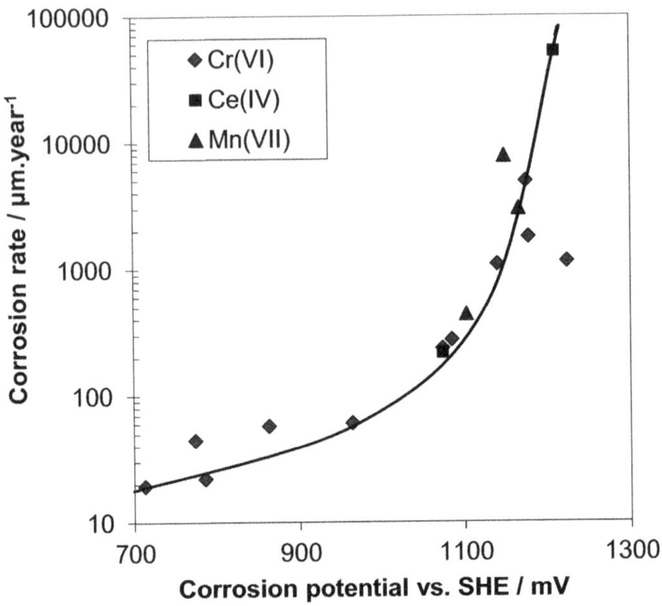

Figure 2.16. *Relationship between the corrosion rate and the corrosion potential of 1.4306/1.4307 ASS in the presence of different oxidizing ions in 5 mol.L^{-1} HNO_3 at boiling point (Fauvet et al. 2008; Fauvet 2012). For a color version of this figure, see www.iste.co.uk/balbaud/nuclearenergytoday.zip*

From an R&D perspective, handling elements such as Np(VI) and Pu(VI) comes with some very restrictive chemical and radiochemical risks.

Simulants such as V(V) can be used to study the effect of these elements on corrosion. The choice of this oxidizing ion can be explained by its molecular structure and/or the standard potentials of the redox couples close to the ones of the redox couples that involve plutonium and neptunium.

2.5.3. *Halides (chlorides and fluorides)*

2.5.3.1. *Austenitic stainless steels in the presence of chlorides*

The stable species of chlorine is Cl^- in a concentrated nitric acid environment.

Chloride ions have an effect on the corrosion of ASS in nitric acid. The main effects are as follows:

– The corrosion rate increases with the chloride ion concentration (Mirolyubov et al. 1966).

– The corrosion morphology of ASS in a nitric acid environment containing chloride ions is classified as "general corrosion" (Petit et al. 1991; Kolman et al. 1997).

– The corrosion potential decreases when the chloride ion content increases until it reaches the active region (Mirolyubov et al. 1966; Kolman et al. 1997).

– The repassivation of ASS in an environment containing chloride ions is possible when there is little or no renewal (Kolman et al. 1997).

– All the parameters promoting the reduction kinetics of nitric acid will enable the possible repassivation of ASS: increase in the S/V ratio, decrease in the solution agitation, production of corrosion and reduction products (Fe(III), HNO_2, NO_x) (Zhuravlev et al. 1973; Kolman et al. 1997).

– The adsorption of chloride ions on the surface will slow down the reduction kinetics of nitric acid (Mirolyubov et al. 1966; Rozenfeld 1968; Tomashov et al. 1982; Kolman et al. 1997).

2.5.3.2. *Austenitic stainless steels in the presence of fluorides*

The sensitivity of ASS to fluoride ions in a HNO_3/HF environment is less pronounced than for zirconium. ASS undergo depassivation of their oxide layers in this oxidizing/complexing environment.

For 1.4306/1.4307, 1.4335 and 1.4361 ASS in 7 $mol.L^{-1}$ HNO_3 at boiling point (Figure 2.17), the corrosion rates seem to increase considerably from an HF content of about 20 $mg.L^{-1}$. Above this content, the effect of HF on the corrosion rate is less pronounced for 1.4361 ASS, which has a rather high corrosion rate nonetheless.

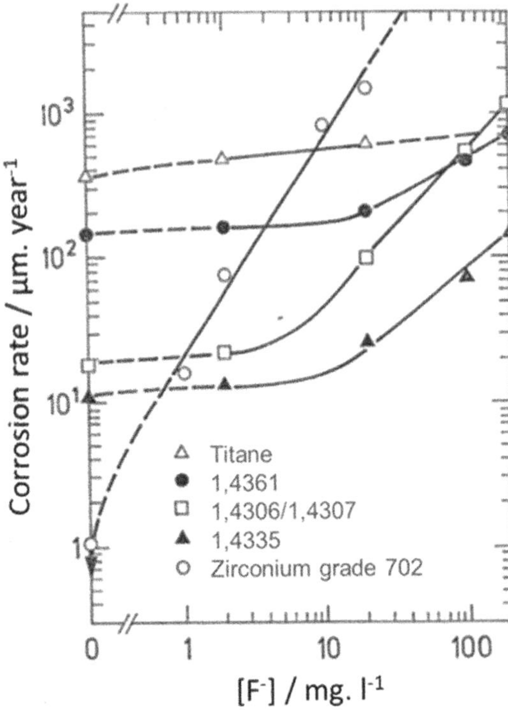

Figure 2.17. *Corrosion rate as a function of the fluoride concentration for 1.4306/1.4307, 1.4335 and 1.4361 ASS grades and zirconium in 7 mol.L^{-1} HNO_3 at boiling point*

2.5.3.3. *Zirconium in the presence of fluorides*

Zirconium is extremely sensitive in a HNO_3/HF environment (Figure 2.17). Corrosion can be described as an oxidation–complexation reaction, in which zirconium oxidizes with a valency of 4 from an anodic viewpoint. It forms complexes with fluorine in solution in the form of ZrF_n^{4-n} (where $1 < n < 6$). From a cathodic viewpoint, the products can result from the reduction of nitric acid (e.g. NO_2, N_2O) or possibly from the proton (H_2).

In these conditions, corrosion is generally uniform. The corrosion kinetics are a function of the quantity of free fluorine, that is, uncomplexed, in solution. In an environment with no renewal, the degradation kinetics decrease (the free fluorine concentration decreases) over time. As corrosion is fast, it can be at least partially limited by transport in the solution, which makes corrosion dependent on the hydrodynamic conditions.

A method that helps limit zirconium corrosion by fluorine involves adding Al or Zr ions in solution, so the fluorine complexes in the solution reduce the concentration of free fluorine. When these ions are added in sufficient quantity, fluorine is rendered practically inert with respect to zirconium.

The general behavior described below can be preceded by a slow corrosion transient related to the presence of an initial protective layer of zirconium oxide (ZrO_2) (Gwinner et al. 2015). This transient can be more or less short depending on the thickness of this layer. The thickness of this layer slowly decreases until it becomes deteriorated.

The chronology of the different processes is illustrated in Figure 2.18.

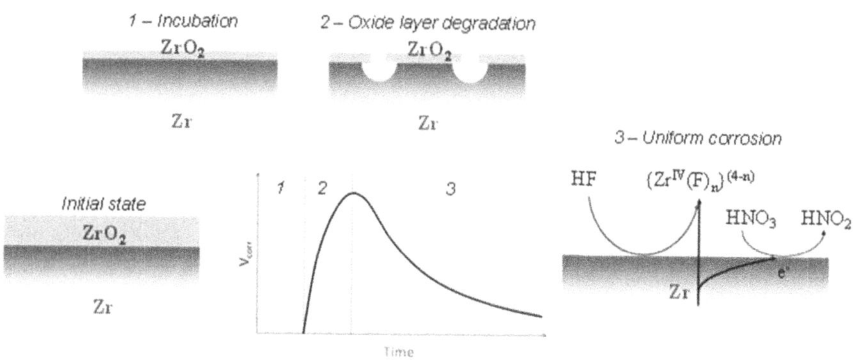

Figure 2.18. *Corrosion of zirconium in a hydrofluoric acid environment. For a color version of this figure, see www.iste.co.uk/balbaud/nuclearenergytoday.zip*

2.5.4. *Temperature and heat flux*

Certain steps of the PUREX process are carried out at high temperature. To do this, the equipment is subject to a heat flux (≤ 10 W.cm^2) that heats the surfaces of the structural material up to temperatures generally exceeding the solution's boiling point temperature. The effects of temperature and heat flux on corrosion depend on the environments and materials in question. For ASS facilities, the temperature seems to be detrimental to the resistance of these materials.

Concerning the thermodynamics of the chemical solution, the increase in the temperature tends to increase the redox potential of the nitric acid medium (Kato et al. 2003a, 2003b). Inversely, the redox potential of most couples involving oxidizing ions tends to decrease when the temperature increases (Kajimura et al. 1987; Nagano 1988b). Consequently, the temperature rise rather tends to stabilize the oxidizing ions in the nitric acid environment. With the temperature rise, the ASS corrosion potential can shift toward its transpassive region and IGC can occur (except for 1.4361 ASS).

Concerning the kinetics, the temperature rise tends to activate the different steps of the corrosion process (Andon et al. 1986; Pinard-Legry et al. 1989; Horn and Schoeller 1990; Kiuchi 1992): reactions in solution, reduction reactions at the interface (nitric acid environment and/or oxidizing ions), and oxidation/dissolution reactions at the interface. For instance, the increase in the equilibrium reaction rates in solution implying the formation of HNO_2 and the decomposition of HNO_2 into NO and NO_2 at boiling point (Kato et al. 2003a, 2003b) can promote autocatalytic reduction mechanisms in a nitric acid environment.

The dissolution rates are governed by an Arrhenius law with a broad region temperature region (from ambient to 120°C). Table 2.4 (Horn and Schoeller 1990) provides the activation energy levels obtained for the three main ASS used in the plant in azeotropic nitric acid. These values are in relatively good agreement with those calculated according to the study (Pinard-Legry et al. 1989) using other concentrations.

ASS grade	Activation energy ($kJ.mol^{-1}$)
1.4306/1.4307	77.2
1.4335	84.8
1.4361	63

Table 2.4. *Activation energy of the dissolution rate of 1.4306/1.4307, 1.4335 and 1.4361 ASS in 14.4 mol.L^{-1} HNO_3 (from ambient temperature to 120°C)*

2.5.5. *Irradiation*

In the case of ASS used in the spent fuel reprocessing-recycling plant, irradiation can have an effect on the anodic (minor effect) or the cathodic behavior (major effect). Some authors demonstrate a slight increase in the passive current under irradiation (Nagai et al. 1998). This is indicative of slightly less effective passivity under irradiation. As for the reduction reactions, irradiation provokes radiolysis of the nitric acid environment during which radicals are formed either from nitric acid (e.g. HNO_2, NO_x and H_2) or from water (e.g. H_2O_2, H_2 and H_3O^+) (Tachimori 1991; Nagai et al. 1998; Whillock et al. 2010). The resulting quantity depends on the dose rate, the dose and the type of irradiation (Nagai et al. 1998; Precek et al. 2010; Garaix et al. 2015; Mincher et al. 2016). The production of HNO_2 or NO_x has antagonist consequences on ASS corrosion, depending on the type of solution. In the case of nitric acid alone, the presence of HNO_2 or NO_x will catalyze the reduction process of nitric acid. This leads to an increase in the ASS corrosion rate (Yamamoto et al. 1996; Nagai et al. 1998). This corrosion rate rises with the dose rate (Nagai et al. 1998). In the case of nitric acid containing oxidizing ions, the presence of HNO_2 or NO_x generally has the effect of reducing these oxidizing species (Tachimori 1991; Yamamoto et al. 1998; Precek et al. 2010; Mincher et al. 2016). As the concentration of oxidizing ions reduces, the corrosion rate also reduces (Yamamoto et al. 1998).

The excellent behavior of zirconium in nitric acid is not modified by irradiation (Whillock et al. 2010).

In conclusion, even if irradiation is classified as one of the relevant parameters, the monitoring of corrosion in industrial plants has not revealed any major role of irradiation in the corrosion of ASS and zirconium.

2.5.6. *Influence of the metallurgical state (welding)*

The elaboration and qualification of welds are important steps. Whether joining the same material or different materials, welds must guarantee good mechanical properties and excellent corrosion resistance. Different processes are used, depending on the materials to be welded so as to avoid the formation of intermetallic compounds, precipitates, hardening phases, Laves phases, etc. (Dey et al. 2009; Shankar et al. 2009; Mengyu et al. 2015; Singh et al. 2018).

Generally speaking, three separate zones can be identified during a welding assembly operation, that is, the fusion zone, the heat-affected zone and the base metal. In the case of ASS, microstructural heterogeneities and composition changes can occur in the fusion and heat-affected zones owing to the temperature–time cycles during the welding operation. These temperature–time cycles can provoke the precipitation of chromium carbides ($M_{23}C_6$) at the grain boundaries (sensitization phenomenon), the formation of secondary phases ($\gamma \rightarrow \delta$ and σ transformation during cooling) and cracks/porosities (Pinard-Legry 1989; Osoba et al. 2017;

Shit et al. 2019). Sensitization – a serious problem during high-temperature processes such as welding – can render ASS sensitive to IGC. Nevertheless, this corrosion morphology can be sidestepped by modifying the ASS composition by reducing the carbon content ([C] < 0.025 wt%) and/or adding stabilizers such as Nb (Pinard-Legry 1989). In a hot, concentrated nitric acid environment, for example, the heat-affected zone of ASS with low carbon contents exhibits no specific corrosion compared with the base metal.

The presence of secondary phases such as delta ferrite (δ ferrite) can have a penalizing effect on the corrosion behavior of welds in concentrated nitric acid. The formation of galvanic cells – due to the difference between the composition (and structure) of secondary phases and austenitic phases – promotes the preferential attack of the delta ferrite phase (Ravi Shankar et al. 2012; Nandakumar et al. 2018). The industrial measures implemented have made it possible to produce welds with corrosion-resistant lifespans equivalent to those of the base materials (Ogawa 1993; Suresh et al. 2017).

As for zirconium and its alloys, the welds show excellent corrosion resistance in nitric acid environments thanks to the formation of a stable, protective layer of ZrO_2 oxide (Shankar et al. 2007). However, dissimilar welding by melting Zr with any kind of ASS proves impossible due to the formation of brittle intermetallic compounds ($ZrFe_2$). Other types of assembly processes such as diffusion brazing or explosion welding offer alternative solutions to avoid/minimize the formation of intermetallic compounds and thus increase the corrosion resistance for heterogeneous junctions (Michel 2021).

2.6. Operating experience: some examples

2.6.1. *Dissolver wheel*

Since the 1990s, the dissolution step in the UP2-800 and UP3 plants has been performed in the wheel dissolvers with the continuous extraction of solutions (Figure 2.19). From a corrosion perspective, the continuous extraction of the dissolution solution allows the maintenance of the chemical balance, leading to more nitrous acid in solution and consequently fewer ions with their oxidizing valencies in solution and a reduction of the redox potential of the solution (see section 2.5.2). The corrosive ability of process liquors can therefore be reduced through design and appropriate operation of the equipment.

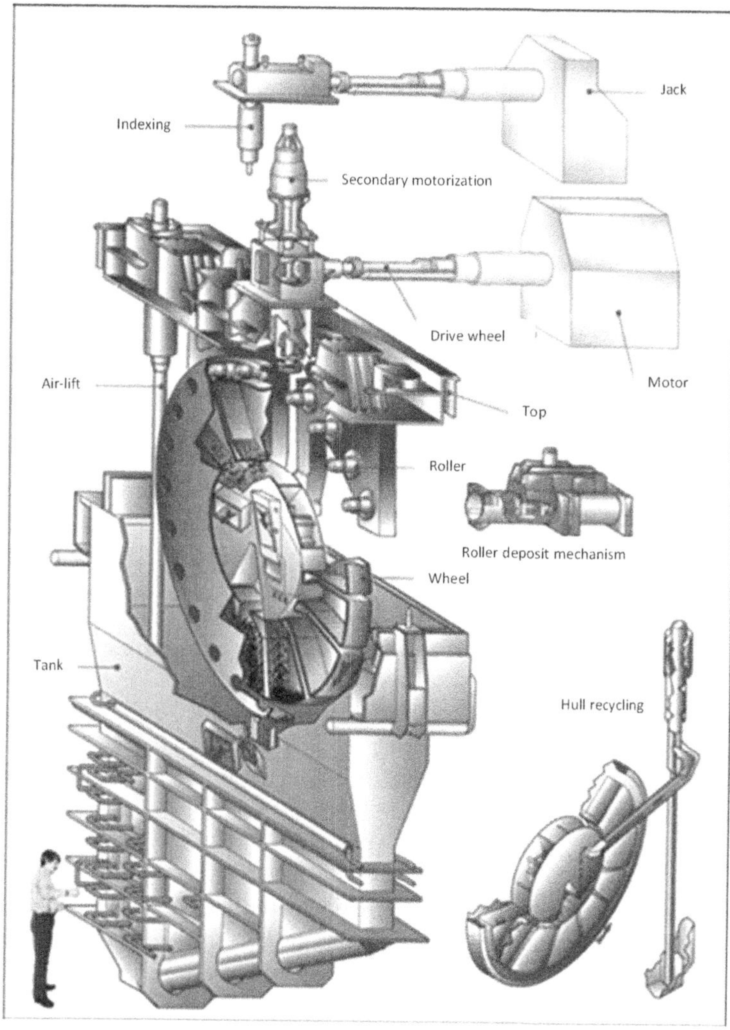

Figure 2.19. *Schematic diagram of the dissolver wheel and its main components at the La Hague plant. For a color version of this figure, see www.iste.co.uk/balbaud/nuclearenergytoday.zip*

As for the equipment inside the dissolver, the materials have mainly been chosen for their corrosion resistance in nitric acid at 90°C. For instance, the dissolver tank has been built from zirconium 702 according to the ASTM standards currently active with specific restrictions on the concentration of some elements, based on recommendations from the Alternative Energies and Atomic Energy Commission

(CEA) (section 2.3.1). This material was chosen owing to its excellent corrosion resistance in nitric acid, including at high temperature and in the presence of oxidizing ions (see section 2.3.1). Considerable attention has been paid to corrosion resistance with respect to nitric condensates (Fauvet 2012). The inside of the dissolver is only partially filled with nitric acid and some immersed parts of internal components can be subjected to nitrous vapor condensation and corrosion due to condensates. This increases the risk of IGC of ASSs. The choice of material for components such as the dissolver wheel rollers, which are subject to the risk of corrosion by nitric condensates and mechanical wear from friction, focused on the 1.4361 ASS containing 18% chromium, 15% nickel and 4% silicon (see section 2.3.1). This grade has the advantage of not developing IGC compared with other austenitic grades under the same test conditions. By design, this ASS is also cold-worked to increase its resistance to friction from the dissolver wheel. Other components, such as the dissolver lid and the wheel hub, which are permanently immersed in the dissolver, are made from 1.4335 ASS according to European standards (see section 2.3.1). This ASS with a low carbon content contains 25 wt% chromium, which makes it extremely resistant to nitric acid.

2.6.2. *Fission product evaporator–concentrator machines*

Fission product evaporators–concentrators are designed to concentrate FP in aqueous nitric acid phase produced during the first extraction cycle. The objective is to reduce their final volume for interim storage upstream of the vitrification operations, and to recover any nitric acid they may contain so it can be recycled in the process (Figure 2.1).

The FP are concentrated in a semi-continuous mode. At the beginning of the evaporation cycle, a formol solution is added at the same time as the solution to undergo concentration. This is to keep the acidity level constant at around 2.5 $mol.L^{-1}$ by denitration by formaldehyde. This denitration process is maintained for the entire duration of the evaporation cycle at boiling temperature. When the expected concentration rate is reached, the supply of the solution containing FP and the formaldehyde solution is stopped, and the evaporator is switched to reflux mode for at least 5 h to completely destroy the residual formaldehyde.

The 1.4361 ASS was chosen as the material for these evaporators in order to resist these conditions (see section 2.3.1). The main advantage of this ASS grade is its low sensitivity to IGC, even in highly oxidizing environments that generally induce IGC in conventional ASS (1.4306/1.4307, 1.4404, 1.4335). This IGC – manifested through the preferential attack of the grain boundaries and then the detachment of these grains (see section 2.4.1.2) – renders the use of conventional

ASS unacceptable in light of the dissolution rates reached in the equipment. Even in severe conditions, the 1.4361 ASS shows no sign of intergranular attack, and its corrosion remains uniform.

The industrial durability of this equipment can be ensured on the condition that a homogeneous, general corrosion rate is maintained in the fission product evaporators–concentrators. These equipment have been designed to operate for a lifespan of 30 years; they were commissioned in 1990 and 1994 for the UP3 and UP2-800 plants, respectively. The aging of these fission product evaporators–concentrators have been monitored within the scope of the lifespan extension of the La Hague plants. However, this equipment was not initially designed to be inspected during its lifetime. A number of innovative developments were necessary to provide solutions to carry out inspections in cells that remain inaccessible to operators due to the ambient radioactivity levels.

These inspections mainly concern:

– Measurement of the equipment wall thickness in areas accessible through existing endoscope ducts located in the adjacent wall just a few meters away. The single-element or phased-array ultrasound transducer is brought into contact with the equipment using a motorized multi-axis arm. The phased-array measurement provides data on the thickness of the boiler wall located under the double-walls heating jacket by measuring through this jacket.

– Measurement of the equipment wall thickness at the entries and exits of the double-wall heating jacket. The ultrasound measurement transducer is fed through the overheated water lines and is able to cover a distance of about 10 m, thanks to an endoscope and drain rod.

– Endoscope video investigation inside boilers. This very unusual investigation was possible using an endoscope that was fed through the drain ejectors, covering a distance of about 10 m. Due to the irradiation in this type of equipment, in situ data processing had to be developed in order to improve the quality of the video footage.

The first measurements of residual thicknesses were completed in 2011. The results of these measurements highlighted a greater loss in the wall thickness than expected from design. A project was therefore implemented to reduce the corrosion rate and thus extend the lifespan of these equipment. At the same time in early 2015, Orano decided to launch its NCPF (new concentration of FP) program, which sets out to replace all of the existing FP evaporator–concentrator equipment in the La Hague plants.

The process to optimize equipment operation has started and aims at increasing the lifespan of these machines. This optimization process is based on R&D knowledge reported in this chapter, and sets out to reduce the general corrosion rate. The following actions were defined:

– increase the frequency of standard rinsing operations designed to reduce the presence of deposits in heated areas and thus limit the ASS surface temperatures;

– limit the operating periods in "total reflux" mode (without denitration with formaldehyde), which are more oxidizing;

– add a fluoride ion complexing agent in specific operating conditions;

– limit the coolant fluid temperature linked to the pressure drop in the double-walls heating jacket owing to the thinner process wall thicknesses due to corrosion.

The NCPF project of construction of new fission product evaporators has benefited from the capitalization of experience feedback. This experience feedback focuses on:

– design, which has benefited from experience feedback on the aging of the first generation of evaporators (definition of corrosion allowance thicknesses, recommendations to limit deposits);

– evaporator fabrication, in particular, the steps involving the procurement of semi-finished products made from 1.4361 ASS and the welding operations. This has benefited from the formalization of experience feedback on the fabrication of the first generation of evaporators;

– experience feedback collected from monitoring the aging of fission product evaporator–concentrator equipment, which has taken advantage of the improved inspectability of the NCPF evaporators, for example, by increasing the fraction of the external surface area that can undergo ultrasonic thickness measurements, and by making new areas inspectable (welds and surfaces under the double-walls heating jacket).

2.6.3. *Zirconium thermosiphons*

At La Hague plants, the PUREX process involves recycling of nitric acid. Indeed, when introduced into the dissolver wheel, the segments of used fuel are dissolved into nitric acid (see section 2.6.1). After dissolution of plutonium, uranium and FP into nitric acid, these elements are separated and purified during further steps of the PUREX process. The acid itself is also purified and concentrated before it returns to the wheel dissolver for further reuse. This recycling process is justified for ecological and economical purposes.

The PUREX process for nuclear fuel reprocessing uses nitric media at different concentrations up to 13.6 $mol.L^{-1}$ and at different temperatures (from room to boiling temperatures). Due to its particularly good behavior in highly oxidizing nitric media with a corrosion rate lower than 1 $\mu m.year^{-1}$ in boiling nitric acid (see section 2.4.2), zirconium was chosen to build the equipment in contact with high concentration nitric solutions at high temperature, and especially in the acid recovery units.

The nitric acid recycling process consists of recovering and condensing nitric vapor at different stages of the plant. Therefore, condensed nitric vapors are concentrated in two types of units, the tritiated acid recovery unit and the non-tritiated acid recovery unit.

For instance, the nitric acid resulting from the tritiated acid recovery unit initially comes from nitric vapors of the fission product evaporators (see section 2.6.2) and it is recycled in the dissolution units (see section 2.6.1). The acid resulting from the non-tritiated recovery units comes from nitric vapors of the purification units and is recycled in the extraction units.

Thermosiphon type evaporators are used in major parts of the acid recycling units, both tritiated and non-tritiated. In UP3 and UP2-800, most of the thermosiphon type evaporators are made of zirconium.

One thermosiphon type evaporator is schematically presented in Figure 2.20. This is made of a boiler and a column. The boiler is a tubular heat exchanger that heats the nitric acid up to its boiling temperature. Boilers are heated up either by water vapor or superheated water, which circulates in the boiler shell; whereas, the acid is heated up in the boiler tubes. These tubes are thus the most sensitive to corrosion since they contain the acid media at the highest temperature and under heat flux.

Throughout the years, Orano developed a non-intrusive ultrasonic method that enables the measurement of the tube thickness without opening the boiler. During maintenance shutdown, ultra-sonic sensors are placed on the outside boiler shell and measure the tube thickness through the external boiler wall.

This method is now widely used at the La Hague plants in order to control the thickness losses and thus the lifetime of this type of thermosiphon evaporators.

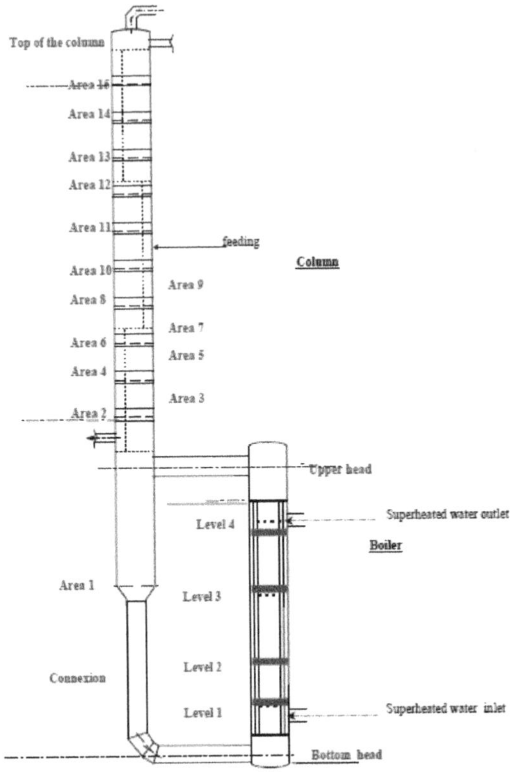

a)

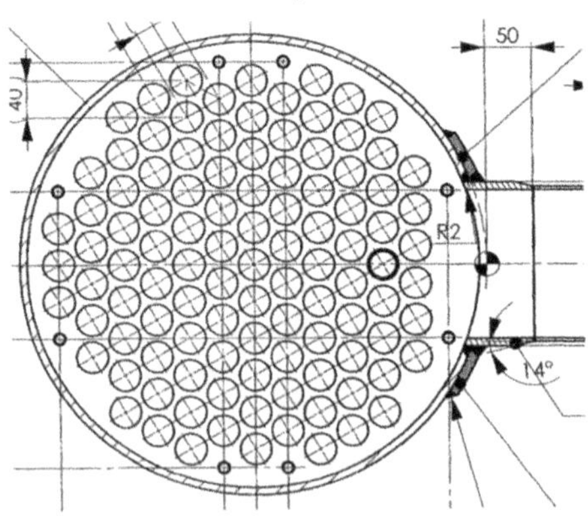

b)

Figure 2.20. *(a) Tritiated acid recovery evaporator. (b) Example of a cross-section of the boiler containing 106 tubes. For a color version of this figure, see www.iste.co.uk/balbaud/nuclearenergytoday.zip*

In UP3 and UP2-800 units, more than 15 zirconium made thermosiphon evaporators are in production since the beginning of the 1990s. All of them are regularly controlled using the non-intrusive method for measuring the tube thickness. Among these equipment, only one thermosiphon has shown a loss of tube thickness, which led to the replacement of the boiler in 2008 after almost two decades of production (Vittoz et al. 2011).

The non-intrusive method enabled anticipation of the replacement of this thermosiphon evaporator of tritiated acid. As illustrated in Figure 2.21, periodic thickness measurement showed a loss of wall thicknesses of the tubes. The boiler was preventatively replaced in 2008, although there was still a significant tube thickness. The purpose was to avoid any risk of leakage.

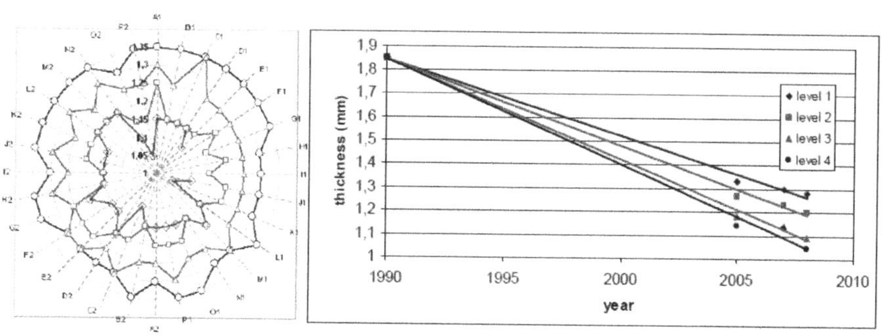

(a) (b)

Figure 2.21. *(a) Tube thicknesses measurements in 2005 at four levels along the height of the boiler. (b) Average thickness evolution. For a color version of this figure, see www.iste.co.uk/balbaud/nuclearenergytoday.zip*

The highest corrosion rate of 44 μm.year^{-1} has been observed on the tubes near the inlet of superheated water. This location corresponds to the highest temperature of the boiler. Moreover, this corrosion rate of zirconium had been attributed to the fluorine content in this specific unit. As described in section 2.5.3.3, zirconium is highly sensitive to fluorine.

Except for units containing traces of fluorine, zirconium used in UP2-800 and UP3 has shown excellent corrosion resistance in nitric acid, even in highly oxidizing conditions.

2.7. New developments

As is the case in many industrial fields, spent nuclear fuel reprocessing–recycling plants are always looking for new solutions to reduce corrosion and thus increase plant performance, while extending the lifespan of its equipment. New approaches to finding innovative and optimized solutions are now being explored:

– In the metallurgy sector, recent developments in material engineering have been made, that is, advanced surface treatment processes and additive manufacturing. Compared with materials elaborated using various different metallurgical techniques (conventional or emerging), research is now focusing on nickel and titanium (Sure et al. 2021) alloys, as well as on high entropy alloys or complex concentrated alloys (Jayaraj et al. 2017). The results of studies are promising. The functionalization of the surface of metallic materials through the use of coatings is a very interesting option for improving their corrosion resistance, mechanical and/or tribological properties (Padhy et al. 2010, 2011; Ningshen et al. 2011; Li et al. 2014; Kazazi et al. 2018; Chabanon et al. 2022). These surface treatments use a variety of technologies (chemical vapor deposition, physical vapor deposition, thermal spray and cold spray) that are increasingly more effective and robust. The possibility of developing dense, adhesive layers with different compositions and microstructures may prove relevant to spent nuclear fuel reprocessing-recycling plants, for example, to maintain rather inaccessible equipment in operational conditions (repair), to improve the corrosion-resistance properties of existing equipment in relation to the changing compositions of environments and to reduce costs with respect to solid materials. In light of the nuclear safety requirements, R&D studies are currently focusing on analyzing the different risks associated with equipment damage and how to manage the aging of such coatings. Furthermore, additive manufacturing currently seems relevant on an industrial level for the optimized fabrication of spare parts. In the future, this technique may also offer solutions to problems of durability in aggressive environments, whether from a design, assembly or innovation perspective (Balbaud-Célérier and Puga 2021). For example, areas at risk of corrosion such as welds and seals between dissimilar materials can be avoided thanks to the possibility of developing one-piece components and materials with composition gradients.

– In the field of modeling, it seems important to develop and use artificial-intelligence tools with the objective of collecting relevant information from databases (data mining), such as from laboratory experiments and/or industrial operating experience. Different tools (algorithms) are being recommended and tested to develop automatic machine learning, focused on material properties (corrosion, mechanics, etc.). For example, different algorithms (k-nearest neighbor, decision tree, gradient boosting decision tree, random forest and support vector machine) have been compared with the purpose of simulating the electrochemical response of copper corrosion (polarization curves and impedance diagrams) within the framework of nuclear waste geological disposal (Gong et al. 2020). Other algorithms such as artificial neural networks have been investigated with the purpose of predicting the degradation in the mechanical properties of materials due to hydrogen (Thankachan et al. 2017). Furthermore, models based on cellular automaton can simulate different types of material corrosion, for example, IGC, with ever-increasing accuracy (Perez-Brokate 2017a, 2017b; Guiso 2020; Guiso et al. 2020, 2021, 2022).

– Finally, these models (and the more classical analytical models) must be able to simulate the actual aging of facilities as a whole (this is called a digital twin) (Polyanchukov et al. 2001; D'Amico et al. 2019; Mnasri et al. 2019). To further enhance the level of efficiency, this approach could be combined with in-service monitoring that is more extensive, accurate, frequent and coupling the effect of chemistry with corrosion on the facilities. Among the corrosion monitoring techniques studied, acoustic emission testing (Bai et al. 2014; Chai et al. 2015) and electrochemical methods (Moody et al. 1984; Suresh et al. 2011; Girija and Kamachi Mudali 2014) are noteworthy.

2.8. References

AFNOR (2014). NF EN 10088-1. Standard. Stainless steels – Part 1: List of stainless steels.

Andon, R.J.L., Pemberton, R.C., Thomas, J.G.N., Harrison, J.M., Worthington, S.E., Shaw, R.D. (1986). Corrosion of stainless-steels under heat-transfer conditions in nitric-acid. *British Corrosion Journal*, 21(2), 119–128.

Andreoletti, G., Chambrette, P., Mercier, A., Moulin, J.-P., Thurin, F., Vittoz, C. (2006). Industrial feedback of intercycle uranyl nitrate concentration in stainless steel 304L evaporator. In *Eurocorr 2006*, Maastricht.

ASTM International (2021). ASTM B551/B551M-12. Standard specification for zirconium and zirconium alloy strip, sheet, and plate.

Bai, W., Chai, M., Li, L., Duan, Q. (2014). Comparative study of acoustic emission characteristics for intergranular corrosion of 316L stainless steel parent material and weldment. In *ASME 2014 Pressure Vessels and Piping Conference*, New York.

Balbaud-Célérier, F. (1998). Mécanisme de corrosion de l'acier inoxydable Z3 CN 18-10 en présence de condensats nitriques. PhD Thesis, Université Pierre et Marie Curie, Paris.

Balbaud-Célérier, F. and Puga B. (2021). Correlation between process parameters and properties of use: Example of corrosion. In *AMETIS: Advanced Manufacturing for Energy and Transportation International School*, Chaâbane, N., Schuster, F. (eds). EDP Sciences, Les Ulis.

Balbaud-Célérier, F., Sanchez, G., Santarini, G., Picard, G. (2000). Cathodic reactions involved in corrosion processes occurring in concentrated nitric acid at 100 degrees C. *Eur. J. Inorg. Chem.*, 4, 665–674.

Balbaud-Célérier, F., Gruet, N., Gwinner, B., Laghoutaris, P. (2020). Corrosion behavior of stainless steels in nitric acid in the context of nuclear fuel reprocessing plants. In *Nuclear Corrosion*, Ritter, S. (ed.). Woodhead Publishing, Duxford/Cambridge/Kidlington.

Beaunier, L., Béranger, G., Charbonnier, J.-C., Cihal, V. (1994). Corrosion intergranulaire. In *Corrosion localisée*, Dabosi, F., Béranger, G., Baroux, B. (eds). EDP Sciences, Les Ulis.

Beavers, J.A., Griess, J.C., Boyd, W.K. (1981). Stress-corrosion cracking of zirconium in nitric-acid. *Corrosion*, 37(5), 292–297.

Benoit, M. (2016). Modélisation de la cinétique de réduction de l'acide nitrique concentré sur acier inoxydable 304L. Doctoral Thesis, Université Pierre et Marie Curie, Paris.

Bourgeois, M. (2000). Retraitement du combustible. Procédés, ingénierie et usines. In *Techniques de l'Ingénieur*, BN3652 v1.

Brunet, S., Coriou, H., Grall, L., Mahieu, C., Pelras, M., Sanatine, O. (1970). Corrosion galvanique d'alliages austénitiques "Fer- Chrome- Nickel" en milieu nitrique. *Les Mémoires Scientifiques de la Revue de Métallurgie*, LXVII(12), 781–787.

Brunet, S., Coriou, H., Grall, L., Mahieu, C., Pelras, M., Sanatine, O. (1971). Corrosion galvanique d'alliages austénitiques "fer-chrome-nickel" en milieu nitrique. Deuxième partie etude du comportement d'alliages austénitiques "fer-chrome-nickel" dans la zone de transpassivité. *Les Mémoires Scientifiques de la Revue de Métallurgie*, LXVIII (7–8), 489–497.

Cabrera, N. and Mott, N.F. (1949). Theory of the oxidation of metals. *Reports on Progress in Physics*, 12(1), 163–184.

Chabanon, A., Michau, A., Schlegel, M., Gündüz, D., Puga, B., Miserque, F., Schuster, F., Maskrot, H., Pareige, C., Cadel, E. et al. (2022). Surface modification of 304L stainless steel and interface engineering by HiPIMS pre-treatment. *Coatings*, 12(6), 727.

Chai, M.Y., Duan, Q., Bai, W.J., Zhang, Z.X., Xie, X.M. (2015). Acoustic emission detection of 316L stainless steel welded joints during intergranular corrosion. *Int. J. Miner. Metall.*, 22(9), 942–949.

Chandra, K., Kain, V., Ganesh, P. (2008). Controlling end-grain corrosion of austenitic stainless steels. *J. Mater. Eng. Perform.*, 17(1), 115–122.

Commissariat à l'énergie atomique (CEA) (2008). *Le traitement-recyclage du combustible nucléaire usé. La séparation des actinides – Application à la gestion des déchets*. Le Moniteur, Paris.

Coriou, H., Hure, J., Plante, G. (1961). Aspect electrochimique de la corrosion d'aciers inoxydables austenitiques en milieu nitrique et en presence de chrome hexavalent. *Electrochim. Acta*, 5(1–2), 105.

Cunat, P.-J. (2000). Aciers inoxydables – Propriétés. Résistance à la corrosion. *Techniques de l'Ingénieur*, M4541 v1.

D'Amico, D., Ekoyuncu, J., Addepalli, S., Smith, C., Keedwell, E., Sibson, J., Penver, S. (2019). Conceptual framework of a digital twin to evaluate the degradation status of complex engineering systems. In *7th CIRP Global Web Conference on Towards Shifted Production Value Stream Patterns through Inference of Data, Models, and Technology (CIRPe)*. Electr Network, Amsterdam.

Darwish, S., Khalil, M.W., Rahim, M.A.A., Ammar, I.A. (1989). The valve metal character of zirconium as inferred from anode potential measurements in mineral acid-solutions. *Materialwissenschaft Und Werkstofftechnik*, 20(9), 299–308.

Dey, H.C., Ashfaq, M., Bhaduri, A.K., Rao, K.P. (2009). Joining of titanium to 304L stainless steel by friction welding. *J. Mater. Process. Technol.*, 209(18–19), 5862–5870.

Emery, A. (2019). Corrosion intergranulaire des aciers inoxydables austénitiques en milieu acide nitrique oxydant. Doctoral Thesis, Université PSL, Paris.

Fauvet, P. (2012). Corrosion issues in nuclear fuel reprocessing plants. In *Nuclear Corrosion Science and Engineering*, Feron, D. (ed.). Woodhead Publishing, Philadelphia.

Fauvet, P., Balbaud, F., Robin, R., Tran, Q.T., Mugnier, A., Espinoux, D. (2008). Corrosion mechanisms of austenitic stainless steels in nitric media used in reprocessing plants. *J. Nucl. Mater.*, 375(1), 52.

Garaix, G., Venault, L., Maurin, J., Blain, G., Fattahi, M., Vigier, N., Moisy, P. (2015). Alpha radiolysis of nitric acid and sodium nitrate with $4He^{2+}$ beam of 13.5MeV energy. *Radiat. Phys. Chem.*, 106, 394–403.

Girija, S. and Kamachi Mudali, U. (2014). Electrochemical noise resistance evaluation of 304L SS in nitric acid and simulated nuclear high level waste. *Corros. Eng. Sci. Technol.*, 49(5), 335–344.

Gong, X., Dong, C., Xu, J., Wang, L., Li, X. (2020). Machine learning assistance for electrochemical curve simulation of corrosion and its application. *Mater. Corr.*, 71(3), 474–484.

Guiso, S. (2020). Modélisation de la corrosion intergranulaire par automates cellulaires. Doctoral Thesis, Université PSL, Paris.

Guiso, S., di Caprio, D., de Lamare, J., Gwinner, B. (2020). Intergranular corrosion: Comparison between experiments and cellular automata. *Corr. Sci.*, 177, 108953.

Guiso, S., di Caprio, D., de Lamare, J., Gwinner, B. (2021). Influence of the grid cell geometry on 3D cellular automata behavior in intergranular corrosion. *J. Comput. Sci. Technol.*, 53, 101322.

Guiso, S., Brijou-Mokrani, N., de Lamare, J., Di Caprio, D., Gwinner, B., Lorentz, V., Miserque, F. (2022). Intergranular corrosion in evolving media: Experiment and modeling by cellular automata. *Corr. Sci.*, 205, 110457.

Gwinner, B., Badji, H., Benoit, M., Brijou-Mokrani, N., Fauvet, P., Gruet, N., Laghoutaris, P., Miserque, F., Robin, R., Tabarant, M. (2015). Corrosion of zirconium in the context of the spent nuclear fuel reprocessing plant. In *Global 2015*, Paris.

Gwinner, B., Auroy, M., Balbaud-Célérier, F., Fauvet, P., Larabi-Gruet, N. Laghoutaris, P., Robin, R. (2016). Towards a reliable determination of the intergranular corrosion rate of austenitic stainless steel in oxidizing media. *Corr. Sci.*, 107, 60–75.

Gwinner, B., Borgard, J. M., Dumonteil, E., Zoia, A. (2017). A stereological approach for measuring the groove angles of intergranular corrosion. *Corr. Sci.*, 115, 1–7.

Horn, E.M. and Schoeller, K. (1990). Corrosion resistance of austenitic and ferritic stainless alloys in 20 to 75% nitric acid as a function of temperature and concentration. *Mater. Corr. Werkst. Korros.*, 41(3), 97–112.

Ishijima, Y., Kato, C., Motooka, T., Yamamoto, M., Kano, Y., Ebina, T. (2013). Stress corrosion cracking behavior of zirconium in boiling nitric acid solutions at oxide formation potentials. *Mater. Trans.*, 54(6), 1001–1005.

Jayaraj, J., Thinaharan, C., Ningshen, S., Mallika, C., Kamachi Mudali, U. (2017). Corrosion behavior and surface film characterization of TaNbHfZrTi high entropy alloy in aggressive nitric acid medium. *Intermetallics*, 89, 123–132.

Jeng, J.-Y., Quayle, B.E., Modern, P.J., Steen, W.M. (1993). Prevention and repair of "end grain corrosion" of type 347 stainless steel in hot nitric acid using laser surface remelting. In *Surface Engineering, Volume III: Process Technology and Surface Analysis*, Datta, P.K. and Gray, J.S. (eds). Royal Society of Chemistry, Cambridge.

Kain, V., Chouthai, S.S., Gadiyar, H.S. (1992). Performance of AISI 304L stainless steel with exposed end grain in intergranular corrosion tests. *Brit. Corr. J.*, 27(1), 59–65.

Kajimura, H., Morikawa, H., Nagano, H. (1987). Formation mechanism of Cr6+ ions and their accelerating effet on the corrosion of stainless steel in high temperature concentrated nitric acid. *Boshoku Gijutsu*, 36, 636–642.

Kaneko, M. and Abe, S. (1995). The mechanism of end grain corrosion of stainless steel in nitric acid solutions and its prevention. *Tetsu To Hagane*, 81(8), 857–861.

Kato, C., Yano, M., Kiuchi, K., Sugimoto, K. (2003a). Effects of heat-transfer on corrosion of zirconium in a boiling nitric acid solution. *Corr. Engineering*, 52, 53–67.

Kato, C., Kiuchi, K., Sugimoto, K. (2003b). Thermodynamic study on redox reactions of boiling nitric acid solutions. *Corr. Eng.*, 52(1), 69–85.

Kazazi, M., Haghighi, M., Yarali, D., Zaynolabedini, M. (2018). Improving corrosion resistance of 316L austenitic stainless steel using ZrO_2 sol-gel coating in nitric acid solution. *J. Mater. Eng. Perform.*, 27(3), 1093–1102.

Khan, S., Chandra, K., Kain, V., Reddy, A.V.R. (2012). End grain corrosion: Establishing microstructural causes and preventive steps. In *Textures of Materials, Pts 1 and 2*, Tewari, A., Suwas, S, Srivastava, D, Samajdar, I, Haldar, A. (eds). Materials Science Forum, Mumbai.

Kiuchi, K. (1992). Fundamental studies for life prediction of materials used in spent fuel reprocessing plant. In *Int. Symp. on Advanced Nuclear Energy Research. Roles and Direction of Material Science in Nuclear Technology*, Mito.

Kolman, D.G., Ford, D.K., Butt, D.P., Nelson, T.O. (1997). Corrosion of 304 stainless steel exposed to nitric acid-chloride environments. *Corr. Sci.*, 39(12), 2067–2093.

Lange, R. (2012). Phénomènes de couplage acier 304L – platinoïdes dans les milieux de dissolution des combustibles usés. Doctoral Thesis, Université Pierre et Marie Curie, Paris.

Lange, R., Maisonhaute, E., Robin, R., Vivier, V. (2013). On the kinetics of the nitrate reduction in concentrated nitric acid. *Electrochem. Comm.*, 29(0), 25–28.

Laurent, B. (2017). Corrosion behavior of a silicon enriched austenitic stainless steel in hot and concentrated nitric acid media. Doctoral Thesis, Université PSL, Paris.

Li, C., Chen, W., Jiang, Q., Wang, L., Luo, D. (2014). Corrosion resistance of Ti-based metallic glass coating in concentrated nitric acid. *Mater. Chem. Phys.*, 143(3), 900–903.

Lohrengel, M.M. (1993). Thin anodic oxide layers on aluminum and other valve metals: High-field regime. *Mater. Sci. Eng. R Rep.*, 11(6), 243–294.

Meisterjahn, P., König, U., Schultze, J.W. (1989). Potentiostatic pulse measurements of oxide growth on zirconium. *Electrochim. Acta*, 34(4), 551–561.

Mengyu, C., Can, L., Dongdong, W., Quan, D., Zaoxiao, Z. (2015). Research on intergranular corrosion behavior of welded joints of nuclear grade stainless steel. *Mater. Sci. For.*, 809–810, 390–394.

Michel, A. (2021). Pièces mécaniques soudées – Fontes, alliages de cuivre, de nickel, de titane ou de zirconium. *Techniques de l'ingénieur*, BM5195 V1.

Mincher, B.J., Precek, M., Paulenova, A. (2016). The redox chemistry of neptunium in γ-irradiated aqueous nitric acid in the presence of an organic phase. *J. Radioanal. Nucl. Chem.*, 308(3), 1005–1009.

Mirolyubov, E.N., Kazakov, A.P., Kurtepov, M.M. (1966). In *Corrosion of Metal and Alloys*, Volume 2, Tomashov, N.D. and Mirolyubev, E.N. (eds). IPST, Jerusalem.

Mnasri, H., Wassar, T., Franchek, M.A., Marotta, E. (2019). Data-driven modeling of carbon dioxide corrosion for integrity management application. *SPE Prod.*, 34(4), 667–677.

Moody, J.R., Qin, X.P., Strutt, J.E. (1984). The application of a computerized impedance monitoring-system to study the corrosion behavior of 347 stainless-steels in nitric-acid solutions. *J. Electrochem. Soc.*, 131(8), C296.

Mori, M., Terasaki, S., Soma, S., Hara, N., Sugimoto, K. (2003). Galvanic corrosion behavior of dissimilar metal joint Zr/Ta/R-SUS 304 ULC in hot nitric acid. *Corr. Eng.*, 52, 35–51.

Motooka, T. and Kiuchi, K. (2002). Effect of neptunium ions on corrosion of stainless steel in nitric acid solution. *J. Nucl. Sci. Technol.*, 39(3), 367–370.

Nagai, T., Takeuchi, R., Takeda, S., Yamamoto, T.A., Tsukui, S., Okamoto, S. (1998). Corrosion behavior of stainless steel in nitric acid solution under gamma-ray irradiation. *J. Nucl. Sci. Technol.*, 35(7), 502–507.

Nagano, H. (1988a). Corrosion of stainless steels in highly oxidizing nitric acid. *Corr. Eng.*, 37(5), 301–310.

Nagano, H. (1988b). Stainless-steel corrosion in concentrated nitric acid. *Corr. Eng.*, 37, 263–274.

Nandakumar, T., Suresh, G., Mudali, U. (2018). Effect of microstructure of 304L SS and thermal aging of 304L SS weld on the corrosion behavior in simulated high level waste. *Trans. Indian Inst. Met.*, 71(1), 167–176.

Ningshen, S., Kamachi Mudali, U., Krishnan, R., Raj, B. (2011). Corrosion behavior of Zr-based metallic glass coating on type 304L stainless steel by pulsed laser deposition method. *Surf. Coat. Technol.*, 205(15), 3961–3966.

Ogawa, T. (1993). Nitric acid corrosion behaviour and hot cracking resistance of austenitic stainless steels weldments. In *1st European Stainless Steel Conference*. Associazione Italiana di Metallurgia, Milan.

Okubo, M., Shintani, M., Ishimaru, H., Tokunaga, K., Nagayama, M., Itoh, A. (1987). Demonstration tests on corrosion resistance of equipment for spent fuel reprocessing process. In *SFEN International Conference on Nuclear Fuel Reprocessing and Waste Management 'RECOD 87'*. Société Française d'Energie Nucleaire (SFEN). Paris.

Osoba, L., Okeniyi, J., Pogoson, B., Fasuba, O. (2017). Effects of single pass and multipass welding on austenitic stainless steel corrosion in aggressive environments. *Gazi Univ. J. Sci.*, 30(4), 514–529.

Ozawa, M., Yamamura, O., Gonda, K. (1985). Corrosion Of type-304l stainless-steel in concentrated plutonium nitrate solution. *J. Nucl. Sci. Technol.*, 22(1), 68–69.

Padhy, N., Kamachi Mudali, U., Raj, B., Kamal, S., Chandra, R. (2010). Corrosion performance of TiO_2 coated type 304L stainless steel in nitric acid medium. *Surf. Coat. Technol.*, 204(16–17), 2782–2788.

Padhy, N., Kamachi Mudali, U.K., Raj, B., Chawla, V., Chandra, R. (2011). Corrosion behaviour of single (Ti) and duplex (Ti-TiO_2) coating on 304L stainless steel in nitric acid medium. *Mater. Chem. Phys.*, 130(3), 962–972.

Pascal, P. (1962). *Nouveau traité de chimie minérale, volume 15, part 3. Transuraniens.* Masson, Paris.

Perez-Brokate, C.F., di Caprio, D., Féron, D., de Lamare, J., Chaussé, A. (2017a). Pitting corrosion modelling by means of a stochastic cellular automata-based model. *Corros. Eng. Sci. Technol.*, 52(8), 605–610.

Perez-Brokate, C.F., di Caprio, D., Féron, D., de Lamare, J., Chaussé, A. (2017b). Probabilistic cellular automata model of generalised corrosion, transition to localised corrosion. *Corros. Eng. Sci. Technol.*, 52, 186–193.

Petit, M.C., Desjardins, D., Puiggali, M., El Kheloui, A., Clement, C. (1991). The electrochemical behaviour of stainless steels in hot nitric solutions as related to the SCC phenomenon. *Corr. Sci.*, 32(12), 1315.

Pinard-Legry, G. (1989). Corrosion des matériaux métalliques. *Techniques de l'Ingénieur*, Saint Denis.

Pinard-Legry, G., Pelras, M., Turluer, G. (1989). Corrosion resistance of metallic materials for use in nuclear fuel reprocessing. In *A Working Party Report on Corrosion in the Nuclear Industry*. EFC Pub., The Institute of Metals, London.

Polyanchukov, V.G. (2001). Modern corrosion monitoring of complex systems in acid media – The bridge into the 21st century. *Mater. Corr. – Werkst. Korr.*, 52(2), 117–123.

Precek, M., Paulenova, A., Tkac, P., Knapp, N. (2010). Effect of gamma irradiation on the oxidation state of neptunium in nitric acid in the presence of selected scavengers. *Sep. Sci. Technol.*, 45(12–13), 1699–1705.

Ravi Shankar, A., Niyanth, S., Vasudevan, M., Kamachi Mudali, U. (2012). Microstructural characterization and corrosion bevavior of activated flux gas tungsten arc-welded and multipass gas tungsten arc-welded stainless steel weld joints in nitric acid. *Corrosion*, 68(8), 762–773.

Razygraev, V.P. (1990). Kinetiks of the reduction of nitric acid on stainless steels and alloys. *Zashch. Met.*, 26(1), 54–60.

Razygraev, V.P., Lebedeva, M.V., Kabakchi, S.A., Ponomareva, E.Y., Balovneva, R.S., Lobanova, L.P. (1988). Features of cathode reduction of boiling solutions of nitric acid on a platinum electrode. *Zh. Prikl. Khim.*, 61, 71–79.

Robin, R., Andreoletti, G., Fauvet, P., Terlain, A. (2004). Effect of noble metals on the corrosion of AISI 316L stainless steel in nitric acid. In *ATALANTE 2004*, Nîmes.

Robin, R., Miserque, F., Spagnol, V. (2008). Correlation between composition of passive layer and corrosion behavior of high Si-containing austenitic stainless steels in nitric acid. *J. Nucl. Mater.*, 375(1), 65–71.

Rozenfeld, I.L. (1968). Influence of chloride ions on the rate of dissolution of stainless steel 1Kh18N9T in nitric acid during cathodic polarization. *Zashch. Met.*, 4(4), 367–370.

Sasahira, A., Takahashi, M., Nakamura, T., Izumida, T., Fukasawa, T., Kawamura, F., Ikeda, T. (1991). Gamma ray irradiation behavior of fission products in nitric acid solution. In *Third International Conference on Nuclear Fuel Reprocessing and Waste Management*, Sendai.

Schmid, G. (1959). Die autokatalytische Natur de kathodischen Reduktion von Salpetersäure zu salpetriger Säure. I.Zur Deutung stationärer Strom-Spannungs-Kurven. *Z. Elektrochem. Angew. Phys. Chem.*, 63(9), 1183–1188.

Schmid, G. (1961). Die autokatalytische Natur der kathodischen Reduktion von Salpetersäure zu salpetriger Säure. III. Mathematische Behandlung einer autokatalytischen Elektrodenreaktion 1. Ordnung. *Z. Elektrochem. Angew. Phys. Chem.*, 65(6), 531–534.

Schmid, G. (1964). Die autokatalytische Natur der kathodischen Reduktion von Salpetersäure zu salpetri¬ger Säure. IV. Der potentiostatische Einschaltvorgang. *Ber. Bunsengesellschaft Physik Chem.*, 68(7), 677–688.

Schmid, G. (1969). Das Verhalten von salpetriger und Salpetersäure an der rotierenden Scheibenelektrode. I. HNO_2 in schwefelsauren Lösungen ohne HNO_3-Zusatz. *Ber. Bunsengesellschaft Physik Chem.*, 73(2), 189–199.

Schmid, G. and Delfs, J. (1959). Die autokatalytische Natur der kathodischen Reduktion von Salpetersäure zu salpetriger Säure. II.Der galvanostatische Einschaltvorgang. *Z. Elektrochem. Angew. Phys. Chem.*, 63(9), 1177–1182.

Schosger, J.-P. (1996). Contribution à la connaissance du comportement de l'acier Z3 CN 18-10 dans l'acide nitrique concentré, chaud et confiné. Doctoral Thesis, Institut National Polytechnique de Toulouse, Toulouse.

Shankar, A.R., Raju, V. R., Rao, M. N., Mudali, U. K., Khatak, H. S., Raj, B. (2007). Corrosion of Zircaloy-4 and its welds in nitric acid medium. *Corr. Sci.*, 49(9), 3527–3538.

Shankar, A.R., Babu, S. S., Ashfaq, M., Mudali, U. K., Rao, K. P., Saibaba, N., Raj, B. (2009). Dissimilar joining of Zircaloy-4 to type 304L stainless steel by friction welding process. *J. Mater. Eng. Perform.*, 18(9), 1272–1279.

Shit, G., Kuppusamy, M.V., Ningshen, S. (2019). Corrosion resistance behavior of GTAW welded AISI type 304L stainless steel. *Trans. Indian Inst. Met.*, 72, 2981–2995.

Sicsic, D., Balbaud-Celerier, F., Tribollet, B. (2014). Mechanism of nitric acid reduction and kinetic modelling. *Eur. J. Inorg. Chem.*, 36, 6174–6184.

Singh, V., Singh, V., Mohan, N.A. (2018). Study of corrosion behaviour of SS-316 cladding deposited by shielded metal arc welding. In *IOP Conference Series: Materials Science and Engineering*, Kattankulathur.

Sure, J., Vishnu, D.S.M., Kumar, R.V., Mudali, U.K., Schwandt, C. (2021). Corrosion performance of electrochemically prepared Ti-5Ta-2Nb alloy in concentrated nitric acid. *Mater. Today Commun.*, 26, 101786.

Suresh, G., Mudali, U.K., Raj, B. (2011). Corrosion monitoring of type 304L stainless steel in nuclear near-high level waste by electrochemical noise. *J. Appl. Electrochem.*, 41(8), 973–981.

Suresh, G., Kishor, P.S.V.R.A., Dasgupta, A., Upadhyay, B.N., Mallika, C., Mudali, U.K. (2017). Microstructure and corrosion behavior of laser melted 304L SS weldment in nitric acid medium. *J. Mater. Eng. Perform.*, 26(2), 773–782.

Tachimori, S. (1991). Numerical-simulation for chemical-reactions of actinide elements in aqueous nitric-acid solution. *J. Nucl. Sci. Technol.*, 28(3), 218–227.

Takeda, S. (1996). On the role of plutonium under the dissolution of 304 ULC stainless steel in plutonium nitrate solutions. *Zair. Kankyo/ Corros. Eng.*, 45(11), 662–666.

Takeda, S., Nagai, T., Yasu, S., Koizumi, T. (1995). Corrosion performance of several metals in plutonium nitrate solution. *Zairyo-to-Kankyo*, 44, 24–29.

Takeuchi, M. and Whillock, G.O.H. (2004). Effect of endgrain attack on corrosion of 18Cr-10Ni austenitic stainless steel in simulated dissolver liquor. *J. Nucl. Sci. Technol.*, 41, 702–708.

Tcharkhtchi-Gillard, E., Benoit, M., Clavier, P., Gwinner, B., Miserque, F., Vivier, V. (2016). Kinetics of the oxidation of stainless steel in hot and concentrated nitric acid in the passive and transpassive domains. *Corr. Sci.*, 107, 182–192.

Thankachan, T., Prakash, K.S., David Pleass, C., Rammasamy, D., Prabakaran, B., Jothi, S. (2017). Artificial neural network to predict the degraded mechanical properties of metallic materials due to the presence of hydrogen. *Int. J. Hydrogen Energy.*, 42(47), 28612–28621.

Tomashov, N.D., Chernova, G.P., Rutten, M.Y. (1982). Electrochemical-behavior of stainless-steels in nitric-acid solutions with added ferric-chloride and hydrochloric-acid. *Prot. Met.*, 18(6), 671–678.

Vervey, E.J.W. (1935). Electrolytic conduction of a solid insulator at high fields. The formation of the anodic oxide film on aluminium. *Physica*, 2, 1059.

Vetter, K.J. (1950a). Die Abhängigkeit der Anfangsneigung der Stromdichte-Potentialkurve von den Konzentrationen,Rückschlüsse auf den potentilbestimmenden Vorgang bei Redoxsystemen. *Z. Phys. Chem.*, 194, 284–296.

Vetter, K.J. (1950b). Uber den enstellungsmechanismus des HNO_2/HNO_3 redoxpotentials. *Z. Phys. Chem.*, 194, 199–206.

Vetter, K.J. (1959). Entgegnung auf die vorstehende Arbeit von G.Schmid über "Die autokatalytische Natur der kathodischen Reduktion von Salpetersäure zu salpetriger Säure". *Z. Elektrochem.*, 63(9), 1189–1191.

Vittoz, C., Fauvet, P., Fieulaine, B. (2011). Zirconium corrosion in nitric acid containing traces of hydrofluoric acid. In *EUROCORR 2011*, Stockholm.

Weast, R.C. (1988–1989). *Handbook of Chemistry and Physics*, 69th edition. CRC Press, Boca Raton, FL.

Whillock, G.O.H., Dunnett, B.F., Takeuchi, M. (2005). Techniques for measuring the end-grain corrosion resistance of austenitic stainless steels. *Corrosion*, 61(1), 58–67.

Whillock, G.O.H., Worthington, S.E., Richardson, T.J.A. (2010). Corrosion in nitric acid. In *Shreir's Corrosion*, Richardson, T.J.A. (ed.). Elsevier, Oxford.

Yamamoto, T., Tsukui, S., Okamoto, S., Nagai, T., Takeuchi, M., Takeda, S., Tanaka, Y. (1996). Gamma-ray irradiation effect on corrosion rates of stainless steel, Ti and Ti-5Ta in boiling 9N nitric acid. *J. Nucl. Mater.*, 228(2), 162–167.

Yamamoto, T., Tsukui, S., Okamoto, S., Nagai, T., Takeuchi, M., Takeda, S., Tanaka, Y. (1998). Gamma-ray irradiation effects on corrosion rates of stainless steel in boiling nitric acid containing ionic additives. *J. Nucl. Sci. Technol.*, 35(5), 353–356.

Yau, T.L. (1988). Factors affecting the stress-corrosion cracking susceptibility of zirconium in 90-percent nitric-acid. *Corrosion*, 44(10), 745–749.

Zhuravlev, V.K., Kurtepov, M.M., Oreshkin, V.I. (1973). Effect of agitation on the corrosion and electrochemical behavior of Kh18N10T stainless steel in nitric acid solutions with a chlorine ion addition. *Zashch. Met.*, 8(2), 183–187.

3

Materials and Processes for Geological Disposal of High-level Waste and Spent Fuel

Fraser KING
Integrity Corrosion Consulting Ltd, Nanaimo, BC, Canada

3.1. Introduction

The canister is an important part of the multi-barrier system that has been widely adopted for the deep geological disposal of heat-generating nuclear waste[1]. Since the canister is the only absolute barrier in the system, with all other barriers being permeable to some degree, and since the concept of a "container" is meaningful to both the expert and the lay person, the canister tends to draw much attention. However, the importance of the canister as a barrier depends on the properties of the other

For a color version of all figures in this chapter, see www.iste.co.uk/balbaud/nuclearenergytoday.zip.

1 The term canister is used here to describe the primary barrier. It can also referred to as a container, waste package, or overpack in various international waste management programmes. Heat-generating waste refers to both immobilized high-level waste from recycling (HLW) and the direct disposal of spent fuel (SF).

Materials and Processes for Nuclear Energy Today and in the Future, coordinated by Fanny BALBAUD-CÉLÉRIER and Céline CABET.

barriers, especially that of the host rock. For low-permeability sedimentary rock, the canister lifetime has little impact on the overall safety of the system, whereas for permeable crystalline rock, the canister may be the primary barrier in the entire system.

Topic/canister material	References
Repository environment and evolution	Andra 2005; Shoesmith 2006; Johnson and King 2008; Diomidis and Johnson 2014; King et al. 2017; King 2020
Copper	King et al. 2010; King 2013; Scully et al. 2016; Hänninen et al. 2017; Hall 2021; Posiva 2021a; Diomidis and King 2022
Carbon steel/cast iron	Marsh and Taylor 1988; King 2008; Féron et al. 2009; Kursten 2009; Smart 2009; Kursten et al. 2016; Crusset et al. 2017; JNC 2000, 2021
Titanium alloys	JNC 2000; Hua and Gordon 2004; Hua et al. 2005; Shoesmith 2006; DOE 2008
Nickel alloys and stainless steels	Hua and Gordon 2004; DOE 2008; Payer et al. 2019
Ceramics	Baroux 2013; Holdsworth 2013; Kerber and Knorr 2013; Holdsworth et al. 2014, 2018
Multiple materials	Andra 2001; Kursten et al. 2003; ONDRAF/NIRAS 2004; Shoesmith 2006; Bennett and Gens 2008; Landolt et al. 2009; King and Padovani 2011; King 2013, 2017, 2020; King et al. 2016; RWM 2016; Abdelouas et al. 2022

Table 3.1. *Sources of additional information on the corrosion behavior of HLW/SF canister materials*

The design and performance of HLW/SF canisters have been investigated since the 1970s. A number of reviews of the subject have been published, to which the interested reader is referred for further details (Table 3.1). Figure 3.1 shows a number of repository designs from various international deep geological repository programmes and illustrates the relationship between the canister and the other engineered and natural barriers in the system. Figure 3.2 provides a more-detailed view of a number of different canister designs from the same waste management programs highlighted in the previous figure. These figures give an indication of the differences, and the many similarities, between the different proposed repository designs. Table 3.2 summarizes the current status of a number of international waste management programs, with an indication of the host rock type and reference canister design for each.

Due to space limitations, the focus here is on the results of more recent studies and the current trends and issues. The present discussion is limited to the design and performance of metallic HLW/SF canisters. Canisters fabricated from ceramics are excluded, as are containers for the disposal or storage of low and intermediate level waste (LILW).

3.2. Canister requirements

A useful starting point is to consider the basic requirements for the canister. These may be formalized in terms of "safety functions" and "performance targets" or may simply be a list of desired properties.

For example, a basic safety function for the canister is that it provides a corrosion barrier to prevent groundwater (or, more correctly, buffer pore water) from reaching the HLW/SF. The corresponding "safety function indicator" or performance target would then be that the canister wall thickness should be greater than zero for the required service life of the canister (Posiva 2021b).

Other examples of canister requirements include the ability to:

– maintain sub-criticality of the spent fuel;

– limit the external gamma dose rate to a level that does not affect the corrosion rate;

– conduct the decay heat of the HLW/SF;

– maintain mechanical stability under postulated external loads in the repository.

Each national nuclear waste management program will have their own list of canister requirements (see, for example, SKB 2011; Patel et al. 2012; Posiva 2021b).

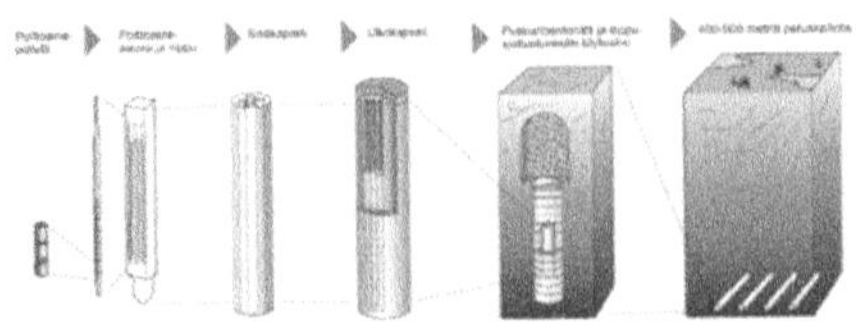

(a) Multi-barrier concept of KBS-3 system (www.posiva.fi)

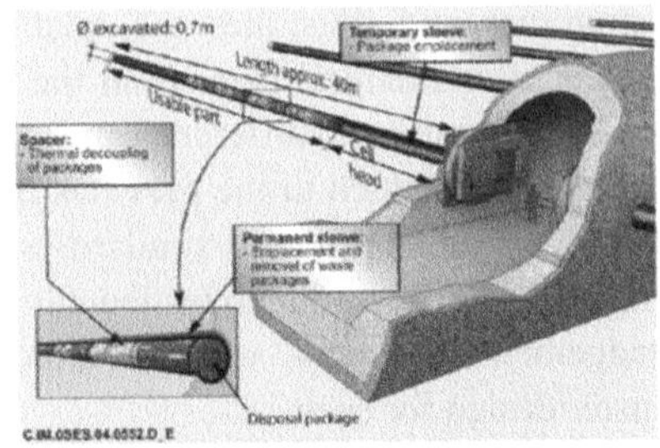

(b) Disposal cell for HLW canisters (Andra 2005)

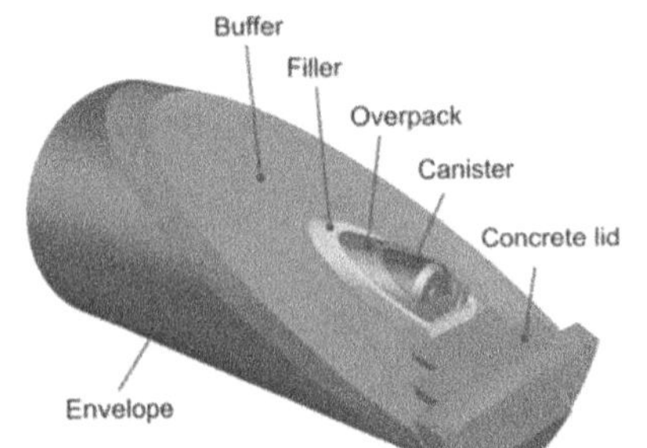

(c) Engineered barrier design for Belgian supercontainer concept (ONDRAF/NIRAS 2013)

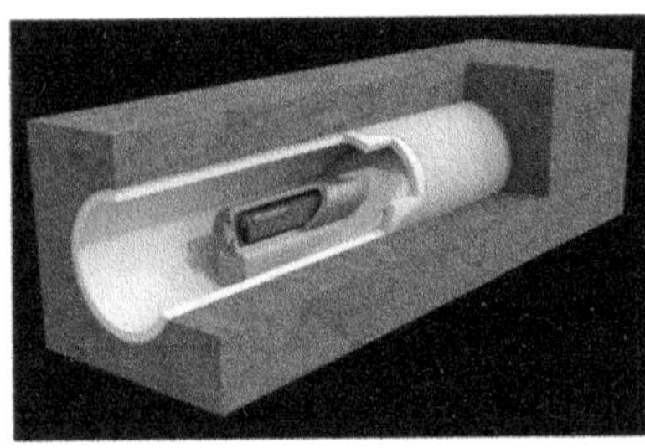

(d) Nagra concept for horizontal emplacement of HLW and SF canisters (www.nagra.ch)

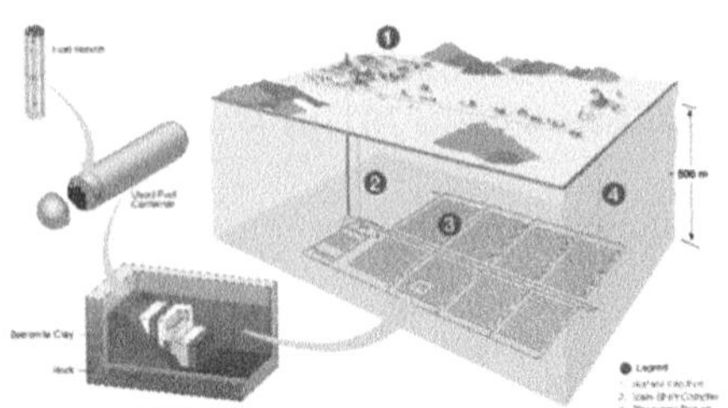

(e) Buffer box arrangement for a deep geological repository in Canada (www.nwmo.ca)

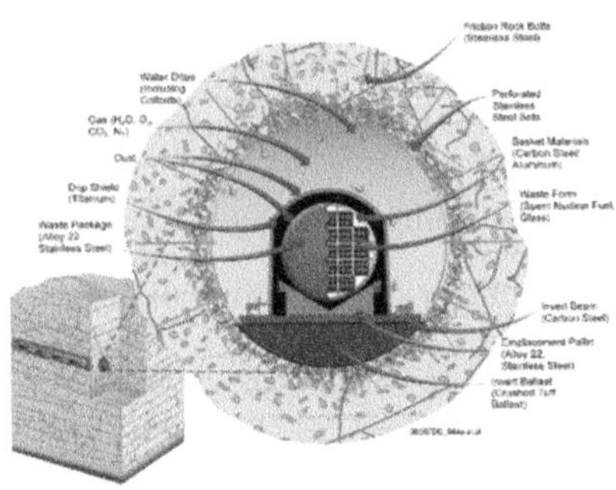

(f) Waste packages and drip shield layout for Yucca Mountain repository (DOE 2008)

Figure 3.1. *Various international multi-barrier concepts for the disposal of HLW/SF. Reproduced with permission of Posiva Oy, Andra, ONDRAF/NIRAS, Nagra, NWMO and US Department of Energy*

(a) Component parts of KBS-3 copper-cast iron canister (www.posiva-fi)

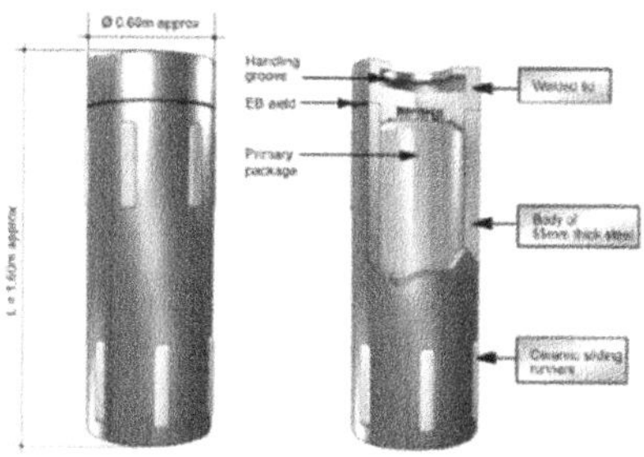

(b) Schematic of French carbon steel canister for HLW (Andra 2005)

(c) Half-scale mock-up of Belgian supercontainer (ONDRAF/NIRAS 2013)

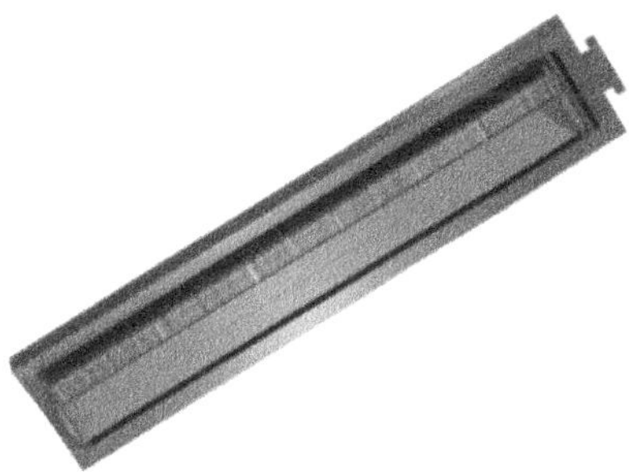

(d) Schematic of Swiss carbon steel canister for SF (Diomidis et al. 2016)

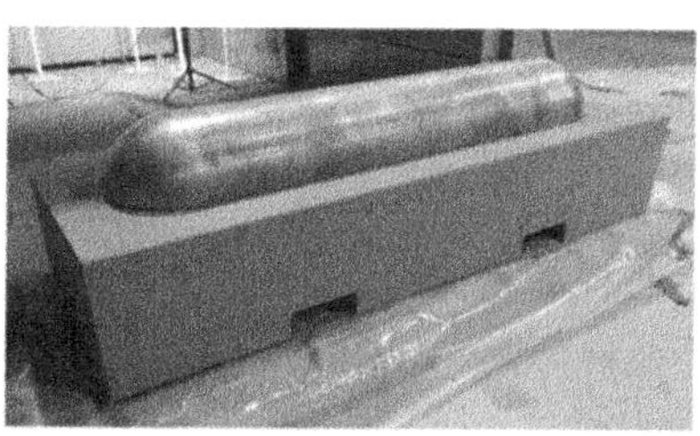

(e) Copper-coated used fuel container with bottom half of buffer box (www.nwmo.ca)

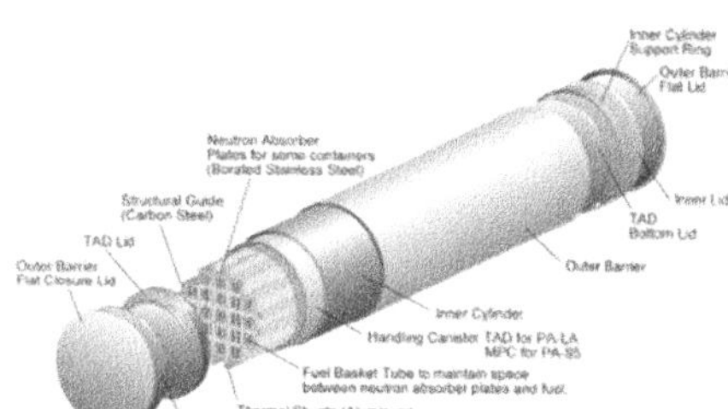

(f) Components parts of Alloy 22 waste package (DOE 2008)

Figure 3.2. *Details of various HLW/SF canisters developed by different international waste management organizations. Reproduced with permission of Posiva Oy, Andra, ONDRAF/NIRAS, Nagra, NWMO and US Department of Energy*

	Host rock	Reference canister concept	Comments	Reference
Belgium	Boom clay	Carbon steel overpack with cementitious buffer	"Supercontainer" design with integrated canister and buffer to be emplaced as a single unit.	ONDRAF/NIRAS 2013; Bel et al. 2006; Kursten et al. 2016, 2021
Canada	Crystalline or sedimentary	Copper-coated carbon steel with compacted bentonite buffer	Direct disposal of SF. Canister and buffer material assembled into a "buffer box" which is then emplaced underground. Two potential sites are under investigation, with site selection expected by 2024.	Hall et al. 2021; Keech et al. 2021
China	Crystalline	Carbon steel with compacted bentonite buffer	Underground research laboratory under construction at Beishan, northwest China.	Chen et al. 2017; Wei et al. 2021
Czech Republic	Crystalline	Outer carbon steel case with inner Type 316L stainless steel container. Compacted bentonite buffer material	Horizontal emplacement of canisters in long boreholes. Outer copper or titanium alloy shell with internal carbon steel structural support has been proposed as an alternative canister design. Site selection in progress.	Pospiskova et al. 2017
Finland	Crystalline	Copper-cast iron canister with compacted bentonite buffer	Disposal system based on KBS-3 methodology. Operating license application for Olkiluoto site submitted in December 2021. Operations could start as early as 2025.	Salonen et al. 2021; Posiva 2021b
France	Claystone	Carbon steel canister in un-backfilled horizontal boreholes	Boreholes are not backfilled to permit retrieval of waste for a period of several hundred years. Current plan is to reprocess but disposal of SF is possible if the fuel is classified as a waste. Construction license application to be submitted in 2022.	Crusset et al. 2017
Germany	To be determined	To be determined	Site identification and selection process in progress.	https://www.bge.de/en/repository-search/siting-regions/

	Host rock	Reference canister concept	Comments	Reference
Japan	To be determined	Carbon steel with compacted bentonite buffer	Copper-coated carbon steel canister as an alternative to the reference concept.	Ogawa et al. 2017, 2021; Suzuki et al. 2021
Korea	To be determined	Cu-coated steel or Cu-cast iron canister with compacted bentonite buffer	Korean program at a generic stage with no site or host rock type yet identified. Canister development studies have focused on either a KBS-3 or copper-coated steel design.	Choi et al. 2013; Kook et al. 2017
The Netherlands	–	–	Because of the relatively small amount of waste produced, the Dutch government has opted for long-term storage (at least 100 years) rather than deep geological disposal.	https://www.covra.nl/en/
Russia	Granite	Stainless steel canister with highly compacted bentonite	A site has been selected in the Nizhnekansky granitoid massif, located near Zheleznogorsk (Krasnoyarsk region). Details of the programme are unclear, especially the proposed use of a stainless steel canister for the disposal of heat-generating HLW.	Petrov et al. 2019; https://www.oecd-nea.org/jcms/pl_33753/russian-federation-report-web
Spain	To be determined	Metallic canister with bentonite or cementitious buffer	Centralized Temporary Storage (CTS) facility to be constructed for SF/HLW until final deep geological repository is available.	Ministry of Industry, Tourism, and Commerce (2006)
Sweden	Crystalline	Copper-cast iron canister with compacted bentonite buffer	Disposal system based on KBS-3 methodology. Construction license application for Forsmark site approved in 2022.	SKB 2010a, 2010b; SKB 2011
Switzerland	Opalinus Clay	Carbon steel in bentonite-backfilled disposal tunnels	Copper-coated carbon steel canister as an alternative to the reference concept. General license application for preferred site to be submitted by the end of 2024.	Diomidis et al. 2017; Reddy et al. 2021a, 2021b; Diomidis and King 2022

	Host rock	Reference canister concept	Comments	Reference
Taiwan	Crystalline	Copper-cast iron canister with compacted bentonite buffer	Disposal system based on KBS-3 methodology.	Hung et al. 2017; Taipower 2022
United Kingdom	To be determined	To be determined	The program is at the stage of site identification and, therefore, canister issues have been generic in nature to date.	King and Padovani 2011; King et al. 2016
United States	Volcanic tuff	Alloy 22 waste package/Grade-7 Ti drip shield	While the license application has been suspended, Yucca Mountain remains the designated host site according to the 1987 Nuclear Waste Policy Act. The only unsaturated repository site currently under consideration.	DOE 2008; Hua and Gordon 2004; Rechard and Voegele 2014

Table 3.2. *Summary of national nuclear waste disposal programs considering deep geological disposal of HLW/SF*

3.3. Choice of material, canister design and canister fabrication

Once the canister requirements and safety functions are defined, it is then possible to select a material and design that can fulfill those requirements. The selected design must also be able to be fabricated with existing, or reasonably foreseeable, technologies.

Table 3.2 summarizes a number of canister materials and designs that have been selected by various national waste management programs. There are a number of factors that determine the selection of the canister material and design, as shown below (King 2012).

– The desired canister lifetime: with copper or a passive material preferred if lifetimes > 100,000 years are required.

– The impact of the canister on the other engineered and natural barriers: for example, under anaerobic conditions carbon steel corrodes with the formation of Fe(II) and H_2. Ferrous ions can react with bentonite clay buffer materials, resulting in a loss of the desired buffer swelling pressure. Hydrogen generation can lead to a build up of pressure in the repository which may impact the integrity of both the buffer material and of the host rock itself, particularly in the case of low-permeability sedimentary rocks.

– The nature of the environmental conditions in the repository: some environmental conditions may preclude the use of certain materials; for example, copper may be unsuitable for sites with high sulfide concentrations in the groundwater. However, as discussed below, it is important to remember that the near-surface environment may be different from that in the host rock and it is the former that determines the corrosion behavior of the canister.

– The amount of waste/resource usage/cost: for programs with large amounts of HLW/SF that will require, for example, several tens of thousands of canisters, the cost and availability of a particular alloy may be a determining factor.

– Program-specific or regulatory requirements: for example, if the canisters have to be retrievable for a certain length of time, then this requirement will affect the design of both the overall engineered barrier system (e.g. the disposal room cannot be backfilled) and of the canister itself (e.g. an alloy must be selected which provides adequate long-term performance in the absence of a buffer or backfill material).

From the viewpoint of the long-term corrosion performance, the range of candidate canister materials can be divided into "corrosion-allowance" and "corrosion-resistant" alloys. Corrosion-allowance materials are those that corrode uniformly (or actively), with little or no tendency toward localized corrosion or environmentally assisted cracking (EAC), such as copper or carbon steel. Corrosion-resistant materials are those protected by a passive film, such as titanium- and nickel-based alloys, and which corrode uniformly at very slow rates but which can be susceptible to localized film breakdown under certain environmental conditions. Some materials can fall into both groups. For example, carbon steel may be active or passive depending on the nature of the buffer material (being active in contact with compacted bentonite but passive in the presence of a cementitious backfill).

As well as having suitable corrosion properties, the canister must also be structurally stable during handling and under postclosure conditions. Ductile materials, such as copper, require an internal structural support (typically provided by an inner carbon steel or cast iron vessel), whereas stronger materials, such as carbon steel, may not.

These dual requirements of resistance to both chemical (corrosion) and mechanical loads, coupled with the properties of the various candidate alloys, lead to the two major types of canister design: single-shell and dual-shell (or dual-wall). In dual-wall designs, the outer shell provides corrosion resistance and the inner shell or vessel provides the structural support. These may be physically separate shells, as in the KBS-3 copper-cast iron design (SKB 2011) or intimately bonded, as in the copper-coated carbon steel design being developed in Canada and elsewhere (Hall et al. 2021). As well as copper, a number of metallic and ceramic coating materials have been considered (Holdsworth et al. 2018; Abdelouas et al. 2022). In single-shell canister designs, the one material provides both corrosion resistance and structural integrity, as in the Nagra carbon steel canister (Patel et al. 2012).

Other design considerations include the use of multi-purpose canisters or "supercontainers" comprising both the canister and associated buffer material. Multi-purpose containers (MPC) have been considered for use in some countries (IAEA 2000), for example in the United States (Rechard and Voegele 2014; Hardin et al. 2015). In the MPC concept, the same canister would be used for several functions, including interim storage, transportation and disposal. Despite the obvious advantages in terms of efficiency and avoiding the need to handle the SF/HLW repeatedly, there has been limited uptake of the MPC concept, in large part because of the difficulties associated with having to relicense the canister for the different functions. The supercontainer concept, in which the canister and buffer material are prefabricated into a single unit above ground and then moved underground and emplaced, has been more widely promoted. The Belgian Supercontainer, comprising a carbon steel overpack and cementitious backfill surrounded by a stainless steel liner (Figures 3.1(c) and 3.2(c)), would be fabricated in the surface-shielded facilities at the repository site and then the entire assembly would be maneuvered into the final repository (ONDRAF/NIRAS 2013). The Canadian buffer box assembly is another example of the supercontainer concept (Figures 3.1(e) and 3.2(e)). In this case, the copper-coated carbon steel container would be located within two shaped compacted buffer blocks. Because of the relatively small size of the container, which in turn is possible because of the small size of CANDU fuel bundles, the entire buffer box assembly can be handled with conventional fort lift technology.

Canister fabrication generally employs conventional manufacturing technology, welding techniques and inspection technology (King et al. 2016). Apart from the need to conduct the loading, sealing and inspection processes in a hot cell or shielded facility, the fabrication and welding of SF/HLW canisters can generally be achieved using conventional methods widely used for the manufacture of similarly sized and shaped components for other industries. This is not to say, however, that significant development work has not been required; for example, in areas such as the manufacturer of thick-walled (5 cm) copper vessels with suitable mechanical and microstructural properties (SKB 2010c; Raiko 2013), or high-quality, low porosity copper coatings (Giallonardo et al. 2017), or methods for adequate thermal stress relief of closure welds in thick carbon steel vessels that avoid exceeding the upper temperature limits for the SF or HLW containers inside the sealed canister (Patel et al. 2012).

3.4. Environmental conditions

As for other barriers and processes in disposal system, it is necessary to predict the long-term performance of the canister, for which a detailed understanding of the nature of the mechanical and chemical environments is essential. Tables 3.3 and 3.4 summarize the most important environmental factors that affect the mechanical and corrosion performance of the canisters, respectively. Details of the precise environmental conditions to which the canisters will be exposed can be found in the references cited in Tables 3.1 and 3.2, with indicative ranges provided in Tables 3.3 and 3.4.

Although the discussion here considers mechanical and chemical loads separately, these loads sometimes act together to affect the joint mechanical–corrosion performance of the canister; for example, as in the case of a tensile stress resulting from an external load in combination with a permissive chemical environment leading to EAC.

Although the precise environmental conditions will vary from design-to-design and from site-to-site, there are a number of common features of the environmental conditions shared by (virtually) all disposal systems. Firstly, it is well recognized that the nature of the post-closure near-field environment changes with time (Andra 2005; Diomidis and Johnson 2014; King et al. 2017). This evolution is usually characterized as a transition from an early warm, aerobic, unsaturated phase to a long-term cool, anoxic, saturated period. In terms of the relative durations of these phases, if the initially trapped atmospheric O_2 is consumed in a period of, say, 10 years, the latter anoxic period would represent 99.999% of the service life of a canister designed to remain intact for a period of 1 million years.

Environmental parameter	Effect	Indicative values
Residual stress	If not relieved by some form of post-weld heat treatment, tensile or compressive residual stresses could be present during the unsaturated phase. Surface tensile residual stress could lead to environmentally-assisted cracking under unsaturated conditions, but the residual stress will be counteracted by hydrostatic and/or lithostatic loads as the repository saturates.	The level of residual stress will depend on the fabrication method, with an overall balance of tensile and compressive stresses. Conservatively, the maximum tensile residual stress can be assumed to be equal to the yield strength of the material.
Buffer swelling pressure	Bentonite-based buffer materials develop a swelling pressure when confined and upon water saturation, with the swelling pressure dependent (among other factors) on the smectite (montmorillonite) content. The swelling pressure may be isostatic or asymmetric, in the case of uneven density distribution or nonuniform wetting. Asymmetric swelling can result in a bending load on the canister and a tensile stress component, as opposed to the compressive surface stresses that develop due to uniform swelling.	Typically, in the range 2–10 MPa, depending on buffer density and composition (smectite content), porewater salinity and temperature.
Hydrostatic pressure	For repositories located in the saturated zone, a hydrostatic pressure will develop due to the water column extending from the surface to repository depth. The height of this water column may be extended during future glaciations if a continuous water path exists from the surface to the base of the icesheet. The hydrostatic pressure will be isostatic.	The hydrostatic pressure amounts to 1 MPa for every 100 m depth of the repository (below ground level or below the surface of the icesheet). In some countries, icesheets of up to 4 km in thickness are considered, resulting in a maximum total hydrostatic pressure of 40–50 MPa.
Lithostatic loads	In lower strength rocks, such as claystones and salt deposits, an isostatic or asymmetric lithostatic load can be exerted on the canister.	Depends on the depth of the repository, the nature of the host rock and the level of in situ stress.
Shear loads	In the event of seismic-induced motion across a fracture that intersects the canister located in the repository, a shear load could be transferred to the canister (depending on the properties of the surrounding buffer material, if any). If the shear load is high enough, the canister could be plastically deformed, with a resulting residual stress.	Depends on the extent of the movement across the fracture, the properties of the buffer material (if any), and the canister design.
Impact loads	Impact loads are possible during handling and could result in localized deformation, cold work or, in the extreme, canister breaching.	Depending upon the severity of the impact, the canister may be reworked and/or scrapped.
Elevated temperature	The increase in canister temperature will affect material properties and creep rates. Thermal stresses may develop due to differential thermal expansion, depending on the canister design.	A maximum canister surface temperature of 100°C is specified in many programs (up to 140°C), with higher internal temperatures.

Table 3.3. *Environmental factors that affect the long-term mechanical performance of HLW/SF canisters*

Environmental parameter	Effect	Indicative values
Groundwater composition	Deep groundwaters tend to be saline, with Cl^-, SO_4^{2-}, and HCO_3^- as the predominant anions and Na^+ and Ca^{2+} as the predominant cations. Depending on the canister material, the groundwater composition may affect the balance between localized and general corrosion. Deep groundwaters are generally anoxic and may contain sulfide, the latter species being of particular concern for copper corrosion.	Salinity varies significantly from site to site, with deep groundwaters varying from fresh (total dissolved solids TDS of < 1 g/L), to brackish (TDS ~ 10's g/L), to saline (TDS as high as 300 g/L).
Buffer porewater composition	The canister is in direct contact with the buffer porewater, rather than the groundwater itself. The alkaline pH of cementitious porewater will passivate carbon steel. Bentonite buffer porewater exhibits a pH in the range of 7–8 and is buffered by the presence of calcite. Ion exchange and reactions involving accessory minerals (such as gypsum and various Fe(II)/Fe(III)-containing phases) will modify the composition of the incoming groundwater.	The pH will range from 7–8 to > 12 for cementitious buffers. The ionic strength of the porewater will depend on the content of reactive accessory minerals, the initial porewater composition and the salinity of the incoming groundwater. Over time, the buffer porewater composition will resemble that of the groundwater.
Elevated temperature	Elevated temperatures will accelerate the rates of thermally activated electrochemical, chemical, and mass-transport processes. Some solid phases exhibit retrograde solubility (e.g. gypsum) and are less soluble at higher temperatures.	A maximum canister surface temperature of 100°C is specified in many programs (up to 140°C), with higher internal temperatures. The maximum temperature is generally attained approximately 10 years post-closure, with the major part of the thermal pulse lasting several thousand years, and ambient temperature reestablished after a few tens of thousands of years.
Availability of oxidants/redox conditions	Because deep groundwaters are anoxic, the availability of O_2 is limited to that initially trapped in the pores of the buffer and backfill materials used to seal the repository. The oxic phase will be of relatively short duration: as little as a few weeks or months following closure. Under anoxic conditions, the evolution of H_2 is the predominant cathodic process, either from the reduction of the limitless H_2O in the saturated zone or, in the case of copper, due to the presence of sulfide.	The initial inventory of trapped O_2 depends on the repository design, but generally is in the range 1–10 moles per m^2 canister surface area. For most materials, this O_2 inventory amounts to a depth of corrosion of 30–300 μm if distributed uniformly over the canister surface. Sulfide may be present in the groundwater (typically < 1 mg/L) but can also be produced by microbial sulfate reduction.

Radiation	Decay of the SF/HLW results in an external γ- (and neutron) radiation field which will result in the formation of oxidizing and reducing molecular and radical radiolysis products. These radiolytic oxidants may support additional general corrosion, but the absorbed dose rates are insufficient to promote localized corrosion or environmentally assisted cracking due to the ennoblement of the corrosion potential E_{CORR}.	Because of the use of thick-walled canister designs, the maximum dose rates are of the order of tens of mGy/hr, with the exception being the Canadian thin-walled copper-coated canister design with a maximum dose rate of ~ 1 Gy/hr. Because Cs-137 is the principle γ-emitter in SF (half-life 30 years), the period of highest dose rate is generally taken to be the first 300 years post-closure, but the in-growth of γ-emitters at long times does result in the accumulation of a significant total dose throughout the canister lifetime.
Saturation	Buffer and backfill materials are emplaced in the partially saturated condition and slowly saturate with incoming groundwater. At early times, the material closest to the canister surface will further desiccate due to the heat flux from the decay of the SF/HLW. Therefore, there will be an initial period of unsaturated conditions during which the canister surface may be too dry to support aqueous corrosion processes. Initial wetting of the canister surface may be nonuniform, increasing the possibility of localized corrosion.	Repository saturation times will depend on the heat loading and the local hydraulic conductivity of the host rock. In many repository designs, complete saturation of the buffer and backfill materials is expected within a period of a few decades postclosure. However, in some low-permeability sedimentary formations, saturation may take tens or hundreds of thousands of years. In salt formations, or course, the nearfield is permanently unsaturated, with the exception of pockets of brine that may be present in the formation.
Microbial activity	Most repositories are designed to prevent microbial activity in the immediate vicinity of the canister surface. This is achieved using either highly compacted bentonite or alkaline cementitious backfills, both of which suppress microbial activity. Thus, biofilm formation on the canister surface is not expected. Microbial activity further away from the canister is possible; the most significant effect potentially being the microbial reduction of sulfate to sulfide.	The extent of microbial sulfate reduction may be limited by the availability of electron donor(s) (generally, either dissolve organic carbon or H_2) or electron acceptor, (i.e. sulfate, the main source of which is gypsum in bentonite).
Mass transport	By design, the repository near-field is characterized by restrictive mass-transport conditions. This will limit the transport of corrosive species to the canister surface, as well as the removal of corrosion products. As a result, the environmental conditions at the canister surface, which determine the corrosion behaviour, can be quite different from those further form the canister.	Effective diffusivities of species in compacted bentonite buffer material are typically 2–3 orders of magnitude lower than in bulk solution.

Table 3.4. *Environmental factors that affect the long-term corrosion performance of HLW/SF canisters*

This evolution of the environment has important implications for the corrosion behavior of the canister and the justification of long-term predictions. For most metals, the faster, more-localized (and, perhaps, less predictable) forms of corrosion tend to occur under warm, oxic conditions. For example, the crevice corrosion and pitting of passive alloys is promoted by higher temperatures and more-oxidizing conditions. In contrast, corrosion processes under cooler, anoxic conditions tend to be more-uniform in nature since there is insufficient electrochemical driving force to sustain the separated anodic and cathodic processes characteristic of localized corrosion. Such uniform corrosion processes tend to be easier to predict since the rates can be experimentally determined or predicted based on mass-transport considerations. Thus, the challenge of making predictions of the long-term corrosion performance over periods of hundreds of thousands of years is greatly simplified if the less-predictable localized corrosion processes are limited to only the initial stages in the evolution of the repository environment. It even becomes possible to conduct long-term experiments over meaningful timescales if, say, the period of pitting or crevice corrosion is limited to the first few years of the repository evolution.

Another important aspect of the early environmental evolution is that the near-surface environment is unsaturated. Even for repositories that saturate within a decade or so, the canister surface may be too dry initially to support aqueous corrosion processes. During this period, a fraction of the initially trapped O_2 will be consumed by processes other than canister corrosion, such as the oxidation of mineral phases in the buffer, backfill and host rock or by aerobic microbial activity in regions of the repository where such activity is possible. As a result, the already limited inventory of atmospheric O_2 will be further reduced and be unavailable to support rapid and/or localized corrosion of the canister. On the other hand, once the canister surface does wet sufficiently to start corroding, the coverage of adsorbed or deliquesced surface H_2O may be nonuniform. This nonuniform wetting could promote localized corrosion by facilitating the spatial separation of anodic and cathodic processes, the latter occurring around the periphery of surface H_2O droplets where the access to atmospheric O_2 is greatest.

In addition to the trapped atmospheric O_2, γ-irradiation of the humid or saturated buffer pore environment may also produce oxidizing species in the near field. As noted in Table 3.4, the magnitude of the external γ-fields (typically a few mGy/h to a maximum, for some canister designs, of ~ 1 Gy/h) is generally too small for this to be a significant source of oxidants. However, the debate continues concerning whether the response of the canister to irradiation is determined by the dose rate or the total absorbed dose (Abdelouas et al. 2022). Although the dose rates are modest (at least in comparison to those experienced in-reactor), the accumulated total dose over a period of 1 million years is not insignificant (of the order of a MGy), in part due to the ingrowth of γ-emitters during the decay of spent fuel (Morco et al. 2017).

It is not only the nature of the chemical environment that evolves over time, but also the magnitude and form of the mechanical loads. After emplacement but prior to the development of the external loads (buffer swelling, hydrostatic and lithostatic pressures), the canister is subject only to the residual stress from fabrication, if any. From a canister performance perspective, residual stress is of most concern if there is a tensile component at the surface, as this could then support some form of EAC in the presence of suitable environmental conditions. The effect of residual stress is counteracted by the imposition of the external load(s). External loading will tend to impart compressive stresses on the canister if the loads are isostatic. However, if the loads are asymmetric, as in the case of uneven buffer swelling, bending stresses may be applied. However, unless there is a permanent asymmetric lithostatic pressure, the external loading will tend to become isostatic over time.

In summary, both the chemical and mechanical environment can be expected to evolve over time following the initial emplacement of the canisters. This evolution has important implications for the post-closure performance of the canister, since the corrosion and mechanical degradation mechanisms to which the canister will be subjected will also change over time.

3.5. Canister performance

3.5.1. *Corrosion performance*

There are many publications describing the corrosion behavior of specific canister materials (Table 3.1). Here, the approach is instead to frame the discussion in terms of different corrosion processes. As elsewhere in this chapter, issues of current importance are highlighted.

3.5.1.1. *Unsaturated corrosion*

It is important to recognize that there will be a period of unsaturated conditions in the repository for a period of time after emplacement of the canisters. The unsaturated period is an important, although until recently rarely considered, phase in the evolution of the corrosion performance. Conditions may be such that corrosion either does not occur at all if the surface is too dry to sustain aqueous electrochemical processes (on the presumption that the rate of oxidation of the canister materials in dry air is inconsequential at the temperatures of interest, i.e. $< 150°C$) or could be localized if the surface is non-uniformly wetted.

If the initially trapped atmospheric O_2 is consumed prior to the wetting of the canister surface, then there may be no period of oxic corrosion. There is evidence from some full-scale tests that the initially trapped O_2 is "consumed" within a matter of days or weeks following backfilling of the disposal tunnels (Giroud et al. 2018). The process or processes responsible for the loss of O_2 from the FE-G experiment are still somewhat uncertain, and it is possible that the (temporary) disappearance of oxygen is due to a reversible sorption process and that the O_2 may be desorbed at a later stage. However, if the O_2 has been irreversibly sequestered then it opens up the possibility of the use of materials such as stainless steel that have excellent performance under anaerobic conditions but which are generally not considered for the disposal of heat-generating wastes because of the concern over localized corrosion and stress corrosion cracking (SCC) under aerobic conditions.

However, if the canister surface does wet during the oxic, unsaturated phase, then corrosion may be localized and rapid. Formation of droplets of a deliquesced salt solution, or other regions of nonuniform wetting, will naturally result in the spatial separation of anodic and cathodic processes via the Evans droplet mechanism. Greater access of O_2 to the periphery of the wetted region results in this location becoming the cathode, which can then support anodic dissolution in the center of the droplet. Rapid diffusion of O_2 through unsaturated buffer and backfill materials could sustain the cathode and result in deep local penetration. This scenario is currently under investigation for copper, but is of concern for all canister materials. While this mechanism is unlikely to lead to through-wall penetration for the 5-cm-thick KBS-3 canister, it is of concern for the copper-coated design where the corrosion allowance is of the order of a few mm.

The other important consideration under unsaturated conditions is that any aqueous phase contacting the (hot) canister surface will be subject to evaporation, leading to concentration of the electrolyte. Evaporative conditions can lead to environments that are more aggressive than those encountered under isothermal conditions at elevated temperature. The effect of evaporation on the evolution of the aqueous phase on the canister surface was extensively studied in the Yucca Mountain Program (YMP) due to the possibility of seepage water dripping onto the canister in the un-backfilled drifts (DOE 2008).

3.5.1.2. *Uniform corrosion*

Once the surface of the canister becomes uniformly wetted, corrosion will be more-uniform in nature. All canister materials will be subject to uniform corrosion, albeit at vastly different rates. Active materials, such as carbon steel in contact with bentonite buffer, will corrode uniformly at a rate of the order of μm/year. The corrosion rate will be higher under oxic conditions and may be limited by the rate of supply of O_2. However, since carbon steel is a base material, corrosion will continue under anaerobic conditions indefinitely (accompanied by the evolution of H_2), although at a lower rate. Passive materials, such as titanium and nickel alloys and stainless steels, corrode much more slowly (of the order of 1–10 nm/year) due to the presence of a stable passive film. Carbon steel is also passive in cementitious backfill, with a correspondingly low corrosion rate. Copper, although typically classified as an active material, is also near-noble and will only corrode in the presence of suitable oxidants, such as the initially trapped O_2. However, if the activity of dissolved copper is reduced sufficiently by the presence of strong complexing species, the metal becomes thermodynamically unstable in H_2O and will corrode with the evolution of H_2. This situation arises in the presence of sulfide (since the solubility product of Cu_2S is of the order of 10^{-46}), and for this reason there has been extensive study of the corrosion of copper canisters by HS^- (Chen et al. 2014; Guo et al. 2021; Posiva 2021a). Chloride ions also complex Cu(I), although not sufficiently strongly that copper dissolution is likely to be significant at the Cl^- concentration and pH typical of repository conditions (Lilja et al. 2021). There has also been an ongoing debate about whether copper corrodes in pure, O_2-free H_2O, but the consensus now is that this is not a phenomenon of concern, even though the original observations cannot be explained (Hedin et al. 2018).

Methods used for the prediction of the rate and/or extent of uniform corrosion are discussed in section 3.6.

3.5.1.3. *Localized corrosion*

Canisters are generally designed with a sufficient wall thickness that uniform corrosion will not lead to early failure of the canister. Localized forms of attack are of a larger concern, such as pitting or crevice corrosion. From an electrochemical point of view, the coupling of a large cathode to a small anode can lead to high rates of localized attack and the possibility of deep penetrations ahead of the uniform corrosion front. Localized corrosion is of most concern for passive materials that are generally used in thinner sections than corrosion allowance canister materials. Pits form on bold, exposed metal surfaces, whereas crevice corrosion is associated with occluded geometries that lead to differential environmental conditions within and outside of the crevice.

The lifecycle of a pit or crevice can be divided into initiation (birth), propagation (growth) and stifling (death) stages. The classic electrochemical criterion for initiation is that the corrosion potential E_{CORR} exceeds the passive film breakdown potential E_B. Propagation continues provided a critical occluded chemistry is maintained within the pit or crevice by continued dissolution. Eventually, however, the pit or crevice will stifle, either because the environment cannot sustain propagation or because of the natural tendency of the alloy to reform a passive film. Different alloys exhibit different tendencies for the initiation and repassivation of localized corrosion. For example, Ti alloys are immune to pitting under repository conditions as the value of E_B far exceeds the highest value of E_{CORR} that could be attained. However, certain Ti alloys considered as candidate canister materials are susceptible to crevice corrosion that initiates under less aggressive conditions than pitting. There is a wide range of susceptibilities of Ni alloys, which reflects the wide range of alloy compositions for these materials. Increasing Cr, Mo and W contents leads to a decreased susceptibility at the expense of higher costs.

There are two basic arguments that can be used to support the use of a canister material that is susceptible to localized corrosion. One argument is that the material will be immune to initiation under repository conditions, either because the alloy itself is highly resistant or because the near-field environment (in particular, the temperature, Cl^- concentration, and redox conditions) is not sufficiently aggressive to cause E_{CORR} to exceed E_B. Alternatively, it can be argued that while a pit or crevice may initiate, propagation will stifle before the canister wall thickness has been penetrated, or that penetration will only occur at times greater than the target lifetime. Unfortunately, although there have been great advances in our understanding of the birth, growth and death of pits and crevices over the past several decades, it has proven difficult to develop a sufficiently persuasive argument to support the use of canister materials susceptible to localized corrosion. This is one of the main reasons why the majority of international programs have adopted the use of either copper or carbon steel, which are less susceptible to these localized forms of attack.

3.5.1.4. *Environmentally assisted cracking*

As with localized corrosion, EAC mechanisms such as SCC or hydrogen-induced cracking (HIC) can be discussed in terms of initiation, propagation and stifling events. Unlike localized corrosion, however, it is difficult to argue that a propagating crack will not lead to canister failure, such is the typical magnitude of the crack growth rate relative to the required canister lifetime. For this reason, with the exception of a slip-oxidation model proposed for predicting the rate of SCC propagation for Alloy 22 waste packages in the YMP (Andresen et al. 2005), it is generally considered necessary to demonstrate that the canister material is immune to EAC or to assume that canister failure coincides with crack initiation.

To exclude the possibility of SCC, it is common to argue that one or more of the prerequisite factors for cracking are absent, namely, the need for a susceptible material, a supportive chemical environment and sufficient tensile stress. It is usually argued that the necessary environmental conditions will not be present in the repository, such as concentrated carbonate, nitrate or caustic solutions in the case of carbon steel (King 2010) or ammonia, acetate or nitrite in the case of copper (King 2021). Less commonly, it has also been argued that the canister material itself is either not in a suitable metallurgical condition (e.g. the material is too ductile) or has been optimized to reduce the susceptibility to SCC (Crusset et al. 2017). Rarely, is it argued that there will be insufficient tensile stress (or strain), despite the fact that many of the loading conditions in Table 3.3 will result in compressive stresses. This reluctance to argue on the basis of insufficient stress may be because it is difficult to guarantee that there will not be any region of tensile stress on the canister or because of the general lack of threshold stress and threshold stress intensity factor data for relevant SCC systems in the literature.

In the case of hydrogen-related forms of EAC, it is difficult to argue that the environment will not support cracking. For all but the first few years of the canister lifetime, corrosion will be supported by the evolution of H_2 and a significant gas pressure (approximately equivalent to the sum of the hydrostatic and the buffer swelling pressures) could develop at the canister surface, especially in low-permeability host rocks (Diomidis et al. 2016), leading to significant absorbed hydrogen concentrations. Hydrogen can also be produced by the discharge of protons in the acidified occluded environments for materials undergoing localized corrosion, such as the crevice corrosion of unalloyed Ti (Hua et al. 2005). Absorption of a fraction of the discharged H atoms is facilitated in this case by the absence of a passive film within the crevice, which otherwise acts as a barrier to hydrogen entry.

Given that hydrogen production and absorption is inevitable, hydrogen-related degradation is generally treated on the basis of the time to reach a threshold for crack initiation. This threshold may be in the form of a critical absorbed H concentration for the HIC or Ti alloys (Hua et al. 2005) or of a critical hydride layer thickness, again for Ti alloys (Nakayama et al. 2008). For carbon steel, a fracture mechanics approach has been developed to predict the time at which the H-related reduction in toughness of the material reaches the threshold for sub-critical crack growth or fast fracture (King et al. 2014) (the slow crack growth rates in the literature are sufficiently high enough for a sub-critical crack to grow to a critical dimension relatively quickly on a repository timescale). In most cases, canister failure is assumed to be instantaneous once the threshold criterion is reached.

3.5.1.5. *Microbiologically influenced corrosion*

Virtually all engineering materials have been shown to be more or less susceptible to microbiologically influenced corrosion (MIC) (Little et al. 1991). Of the alloys proposed for HLW/SF canisters, only Ti and, possibly, the more corrosion-resistant Ni-Cr-Mo alloys can be claimed to be resistant to MIC. In the former case, the stability of the passive TiO_2 film over a wide range of pH and redox conditions and the absence of multiple oxidation states may explain the relative immunity of Ti alloys to the various effects of microbes. In the case of the highly corrosion-resistant Ni alloys, the general stability of the passive film, and its inherent ability to repassivate should localized breakdown occur, limits the effects of microbes.

The general acceptance that most, if not all, HLW/SF canister materials are susceptible to MIC is based on industrial experience in which these alloys are directly exposed to active microbial colonies. However, the near-field environment is quite inhospitable for microbes and microbial activity at the canister surface is not expected to occur for repository designs involving the use of bentonite or cementitious buffer materials (Aerts 2009; King 2009). The controlling mechanism responsible for the suppression of microbial activity in compacted bentonite is currently uncertain but, from an engineering standpoint, is associated with a threshold dry density variously reported to be in the range 1,400–1,600 kg/m^3. In cementitious backfill, it seems clear that suppression of microbial activity is due to the highly alkaline conditions. In both cases, provided the buffer can be installed with appropriate properties, the resulting suppression of microbial activity at the canister surface precludes the aggressive forms of MIC that occur underneath biofilms that dominate the MIC literature. As a result, microbial activity in more-hospitable locations within the repository will only adversely impact the corrosion behavior of the canisters if corrosive metabolic by-products are able to diffuse to the canister surface. Such remote microbial activity can also be beneficial, since aerobic respiration may be responsible for the rapid consumption of O_2 observed in various field tests.

Of the various forms of microbial activity possible in hospitable regions of the repository, most attention has been paid to the reduction of sulfate to sulfide by sulfate-reducing bacteria (SRB). This emphasis is largely because of the importance of sulfide in the long-term corrosion of copper canisters, but sulfide is also detrimental for carbon steel and Ni-based alloys, in the latter case due to the possibility of depassivation due to the segregation of sulfur to the metal-oxide interface (Marcus 1995). A number of reactive-transport models have been developed to predict the production, transport and consumption of microbially produced sulfide (Idiart et al. 2019; King and Kolář 2019; Pękala et al. 2019). Microbial sulfide production is modeled using Monod kinetics, with the overall rate limited by the availability of either the electron acceptor (i.e. sulfate) or donors (typically assumed to be either organic carbon or H_2). Not all of the sulfide produced is predicted to reach the canister, with reaction of HS^- with Fe(II) and/or Fe(III) species in the buffer being a significant sink.

3.5.1.6. *Effect of radiation*

The magnitude of the external gamma (and neutron) field at the canister surface is small compared with that encountered in-reactor or in waste processing facilities where instances of radiation-induced corrosion (RIC) have been reported. The substantial wall thickness required to withstand the external loads within the repository also provides significant radiation shielding, with maximum external absorbed dose rates typically in the range of 0.1–10 Gy/h (Padovani et al. 2017). Although the dose rate is modest, there is still debate about whether it is the dose rate or the total absorbed dose that is most important in determining the overall impact on corrosion. While, intuitively, it is most likely that the rate of generation of radiolytic oxidants primarily determines the impact on the corrosion kinetics, it is feasible that for a cathodically limited corrosion process the total yield of oxidants determines the extent of corrosion. Determining whether the dose rate or total dose is determining is important because, although the dose rate diminishes to negligible levels after 300 years (equivalent to 10 half-lives of Cs-137), the total dose continues to accumulate throughout the target lifetime of 1 million years for a copper canister, in part because of the in-growth of gamma-emitters as the fuel decays (Morco et al. 2017).

Radiolysis of the near-field environment may have a number of effects on the canister corrosion behavior (Farnan et al. 2018) (the magnitudes of the gamma-field and neutron flux are insufficient to cause embrittlement or other metallurgical changes to the canister materials). Radiolysis will produce a series of oxidizing (and reducing) radiolysis products. Molecular and radical radiolytic species, such as H_2O_2 and $OH^\bullet$, will add to the limited inventory of oxidants within the repository and may support additional uniform corrosion. At sufficiently high dose rates, the E_{CORR} of various materials can be ennobled (shifted in the positive direction), increasing the possibility of localized corrosion (Farnan et al. 2018). However, the degree of ennoblement is minor at repository-relevant dose rates and an increase in susceptibility to pitting or SCC is unlikely. In fact, irradiation has been found to inhibit the crevice corrosion of Grade-2 Ti, possibly by promoting repassivation by the production of oxidizing species within the crevice (King et al. 2002). Indeed, there is evidence that irradiation at low dose rates (< 20 Gy/h) suppresses the uniform corrosion rate of copper (King and Behazin 2021).

Irradiation may also modify the near-field environment (King and Behazin 2021). The generation of nitric acid by the radiolysis of moist air can lead to the formation of nitrate salts, which are believed to promote the surface wetting of copper coupons exposed to irradiated humid air (Ibrahim et al. 2018). The reduction of nitrate could lead to the formation of nitrite, a known SCC agent for copper, as is NH_3 which is another potential N-containing product of the radiolysis of N_2-H_2O atmospheres. The buffer or backfill material close to the container will also be irradiated, but so too will the buffer or backfill material close to the container. In the case of bentonite buffer, $H^•$ radicals and solvated electrons e^-_{aq} may reduce the structural Fe(III) located in the octahedral lattice of the clay to Fe(II) (Gournis et al. 2000), which can then be reoxidized by H_2O_2 (Holmboe et al. 2012). Hydroxyl radicals $OH^•$ can oxidize pyrite present as an accessory mineral in many natural bentonites (Lefticariu et al. 2010).

Overall, however, the effect of gamma irradiation on the corrosion of the canister and on the near-field environment is expected to be minor because of the modest radiation fields involved.

3.5.1.7. *Miscellaneous forms of corrosion*

Specific canister materials and/or designs may be subject to other forms of corrosion. For example, copper-steel (or copper-cast iron) dual-wall canisters may be subject to galvanic corrosion under aerobic conditions if there is a through-wall defect in the outer copper corrosion barrier (Standish et al. 2016). This form of corrosion is of more concern for a copper-coated steel canister than for the thick-walled KBS-3 design, but in both cases the outer copper layer would act as the cathode supporting accelerated corrosion of the exposed underlying carbon steel or cast iron inner vessel. Welds can also be locations of particular susceptibility, although the experience in many waste management programs is that preferential corrosion of welded material is not a major concern.

3.5.2. *Mechanical performance*

Unlike other industrial components, there are no well-accepted codes or standards that specifically relate to the mechanical performance of HLW/SF canisters. However, the general practices adopted in other industries for safety-critical components, such as ASME's (American Society for Mechanical Engineers) Boiler and Pressure Vessel Code (https://www.asme.org/codes-standards/bpvc-standards) can be used as a guide. Generally, the canister is designed to withstand the expected loads with a certain margin of safety, and extreme loads with a lower safety margin. In some cases, the time-dependent degradation of the canister is taken into account, for example, by assuming a reduced wall thickness as a result of corrosion for long-term loading scenarios. Because this is a nuclear application, and since in-service inspection and repair of the canisters is not possible, higher safety margins than would otherwise be considered are appropriate. However, these safety margins should be considered in the context of the overall safety of the system, that is, lower safety margins may be acceptable if the canister is not the primary barrier limiting the dose.

Structural analysis of the mechanical performance is typically performed initially for a perfect, pristine canister. However, all fabricated structures will contain flaws and it is important to take into account the presence of possible defects in the structural analysis. The structural analysis can also be used to determine the critical defect size (as well as shape, orientation and location) for the expected loads, which can then be compared to the available inspection techniques to determine whether a critical defect can be reliably detected.

3.5.2.1. *Plastic collapse*

Plastic collapse, either globally due to wall thinning of the canister or locally if associated with a defect, is characterized by excessive plastic strain due to stresses greater than the yield strength of the material. "Failure" by plastic collapse will occur when the degree of plastic strain exceeds a design-specific limit and may or may not result in a through-wall breach. The likelihood of plastic collapse increases as the load-bearing wall thickness is reduced by corrosion and/or the load increases, for example, as a result of the increase in pressure due to a future glaciation event.

3.5.2.2. *Brittle/ductile fracture*

Fracture results in the propagation of a crack in a material and is associated with a threshold stress or stress intensity factor K_I (the properties of ductile materials are best characterized by the J-integral, but here it is assumed that the behavior is linear elastic). Fast fracture, as characterized by the fracture toughness K_{Ic}, is preceded by a period of slow or sub-critical crack growth until such time that the crack reaches the critical size and the crack growth rate increases significantly. In some industrial applications, it is possible to manage the component lifetime by monitoring the period of slow crack growth and removing the component from service before the critical defect size is reached. This is not possible for HLW/SF canisters and it is necessary to base predictions on the threshold stress intensity factor instead. Both the fracture toughness and the threshold K_I are affected by environmental factors such as temperature, the degree of radiation embrittlement, and the absorption of hydrogen. As noted above, the magnitude of the neutron flux for a SF canister is too low to cause significant radiation embrittlement, but the absorption of H can be significant (see below).

3.5.2.3. *Creep*

In contrast to plastic collapse due to stresses greater than the yield strength, creep is plastic deformation resulting from stresses below yield. Creep is generally associated with high-temperature applications but, even at repository-relevant temperatures, is of concern for copper and Ti alloys (Dutton 1996; Björkblad and Faleskog 2018; Andersson-Östling 2020). Creep rupture will occur when the plastic strain exceeds the creep rupture strain, which is lower than the rupture strain measured under rapid straining conditions. For example, the creep rupture strain of oxygen-free copper can be < 10% (Andersson-Östling 2020), compared with a ductility of 50–60% from tensile loading. For this reason, SKB have developed OFP copper in which 30–100 wppm phosphorus is added to improve the creep ductility. Creep will also result in a redistribution of the stresses in the canister shell, which can both relieve and induce tensile stresses and, thus, impact the SCC behavior (Salonen et al. 2021).

Creep of the outer copper (or Ti) corrosion barrier of a dual-wall canister is only possible, of course, if there is a gap between the outer shell and inner load-bearing component. This is an advantage, therefore, of a coated canister design in which the outer corrosion barrier is applied directly onto a thick-walled steel or cast iron inset that supports the load and which itself is not subject to deformation.

3.5.2.4. *Impact loads*

The only possibility of dynamic (or impact) loading occurs during handling and, in tectonic regions, due to movement across a fault zone that intersects the canister location. Accidents caused by serious handling could damage the canister or its contents, necessitating repair or repackaging of the waste. Accidents caused by minor handling could lead to local plastic deformation, resulting in increased residual stress and an altered microstructure. Criteria are required to determine the level of acceptable handling damage. Shear loading due to a seismic event could result in plastic deformation of the canister if the load is transferred to the canister rather than being absorbed by the buffer material (Jonsson et al. 2018). Plastic deformation will also lead to a residual stress.

3.5.3. *Joint mechanical–corrosion performance*

Generally, the mechanical and corrosion performance of the canister are considered separately. Occasionally, the effect of a reduced wall thickness due to corrosion is taken into account in the structural analyses. Likewise, the distribution and magnitude of tensile stress is sometimes included in the assessment of SCC, although it is common to assume that there will be sufficient stress and instead to argue that the necessary environment will not be present. However, it is good practice to consider the possibility of interactions between the various mechanical and corrosion degradation mechanisms.

The degree of interaction between mechanical and corrosion degradation mechanisms depends on the design of the canister and the materials of construction. Single-shell canister designs, in which the one component provides both structural stability and corrosion resistance, are more likely to be subject to joint mechanical–corrosion degradation than dual-wall designs, in which the outer layer provides corrosion resistance and the inner vessel structural strength. While all materials are subject to uniform corrosion and, hence, a loss of load-bearing wall thickness with time, corrosion allowance materials such as carbon steel will suffer a greater loss of load-bearing capacity than a passive material would. Some materials will experience a loss of ductility or toughness as a result of exposure to the corrosive environment, for example, the absorption of hydrogen during anaerobic corrosion leads to a decrease in fracture toughness of carbon steel and the formation of brittle hydride phases in Ti alloys.

Examples of joint mechanical–corrosion degradation modes that have been considered include:

- SCC of copper (Posiva 2021a; Salonen et al. 2021);
- SCC of carbon steel (Didot et al. 2017; Necib et al. 2017);
- SCC of Alloy 22 (Andresen et al. 2005);
- HIC of carbon steel (King et al. 2014);
- hydride-induced cracking of Ti alloys (Hua et al. 2005; Nakayama et al. 2008);
- thermal aging of Alloy 22 (Rebak et al. 2000).

3.6. Lifetime prediction

As with many other aspects of the disposal system, one of the challenges is making predictions of the corrosion and mechanical performance of the canister over periods of thousands to hundreds of thousands of years. If making the prediction of canisters lifetimes is challenging, then justifying those predictions to both expert and nonexpert audiences is even more challenging. Central to making such justifiable predictions is to be able to demonstrate a thorough understanding of the underlying mechanisms.

It is first necessary to define what is meant by canister "failure". Failure may be defined as the time to completely penetrate the corrosion barrier. But if penetration is the result of a narrow intergranular stress corrosion crack with a tortuous pathway, then it could be argued that the canister still provides a significant mass-transport barrier to the ingress of groundwater and the egress of radionuclides. And what of the internal load-bearing vessel which, while commonly closed with a mechanical fastener rather than a weld, may still act as a significant barrier? Mechanical canister "failure" may not even involve a through-wall penetration if it is defined as a certain degree of plastic deformation.

Regardless of the definition of failure, it is necessary to predict the time-dependent progression of one or more corrosion processes to which the canister will be subject. Table 3.5 summarizes the methods used to predict the long-term performance for the four main corrosion mechanisms of concern (uniform and localized corrosion, EAC and MIC) for each of the four main classes of HLW/SF canister materials (copper, carbon steel/cast iron, Ti alloys and Ni-based alloys). The various methods range from mechanistically based reactive-transport modeling to simple empirical extrapolations. In addition to predicting corrosion damage, some of these methods involve the development of "reasoned arguments" for excluding corrosion processes that are not considered possible under repository conditions.

As noted above, the important aspect of making such predictions is not the mathematical complexity of the model but the ability to demonstrate a thorough mechanistic understanding. It is only through such understanding that predictions over such long timescales can be justified. Other methods for building confidence in the predictions are presented below.

– Archaeological and natural analogues (Neff et al. 2010): for copper and steel/iron canisters, the ability to be able to point to objects that have survived for periods of hundreds or thousands of years (or geological timescales in the case of native copper deposits) can be a persuasive argument, for both expert and nonexpert audiences alike.

– In situ tests: most waste management programs conduct large-scale in situ tests under conditions that closely simulate those in the repository. While corrosion studies are usually limited to post-exposure examination of coupons or simulated canisters, real-time monitoring of different corrosion processes is also possible (Diomidis and King 2020). In situ tests provide an ideal opportunity to study the corrosion behavior during the initial redox transient during which the most severe and difficult to predict forms of corrosion are most likely to occur.

– Alternative models: being able to demonstrate that different models predict similar behavior can also build confidence in corrosion predictions. A simple example is the use of mass-balance arguments and detailed reactive-transport modeling to predict the extent of aerobic corrosion.

	Uniform corrosion	Localized corrosion	EAC	MIC
Copper	Mass-balance or reactive-transport modeling for initially trapped O_2. Reasoned argument or radiolysis modeling for radiation-induced corrosion. Mass-transport modeling for corrosion due to sulfide.	Pitting assessed using pitting factor, extreme-value analysis of maximum pit depth, or probabilistic pitting model for aerobic saturated conditions. Corrosion allowance for surface roughening. Reasoned argument for exclusion of pitting in presence of sulfide.	Reasoned argument for absence of SCC based on joint mechanical–corrosion assessment of environmental and loading conditions.	Reasoned argument based on suppression of microbial activity by highly compacted bentonite. Sulfide-transport model for remote SRB activity.
Carbon steel/iron	Mass-balance or electrochemical modeling for initially trapped O_2. Reasoned argument for excluding radiation-induced corrosion. Empirical corrosion rates or reactive-transport modeling for anaerobic corrosion.	Pitting assessed using depth-dependent pitting factor or extreme-value analysis of maximum pit depth. Cellular automata model for surface roughening.	Reasoned argument for absence of SCC or specification of a resistant alloy. Reasoned argument for absence of HIC based on low absorbed H concentration and use of low-strength alloys. Joint mechanical–corrosion analysis of failure by HIC.	Mass-balance argument based on limitation by either organic carbon or sulfate. Reasoned argument based on suppression of microbial activity by highly compacted bentonite or cementitious buffer.
Ti alloys	Empirical corrosion rates.	Limited propagation argument for Grades 2 and 12 Ti, or use of resistant alloys Grades 7, 16, 29.	HIC based on either critical absorbed H concentration or a critical hydride layer thickness.	Considered to be immune.
Ni alloys	Empirical corrosion rates.	Model based on exceedance of threshold potential, followed by rapid propagation. Reasoned argument for stifling of propagating crevice.	Slip dissolution model for limited crack growth.	Enhancement factor for uniform corrosion.

Table 3.5. *Summary of methods used to predict various corrosion processes for copper, carbon steel, titanium alloys and nickel-based HLW/SF canister materials. Specific examples can be found in King (2014)*

3.7. Outlook

This chapter has provided an overview of the design, fabrication and long-term corrosion and mechanical performance of canisters for the disposal of HLW/SF based on studies carried out over the past 40–50 years. The focus of the chapter has been on the current issues being addressed in various international waste management programs, but we should not lose sight of all of the earlier works that have helped form the basis of our current understanding. References to reviews of earlier works have been provided.

What, then, is the outlook for the future? At the time of writing, Posiva, have applied for an operating license which, if approved, would make Finland the first country in the world to grant such a license. It is hoped that the license could be approved as early as 2025, with actual disposal to begin sometime after that. However, R&D would not stop there and will continue for much of the operational period of up to 100 years. Periodic safety analyses will be required, each based on new and prior knowledge and detailed site information from the construction of the repository itself. Hopefully, we are at the point at which we transition from the long period of R&D to actual implementation of the repository system.

3.8. References

Abdelouas, A., Alonso, U., Bernier-Latmani, R., Bosch, C., Cherkouk, A., Dobrev, D., Fernández, A.M., Finck, N., Havlová, V., Hesketh, J. et al. (2022). Final version as of 15.02.2022 of deliverable D2.1 of the HORIZON 2020 project EURAD. EC grant agreement no: 847593.

Aerts, S. (2009). Effect of geochemical conditions on bacterial activity. SCK CEN Report, SCK•CEN-ER-75.

Andersson-Östling, H.C.M. (2020). Creep testing of copper intended for nuclear waste disposal – Overview of studies from 1985 to 2018. Technical Report, TR-20-06, Swedish Nuclear Fuel and Waste Management Co.

Andra (2001). Materials baseline – Volume 4: Corrosion of metallic materials. Andra Report, C.RP.AMAT.01.060.

Andra (2005). Dossier 2005 argile. Phenomenological evolution of a gelogical repository [Online]. Available at: https://international.andra.fr/sites/international/files/2019-03/5-%20Dossier%202005%20Argile%20Tome%20-%20Phenomenological%20evolution%20of%20a%20geological%20repository.pdf.

Andresen, P.L., Gordon, G.M., Lu, S.C. (2005). The stress-corrosion-cracking model for high-level radioactive-waste packages. *J. of Materials*, 57, 27–30.

Baroux, C. (2013). Summary report of the preliminary feasibility study for ceramic HLW overpacks (2007–2012). Andra Report, CG.NT.ASCM.13.0004.

Bel, J.J.P., Wickham, S.M., Gens, R.M.F. (2006). Development of the supercontainer design for deep geological disposal of high-level heat emitting radioactive waste in Belgium. *MRS Online Proceedings Library*, 932, 1221. doi: 10.1557/PROC-932-122.1.

Bennett, D.G. and Gens, R. (2008). Overview of European concepts for high-level waste and spent fuel disposal with special reference waste container corrosion. *J. Nucl. Mater.*, 379, 1–8.

Björkblad, A. and Faleskog, J. (2018). Evaluation of Cu-OFP creep crack growth and theoretical fracture models for Cu-OFP. Posiva SKB Report 03, Posiva Oy, Swedish Nuclear Fuel and Waste Management Co.

Chen, J., Qin, Z., Shoesmith, D.W. (2014). Key parameters determining structure and properties of sulphide films formed on copper corroding in anoxic sulphide solutions. *Corros. Eng. Sci. Technol.*, 49, 415–419.

Chen, L., Liu, J., Zhang, H.Y., Zong, Z.H., Wang, J. (2017). Site selection of underground research laboratory for geological disposal in China – The application of a new rock mass classification system QHLW. In *Proc. Int. High-level Radioactive Waste Management 2017*, April 9–13, Charlotte. American Nuclear Society, La Grange Park.

Choi, H.-J., Lee, J.Y., Choi, J. (2013). Development of geological disposal systems for spent fuels and high-level radioactive wastes in Korea. *Nucl. Eng. Technol.*, 45(1), 29–40.

Crusset, D., Deydier, V., Necib, S., Gras, J.-M., Combrade, P., Féron, D., Burger, E. (2017). Corrosion of carbon steel components in the French high-level waste programme: Evolution of disposal concept and selection of materials. *Corros. Eng. Sci. Technol.*, 52(Suppl. 1), 17–24.

Didot, A., Herms, E., Féron, D., Chêne, J., Crusset, D. (2017). Stress corrosion cracking in the context of deep geological nuclear disposal – Investigations on P235 and P265 steels. *Corros. Eng. Sci. Technol.*, 52(Suppl. 1), 131–135.

Diomidis, N. and Johnson, L.H. (2014). Materials options and corrosion-related considerations in the design of spent fuel and high-level waste disposal canisters for a deep geological repository in opalinus clay. *JOM*, 66, 461–470.

Diomidis, N. and King, F. (2020). The corrosion of radioactive waste disposal canisters based on in situ tests. In *Nuclear Corrosion: Research, Progress and Challenges*, Ritter, S. (ed.). Woodhead Publishing, Cambridge.

Diomidis, N. and King, F. (2022). Development of copper coated canisters for the disposal of SF and HLW in Switzerland. Technical Report, Nagra NTB 20-01, National Cooperative for the Disposal of Radioactive Waste.

Diomidis, N., Cloet, V., Leupin, O.X., Marschall, P., Poller, A., Stein, M. (2016). Production, consumption and transport of gases in deep geological repositories according to the Swiss disposal concept. Technical Report, Nagra NTB 16-03, National Cooperative for the Disposal of Radioactive Waste.

Diomidis, N., Johnson, L.H., Bastid, P., Allen, C. (2017). Design development of a copper-coated canister for the disposal of spent fuel in a deep geological repository in Opalinus Clay. *Corros. Eng. Sci. Technol.*, 52(Suppl. 1), 31–39.

DOE (2008). Yucca mountain repository license application. U.S. Department of Energy Office of Civilian Radioactive Waste Management, DOE/RW-0573, Rev. 0.

Dutton, R. (1996). A review of the low-temperature creep behaviour of titanium. Atomic Energy of Canada Limited Report, AECL-11544, COG-96-70-I.

Farnan, I, King, F., Roberts, D., Smith, V., Swanton, S., Thetford, R. (2018). Effects of ionising radiation on engineered barrier system performance. Wood Report to RWM, RWM008597.

Féron, D., Crusset, D., Gras, J.-M. (2009). Corrosion issues in the French high-level nuclear waste program. *Corrosion*, 65, 213–223.

Giallonardo, J.D., Keech, P.G., Doyle, D. (2017). Application of copper coatings to used nuclear fuel containers. In *Proc. Int. High Level Radioactive Waste Management 2017*, April 9–13, Charlotte. American Nuclear Society, La Grange Park.

Giroud, N., Tomonaga, Y., Wersin, P., Briggs, S., King, F., Vogt, T., Diomidis, N. (2018). On the fate of oxygen in a spent fuel emplacement drift in Opalinus Clay. *Appl. Geochem.*, 97, 270–278.

Gournis, D., Mantaka-Marketou, A.E., Karakassides, M.A., Petridis, D. (2000). Effect of γ-irradiation on clays and organoclays: A Mössbauer and XRD study. *Phys. Chem. Minerals*, 27, 514–521.

Guo, M., Chen, J., Martino, T., Lilja, C., Johansson, J.A., Behazin, M., Binns, W.J., Keech, P.G., Noël, J.J., Shoesmith, D.W. (2021). The nature of the copper sulfide film grown on copper in aqueous sulfide and chloride solutions. *Materials and Corrosion*, 72, 300–307.

Hall, D.S., Behazin, M., Binns, W.J., Keech, P.G. (2021). An evaluation of corrosion processes affecting copper-coated nuclear waste containers in a deep geological repository. *Prog. Mater. Sci.*, 118, 100766.

Hänninen, H., Forsström, A., Yagodzinskyy, Y. (2017). Copper behaviour in geological nuclear waste disposal. In *Proc. EURCORR 2017/20th International Corrosion Congress*, September 3–7, Prague. European Federation of Corrosion.

Hardin, E., Banerjee, K., Carter, J., Clark, R., Clarity, J., Howard, R., Kalinina, E., Scaglione, J. (2015). Investigations of dual-purpose canister direct disposal feasibility. In *Proc. Waste Management 2015 (WM2015)*, March 15–19, Phoenix.

Hedin, A., Johansson, A.J., Lilja, C., Boman, M., Berastegui, P., Berger, R., Ottosson, M. (2018). Corrosion of copper in pure O_2-free water? *Corros. Sci.*, 137, 1–12. doi: 10.1016/j.corsci.2018.02.008.

Holdsworth, S.R. (2013). Ceramic material solutions for nuclear waste disposal canisters. Nagra Report, NAB 12-45.

Holdsworth, S.R., Graule, T., Mazza, E. (2014). Feasibility evaluation study of candidate canister solutions for the disposal of spent nuclear fuel and high level waste. Report, Nagra NAB 14-90, National Cooperative for the Disposal of Radioactive Waste.

Holdsworth, S.R., King, F., Diomidis, N. (2018). Alternative coating materials as corrosion barriers for SF and HLW disposal canisters. Nagra Report, NAB 18-19.

Holmboe, M., Jonsson, M., Wold, S. (2012). Influence of γ-radiation on the reactivity of montmorillonite towards H_2O_2. *Radiat. Phys. Chem.*, 81, 190–194.

Hua, F. and Gordon, G. (2004). Corrosion behavior of alloy 22 and Ti grade 7 in a nuclear waste repository environment. *Corrosion*, 60(8), 764–777.

Hua, F., Mon, K., Pasupathi, P., Gordon, G., Shoesmith, D. (2005). A review of corrosion of titanium grade 7 and other titanium alloys in nuclear waste repository environments. *Corrosion*, 61, 987–1003.

Hung, C.-C., Wu, Y.-C., King, F. (2017). Corrosion assessment of canister for the disposal of spent nuclear fuel in crystalline rock in Taiwan. *Corros. Eng. Sci. Technol.*, 52(Suppl 1.), 194–199.

IAEA (2000). Multi-purpose container technologies for spent fuel management. Technical document, International Atomic Energy Agency, IAEA-TECDOC-1192.

Ibrahim, B., Zagidulin, D., Behazin, M., Ramamurthy, S., Wren, J.C., Shoesmith, D.W. (2018). The corrosion of copper in irradiated and unirradiated humid air. *Corros. Sci.*, 141, 53–62.

Idiart, A., Coene, E., Bagaria, F., Román-Ross, G., Birgersson, M. (2019). Reactive transport modelling considering transport in interlayer water. New model, sensitivity analyses and results from the Integrated Sulfide Project inter-model comparison exercise. Report, TR-18-07, Swedish Nuclear Fuel and Waste Management Co.

JNC (2000). H12: Project to establish the scientific and technical basis for HLW disposal in Japan, Supporting report 2: Repository design and engineering technology. Technical Report, JNC TN1410 2000-003, Japan Nuclear Cycle Development Institute, Ibaraki.

Johnson, L.H. and King, F. (2008). The effect of the evolution of environmental conditions on the corrosion evolutionary path in a repository for spent fuel and high-level waste in Opalinus Clay. *J. Nucl. Mater.*, 379, 9–15.

Jonsson, M., Emilsson, G., Emilsson, L. (2018). Mechanical design analysis for the canister. Posiva SKB Report 04, Posiva Oy, Swedish Nuclear Fuel and Waste Management Co.

Keech, P.G., Behazin, M., Binns, W.J., Briggs, S. (2021). An update on the copper corrosion program for the long-term management of used nuclear fuel in Canada. *Materials and Corrosion*, 72, 25–31.

Kerber, A. and Knorr, J. (2013). SiC encapsulation of high-level waste for long-term immobilization. *atw. Internationale Zeitschrift fuer Kernenergie*, 58, 8–13.

King, F. (2008). Corrosion of carbon steel under anaerobic conditions in a repository for SF and HLW in Opalinus Clay. Nagra Technical Report 08-12, Wettingen.

King, F. (2009). Microbiologically influenced corrosion of nuclear waste containers. *Corrosion*, 65, 233–251.

King, F. (2010). Stress corrosion cracking of carbon steel used fuel containers in a Canadian deep geological repository in sedimentary rock. Nuclear Waste Management Organization Technical Report, NWMO TR-2010-21.

King, F. (2012). Factors in the selection of container materials for the disposal of HLW/SF. *MRS Proceedings*, 1475, 241–252.

King, F. (2013). Container materials for the storage and disposal of nuclear waste. *Corrosion*, 69, 986–1011.

King, F. (2014). Predicting the lifetimes of nuclear waste containers. *J. Mater.*, 66, 526–537.

King, F. (2017). Nuclear waste canister materials: Corrosion behaviour and long-term performance in geological repository systems. In *Geological Repository Systems for Safe Disposal of Spent Nuclear Fuels and Radioactive Waste*, Apted, M.J. and Ahn, J. (eds). Elsevier Ltd, Amsterdam.

King, F. (2020). Canister materials for the disposal of nuclear waste. In *Comprehensive Nuclear Materials*, 2nd edition, Konings, R. and Stoller, R.E. (eds). Elsevier, Amsterdam.

King, F. (2021). Assessment of the stress corrosion cracking of copper canisters. Posiva Working Report, WR-2021-11, Posiva Oy, Eurajoki.

King, F. and Behazin, M. (2021). A review of the effect of irradiation on the corrosion of copper-coated used fuel containers. *Corros. Mater. Degrad.*, 2, 678–707.

King, F. and Kolář, M. (2019). Copper sulfide model (CSM) – Model improvements, sensitivity analyses, and results from the integrated sulfide project inter-model comparison exercise. SKB Technical Report, TR-18-08, Swedish Nuclear Fuel and Waste Management Co.

King, F. and Padovani, C. (2011). Review of the corrosion performance of selected canister materials for disposal of UK HLW and/or spent fuel. *Corros. Eng. Sci. Technol.*, 46(2), 82–90.

King, F., Ikeda, B.M., Shoesmith, D.W. (2002). Radiation effects and the modelling of the corrosion of nuclear waste disposal containers. In *Effects of Radiation and Environmental Factors on the Durability of Materials in Spent Fuel Storage and Disposal*. International Atomic Energy Agency Report, IAEA-TECDOC-1316.

King, F., Lilja, C., Pedersen, K., Pitkänen, P., Vähänen, M. (2010). An update of the state-of-the-art report on the corrosion of copper under expected conditions in a deep geologic repository. Swedish Nuclear Fuel and Waste Management Co. Technical Report, TR-10-67 and Posiva Oy Report, POSIVA 2011-01.

King, F., Lilja, C., Vähänen, M. (2013). Progress in the understanding of the long-term corrosion behaviour of copper cansiters. *J. Nucl. Mater.*, 438, 228–237.

King, F., Burt, D., Ganeshalingam, J., Gardner, P., Sanderson, D., Watson, S., Padovani, C. (2014). Coupled analysis of mechanical- and corrosion-related degradation of carbon steel spent fuel container. *Corros. Eng. Sci. Technol.*, 49, 442–449.

King, F., Sanderson, D., Watson, S. (2016). Durability of high-level waste and spent fuel disposal containers – An overview of the combined effect of chemical and mechanical degradation mechanisms. AMEC Report, 17697/TR/03 [Online]. Available at https://rwm.nda.gov.uk/publication/durability-of-high-level-waste-and-spent-fuel-disposal-containers/.

King, F., Hall, D.S., Keech, P.G. (2017). Nature of the near-field environment in a deep geological repository and the implications for the corrosion behaviour of the container. *Corros. Eng. Sci. Technol.*, 52, 25–30.

Kook, D., Kim, Y., Kessler, J. (2017). Status of spent nuclear fuel storage, transportation, and disposal in the Republic of Korea. In. *Proc. Int. High-level Radioactive Waste Management 2017*, Charlotte, April 9–13. American Nuclear Society, La Grange Park.

Kursten, B. (2019). Uniform corrosion evaluation of C-steel in cementitious environments relevant to the supercontainer design. SCK CEN Report, 26348152 [Online]. Available at: https://publications.sckcen.be/portal/files/6928279/Uniform_corrosion_evaluation_of_C_steel_in_cementitious_environments_relevant_to_the_supercontainer_design.pdf.

Kursten, B., Smailos, E., Azkarate, I., Werme, L., Smart, N.R., Santarini, G. (2003). COBECOMA state-of-the-art document on the corrosion behaviour of container materials. Report, European Commission 5th EURATOM Framework Programme 1998–2002, Brussels.

Kursten, B., Macdonald, D.D., Smart, N.R., Gaggiano, R. (2016). Corrosion issues of carbon steel radioactive waste packages exposed to cementitious materials with respect to the Belgian supercontainer concept. *Corros. Eng. Sci. Technol.*, 52, 11–16.

Kursten, B., Smart, N.R., Senior, N.A., Macdonald, D.D., Caes, S., de Souza, V., Gaggiano, R. (2021). Overview of anaerobic corrosion of carbon steel radioactive waste packages in alkaline media in support of the Belgian supercontainer concept. *Materials and Corrosion*, 72, 32–51.

Landolt, D., Davenport, A., Payer, A., Shoesmith, D. (2009). A review of materials and corrosion issues regarding canisters for disposal of spent fuel and high-level waste in Opalinus clay. Nagra Technical Report 09-02.

Lefticariu, L., Pratt, L.A., LaVerne, J.A., Schimmelmann, A. (2010). Anoxic pyrite oxidation by water radiolysis products – A potential source of biosustaining energy. *Earth Planet. Sci. Lett.*, 292, 57–67.

Lilja, C., King, F., Puigdomenech, I., Pastina, B. (2021). Speciation of copper in high chloride concentrations, in the context of corrosion of copper canisters. *Materials and Corrosion*, 72, 293–299. doi: 10.1002/maco/202011778.

Little, B., Wagner, P., Mansfeld, F. (1991). Microbiologically influenced corrosion of metals and alloys. *Int. Mater. Rev.*, 36, 253–272.

Marcus, P. (1995). Sulfur-assisted corrosion mechanisms and the role of alloyed elements. In *Corrosion Theory and Practice*, Marcus, P. and Oudar, J. (eds). Marcel Dekker, New York.

Marsh, G.P. and Taylor, K.J. (1988). An assessment of carbon steel containers for radioactive waste disposal. *Corros. Sci.*, 28, 289–320. doi: 10.1016/0010-938X(88)90111-4.

Ministry of Industry, Tourism, and Commerce (2006). Sixth General Radioactive Waste Plan (GRWP). Ministerio de Industria, Turismo y Comercio, Madrid.

Morco, R.P., Joseph, J.M., Hall, D.S., Medri, C., Shoesmith, D.W., Wren, J.C. (2017). Modelling of radiolytic production of HNO_3 relevant to corrosion of a used fuel container in deep geologic repository environments. *Corros. Eng. Sci. Technol.*, 52(Suppl. 1), 141–147.

Nakayama, G., Sakakibara, Y., Taniyama, Y., Cho, H., Jintoku, T., Kawakami, S., Takemoto, M. (2008). The long-term behaviors of passivation and hydride layer of commercial grade pure titanium in TRU waste disposal environments. *J. Nucl. Mater.*, 379, 174–180.

Necib, S., Bumbieler, F., Duret-Thual, C., Bulidon, N., Crusset, D., Combrade, P. (2017). Assessment of the resistance to environmentally assisted cracking (EAC) of C-steel casing and overpack in the Cox claystone. *Corros. Eng. Sci. Technol.*, 52(Suppl. 1), 95–100.

Neff, D., Saheb, M., Monnier, J., Perrin, S., Descostes, M., L'Hostis, V., Crusset, D., Millard, A., Dillmann, P. (2010). A review of the archaeological analogue approaches to predict the long-term corrosion behaviour of carbon steel overpack and reinforced concrete structures in the French disposal systems. *J. Nucl. Mater.*, 402, 196–205.

Ogawa, Y., Suzuki, S., Kubota, S., Deguchi, A. (2017). Re-evaluation of the required thickness of the carbon steel overpack for high-level radioactive waste disposal in Japan based on the latest scientific and engineering knowledge. *Corros. Eng. Sci. Technol.*, 52(Suppl. 1), 204–209.

Ogawa, Y., Suzuki, S., Taniguchi, N., Kawasaki, M., Suzuki, H., Takahashi, R. (2021). Corrosion resistance of a cast steel overpack for high-level radioactive waste disposal in Japan. *Materials and Corrosion*, 72, 52–66.

ONDRAF/NIRAS (2004). A review of corrosion and material selection issues pertinent to underground disposal of highly active nuclear waste in Belgium. ONDRAF Report, NIROND 2004-02.

ONDRAF/NIRAS (2013). ONDRAF/NIRAS Research, Development and Demonstration (RD&D) Plan. NIROND-TR 2013-12 E.

Padovani, C., King, F., Lilja, C., Féron, D., Necib, S., Crusset, D., Deydier, V., Diomidis, N., Gaggiano, R., Ahn, T. et al. (2017). The corrosion behaviour of candidate container materials for the disposal of high-level waste and spent fuel – A summary of the state of the art and opportunities for synergies in future R&D. *Corr. Eng. Sci. Tech.*, 52, 227–31. doi: 10.1080/1478422X.2017.1356973.

Patel, R., Punshon, C., Nicholas, J., Bastid, P., Zhou, R., Schneider, C., Bagshaw, N., Howse, D., Hutchinson, E., Asano, R. et al. (2012). Canister design concepts for the disposal of spent fuel and high level waste. Nagra Technical Report, NTB 12-06.

Payer, J.H., Finsterle, S., Apps, J.A., Muller, R.H. (2019). Corrosion performance of engineered barrier system in deep horizontal drillholes. *Energies*, 12, 1491.

Pękala, M., Alt-Epping, P., Wersin, P. (2019). 3D and 1D dual-porosity reactive transport simulations – Model improvements, sensitivity analyses, and results from the integrated sulfide project inter-model comparison exercise. Posiva Working Report, WR 2018-31.

Petrov, V.G., Vlasova, I.E., Rodionova, A.A., Yapaskurt, V.O., Korolev, V.V., Petrov, V.A., Poluektov, V.V., Hammer, J., Kalmmykov, S.N. (2019). Preferential sorption of radionuclides on different mineral phases typical for host rocks at the site of the future Russian high level waste repository. *Appl. Geochem.*, 100, 90–95.

Posiva (2021a). Canister evolution. Posiva Oy Working Report, WR 2021-06.

Posiva (2021b). Safety case for the operating licence application – Design basis. Posiva Oy Report, POSIVA 2021-08.

Pospiskova, L., Dobrev, D., Kouril, M., Stoulil, J., Novikova, D., Kotnour, P., Matal, O. (2017). Czech national programme and disposal canister concept. *Corros. Eng. Sci. Technol.*, 52(Suppl. 1), 6–10.

Raiko, H. (2013). Canister design 2012. Posiva Oy Report, POSIVA 2012-13.

Rebak, R.B., Koon, N.E., Dillman, J.R., Crook, P., Summers, T.S.E. (2000). Influence of aging on microstructure, mechanical properties and corrosion resistance of a Ni-22Cr-13Mo-3W alloy. In *Proc. CORROSION/2000*. NACE International, Houston.

Rechard, R.P. and Voegele, M.D. (2014). Evolution of repository and waste package designs for Yucca Mountain disposal system for spent nuclear fuel and high-level radioactive waste. *Reliability Eng. Systems Safety*, 122, 53–73.

Reddy, B., Padovani, C., Smart, N.R., Rance, A.P., Cook, A., Milodowski, A., Field, L., Kemp, S., Diomidis, N. (2021a). Further results on the in situ anaerobic corrosion of carbon steel exposed in compacted bentonite exposed to natural Opalinus Clay porewater containing native microbial populations. *Materials and Corrosion*, 72, 268–281.

Reddy, B., Padovani, C., Rance, A.P., Smart, N.R., Cook, A., Haynes, H.M., Milodowski, A.E., Field, L.P., Kemp, S.J., Martin, A. et al. (2021b). The anaerobic corrosion of candidate disposal canister materials in compacted bentonite exposed to natural granitic porewater containing native microbial populations. *Materials and Corrosion*, 72, 361–382.

RWM (2016). Geological disposal-waste package evolution status report. NDA Report, DSSC/451/01, Radioactive Waste Management.

Salonen, T., Lamminmäki, T., King, F., Pastina, B. (2021). Status report of the Finnish spent fuel geological repository programme and ongoing studies. *Materials and Corrosion*, 72, 14–24.

Scully, J.R., Féron, D., Hänninen, H. (2016). Review of the NWMO copper corrosion program. Nuclear Waste Management Organization Report, NWMO TR-2016-11, Toronto.

Shoesmith, D.W. (2006). Assessing the corrosion performance of high-level nuclear waste containers. *Corrosion*, 62, 703–722.

SKB (2010a). Fuel and canister process report for the safety assessment SR-site. Swedish Nuclear Fuel and Waste Management Co. Technical Report, SKB TR-10-46.

SKB (2010b). Corrosion calculations report for the safety assessment SR-site. Swedish Nuclear Fuel and Waste Management Co. Technical Report, SKB TR-10-66.

SKB (2010c). Design, production and initial state of the canister. Swedish Nuclear Fuel and Waste Management Co. Technical Report, SKB TR-10-14.

SKB (2011). Long-term safety for the final repository for spent nuclear fuel at Forsmark. Main report of the SR-site project. Swedish Nuclear Fuel and Waste Management Co. Technical Report, SKB TR-11-01.

Smart, N.R. (2009). Corrosion behavior of carbon steel radioactive waste packages: A summary review of Swedish and U.K. Research. *Corrosion*, 65, 195–212.

Standish, T., Chen, J., Jacklin, R., Jakupi, P., Ramamurthy, S., Zagidulin, D., Keech, P., Shoesmith, D. (2016). Corrosion of copper-coated steel high level nuclear waste containers under permanent disposal conditions. *Electrochim. Acta*, 211, 331–342.

Suzuki, S., Ogawa, Y., Giallonardo, J., Keech, P.G. (2021). The design of copper-coating overpack for the high-level radioactive waste disposal concept in Japan. *Materials and Corrosion*, 72, 94–106.

Taiwan Power Company (TPC) (2022). Spent nuclear fuel final disposal program. Candidate site selection and confirmation. Preliminary safety case report on spent nuclear fuel final disposal (Rev. 2). SNFD-B202-IPR-003-003.

Wei, X., Dong, J., Chen, N., Yadav, A.P., Ren, Q., Wei, J., Wang, C., Ma, R., Ke, W. (2021). Effects of bentonite content on the corrosion evolution of low carbon steel in simulated geological disposal environment. *J. Mater. Sci. Technol.*, 66, 46–56.

4

Structural Materials for Sodium Cooled Fast Reactors

Hayato YAMASHITA and Tai ASAYAMA
Japan Atomic Energy Agency, Oarai, Japan

The sodium cooled fast reactor (SFR) is one of the most promising fast breeder reactor (FBR) systems. FBRs are a sustainable energy source that utilizes natural uranium resources efficiently. In addition, major long-lived radioactive fission products, that is, minor actinides (MAs), in spent fuels of nuclear power plants can be used as nuclear fuel in an FBR and reduced through the nuclear fuel cycle. In this context, SFRs are an energy source that conforms to the sustainable development goals adopted at the United Nations Summit in September 2015. This chapter describes the outline of structural materials for SFRs.

4.1. Introduction

Nuclear power plants across the world significantly contribute to a major reduction in greenhouse gas emissions in power generation. To continue and expand this benefit and ensure the reliability and sustainability of energy supply, next generation reactors taking advantage of advanced design concepts and state-of-the-art technologies will need to be developed and deployed. Such new nuclear systems, called Generation IV systems, are being examined and developed as part of a technology roadmap (GIF 2002) within an international forum, the Generation IV International Forum (GIF) (Yvon 2017).

The GIF technology roadmap identified six most promising Generation IV systems. The six systems chosen include the gas-cooled fast reactor system (GFR),

Materials and Processes for Nuclear Energy Today and in the Future,
coordinated by Fanny BALBAUD-CÉLÉRIER and Céline CABET.

the lead-cooled fast reactor systems (LFR), the molten salt-cooled reactor system (MSR), the sodium cooled fast reactor system (SFR), the supercritical water-cooled reactor system (SCWR) and the very-high-temperature reactor system (VHTR). Among these systems, the SFR Generation IV technology builds on the technologies that have been successfully used in the UK, France, Germany, Japan, the United States and the Russian Federation. Therefore, the SFR has a potential to be deployed earlier than the other concepts in GIF by accelerating technology development utilizing proven technologies and experiences on design, construction and operation worldwide (Yvon 2017).

Generally, the SFR consists of primary and secondary sodium systems as well as a water/steam system. The primary sodium system transports heat generated in the reactor core. Thermal energy of the primary sodium is transferred to the secondary sodium through an intermediate heat exchanger (IHX). Thermal energy of the secondary sodium is in turn transferred to the water/steam system through a steam generator (SG) and the steam is used to turn the turbine to generate electricity.

The reason why there is a secondary system in the SFR is to prevent any sodium–water reaction accident from directly affecting the reactor core.

Generally, the primary coolant sodium flows into the reactor core at around 400°C and is heated to a temperature range of 500–550°C in the reactor core. The temperature of the secondary coolant sodium is lower than that of the primary one. Therefore, both the primary and secondary systems do not need to be pressurized because the temperature is sufficiently lower than the boiling point of sodium, that is, 880°C.

The primary system consists of a reactor vessel (RV), a primary circulation pump, an IHX, primary coolant piping and so on. Depending on the arrangement of these components, SFRs are classified into two types, loop type and pool type.

In the loop-type SFR, the RV, the primary circulation pump and the IHX are connected by primary coolant piping. In contrast, in the pool type, the reactor core, the primary circulation pump and the IHX are installed in a large main vessel.

In the loop type, since components are located with space from each other, they are relatively easy to access for maintenance and repair. However, because each component is connected by piping, the overall plant size tends to be large. In contrast, in the pool type, since each component is installed in the main vessel, the size of the plant can be smaller. However, because the main vessel is large, it becomes important to consider the seismic resistance of structures and spatial temperature variations.

For the deployment of SFR, an innovative design enhancing safety and economic efficiency is strongly required (Yvon 2017). For this purpose, it is effective to apply the suitable structural materials for SFR components.

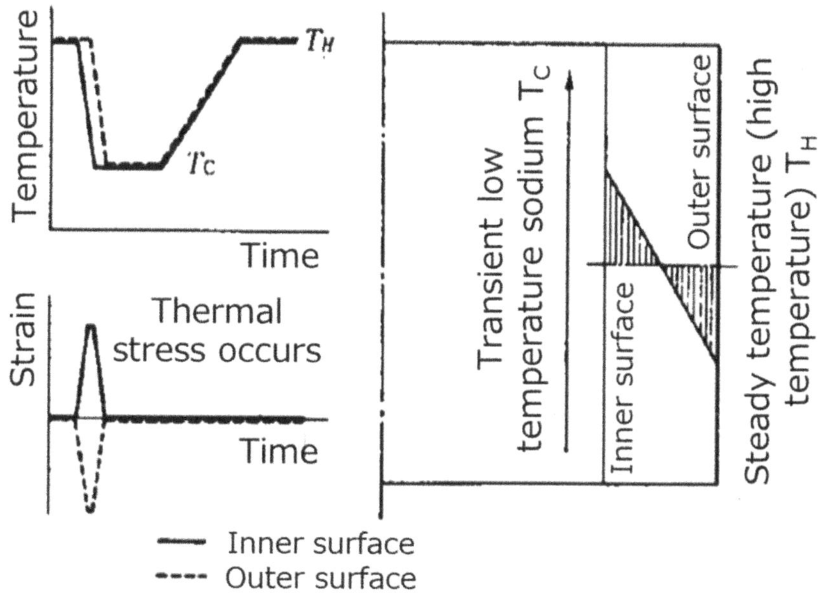

Figure 4.1. *Thermal stress caused by changes of sodium temperature*

Structural materials of SFRs have the following characteristics compared to conventional nuclear power plants such as light water reactors.

– Structural materials are used at elevated temperatures where creep effects could be significant.

– Structural materials are subject to temperature changes caused by thermal transients. Because of excellent heat transfer from sodium to structural materials and low heat conduction within structural materials, large thermal stresses could result from thermal expansion and contraction near the surface region, as shown in Figure 4.1 (Imazu 1986). In the structural designing of SFR components, it is necessary to prevent creep-fatigue failure caused by such cyclic thermal stress.

– Since sodium is liquid under ambient pressure in the operating temperature range of SFR, load caused by internal pressure is insignificant.

– Structural materials contact liquid sodium at elevated temperatures. Therefore, the influence of liquid sodium on material properties must be considered. In addition, since structural materials near the reactor core are exposed to neutron irradiation, the influence of neutron irradiation on material properties cannot be ignored.

This chapter describes the requirements for SFR structural materials, some examples of newly developed structural materials and evaluation methods for the elevated temperature integrity of structural materials for SFRs.

4.2. Requirements for structural materials

Structural materials in SFRs are used in the creep temperature range in the same way as in thermal power plants. However, there is a difference between design requirements in thermal power plants and those in SFRs. Thermal power plant components are designed to withstand mainly internal pressure, while design against cyclic thermal stress (Ozaki and Sasaki 1989) is important for SFR components.

Stresses in SFR components are broadly categorized into load-controlled stresses and deformation-controlled ones, depending on the source of them. As shown in Figure 4.2, for load-controlled stress, high strength material results in superior creep strength. In contrast, for deformation-controlled stresses, high ductile materials with a small ratio of yield stress to tensile strength (i.e. yield ratio) bear smaller stress, σ_L, while high strength material bears larger stress, σ_H. As a result, high strength materials may have inferior creep strength to high ductility materials. Although stress relaxation may occur due to creep deformation at elevated temperatures, creep rupture time could be shorter for high strength materials.

This observation indicates the importance of the balance between strength and ductility in selecting structural materials for components subject to creep fatigue loading caused by cyclic thermal stress (Wada 1986).

SFR structural materials need to have a high-level reliability, since they constitute coolant boundaries. Table 4.1 summarizes structural materials used and/or planned to be used in various SFRs (IAEA 1992; Mannan et al. 2003). In existing SFRs, austenitic stainless steels are widely applied as materials for major components such as reactor vessels (RV), IHX, and piping systems, because of:

– excellent corrosion resistance;

– excellent heat resistance and high creep strength;

– low yield ratio and high ductility which are suitable for deformation-controlled stresses.

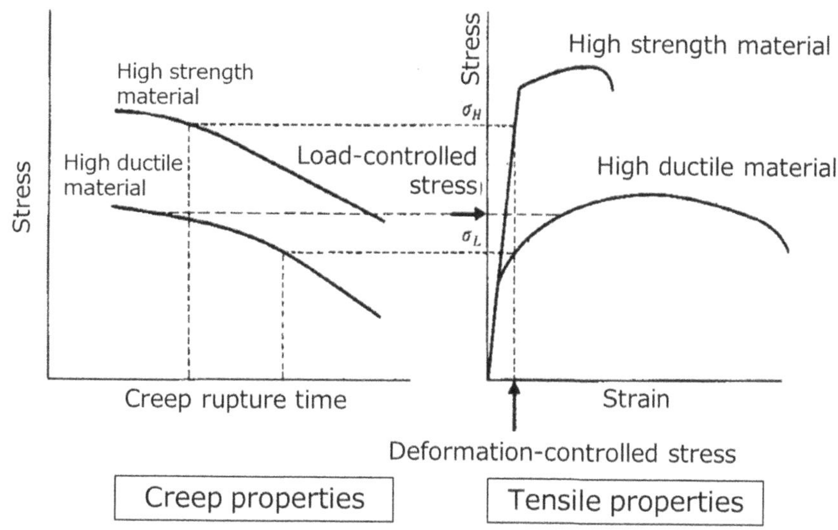

Figure 4.2. *Relationship between the type of stress and material properties*

As generally known, intense chemical reaction takes place when liquid sodium comes into contact with water. Therefore, SGs, which constitute the boundary between liquid sodium and water, are one of the most critical components in SFRs. For the SGs of the existing SFRs, various materials, such as 2.25Cr-1Mo steel, Type 321 stainless steel (SS) and Alloy 800, have been adopted. This is due to the consideration of stress corrosion cracking (SCC) resistance in water and steam environments.

For SFRs in a design phase, several kinds of newly developed materials, such as 316L(N) steel, 316FR steel and modified 9Cr-1Mo steel would be employed. For example, EFR planned to adopt 316L(N) steel for the reactor vessel, IHX and pipes. This steel was developed to improve high-temperature strength and ductility based on the chemical compositions of 316SS. This steel was also used in Super-Phenix. In Japan, fast reactor grade 316 steel (316FR steel) was developed and planned to be employed as the main structural material of primary circuits. This steel was developed to improve creep properties especially in SFR conditions, namely around 550°C. Modified 9Cr-1Mo steel was originally developed in the United States. It demonstrated excellent high temperature strength as compared to conventional 2.25Cr-1Mo steel. This steel was planned to be employed as the material for IHXs, SGs and pipes of JSFR, as well as for the components of KALIMER-600, which was an advanced SFR concept designed in Korea, and for the SGs of EFR. In Japanese prototype SFR “MONJU”, a separate type of SG system was employed. There were an evaporator and a super heater in each loop which had different steam conditions. For the evaporator, 2.25Cr-1Mo steel was adopted considering SCC resistance. For the super heater, 321SS was chosen so as to take advantage of its high-temperature strength. However, separated SG system leads to higher costs. Employing modified 9Cr-1Mo steel, which has both excellent SCC resistance and high-temperature strength, an integrated SG system can be designed with a reduced construction cost.

Country	Reactor	RPV	IHX	Pipes	SG	
					Evaporator	Superheater
Japan	Monju	Type 304 stainless steel	Type 304 stainless steel	Type 304 stainless steel	2.25Cr-1Mo	Type 321 stainless steel
	JSFR*	Type 316FR stainless steel	Mod. 9Cr-1Mo	Mod. 9Cr-1Mo	Mod. 9Cr-1Mo	
France	Phenix	Type 316 stainless steel	Type 316 stainless steel	Type 304 stainless steel & Type 321 stainless steel	2.25Cr-1MoNb	Type 321 stainless steel
	Super-Phenix	Type 316LN stainless steel	Type 316LN stainless steel	Type 316LN stainless steel	Alloy 800	
America	CRBRP	Type 304 stainless steel	Type 304 stainless steel	Type 304 stainless steel & Type 316 stainless steel	2.25Cr-1Mo	2.25Cr-1Mo
Germany	SNR300*	Type 304 stainless steel	Type 304 stainless steel	Type 304 stainless steel	2.25Cr-1MoNbNi	
England	PFR	Type 321 stainless steel	Type 321 stainless steel	Type 321 stainless steel	2.25Cr-1MoNb	Type 316 stainless steel ⇒9Cr-1Mo
Europe	EFR*	Type 316LN stainless steel	Type 316LN stainless steel	Type 316LN stainless steel	Mod. 9Cr-1Mo or 9Cr-1Mo	
Russia	BN350	Type 304 stainless steel (Equivalent)	Type 304 stainless steel (Equivalent)	Type 304 stainless steel (Equivalent)	2.25Cr-1Mo	2.25Cr1Mo (Equivalent)
	BN600	Type 304 stainless steel	Type 304 stainless steel	Type 304 stainless steel	2.25Cr-1Mo	Type 304 stainless steel
	BN800	Type 304 stainless steel	Type 304 stainless steel	Type 304 stainless steel	2.25Cr-1Mo	2.25Cr-1Mo
Korea	KALIMER-600*	Type 316 stainless steel	Mod. 9Cr-1Mo	Mod. 9Cr-1Mo	Mod. 9Cr-1Mo	
India	PFBR	Type 316LN stainless steel	Type 316 stainless steel	Type 304 stainless steel	2.25Cr-1Mo	

*:Design stage

Table 4.1. *List of structural materials for fast reactors in various countries*

4.3. Development of structural materials and evaluation methods

4.3.1. *Creep-fatigue evaluation methods*

Since creep-fatigue caused by cyclic thermal stresses is one of the most important failure modes to be prevented in structural design of SFRs, excellent creep-fatigue properties are required for their structural materials.

To prevent creep-fatigue failure, proper creep-fatigue evaluation methods have to be established. The concept of creep-fatigue evaluation is shown in Figure 4.3. A trapezoidal strain waveform is divided into a strain variation portion (corresponding to start up and shutdown of a plant) and a strain hold period (corresponding to steady state operation of a plant), and the linear damage summation rule is employed for estimating a total damage.

The time fraction rule can be summarized as follows:

– ① Fatigue damage corresponding to a strain range $\Delta\varepsilon_i$ which is repeated for n_i cycles is assumed to be the ratio of n_i to the fatigue life N_{fi} corresponding to $\Delta\varepsilon_i$, that is, n_i/N_{fi}.

– ② Creep damage corresponding to a stress σ_j applied for time t_j is assumed to be the ratio of t_j to creep life t_{Rj} corresponding to σ_j, that is, t_j/t_{Rj}.

– ③ When the summation of fatigue and creep damages reaches a limit value, D, creep fatigue failure is assumed to take place.

$$\sum_i \frac{n_i}{N_{fi}} + \sum_j \frac{t_j}{t_{Rj}} = D \qquad [4.1]$$

If there is only fatigue damage or creep damage, creep-fatigue interaction does not need to be considered. When there is only fatigue damage, it is consistent with the Minor rule shown below.

$$\sum \frac{n_i}{N_{fi}} = 1 \qquad [4.2]$$

When there is only creep damage, it is consistent with the Robinson rule shown below.

$$\sum \frac{t_j}{t_{Rj}} = 1 \qquad [4.3]$$

If both fatigue and creep damages are not negligible, creep-fatigue interaction have to be taken into account, as shown in Figure 4.4.

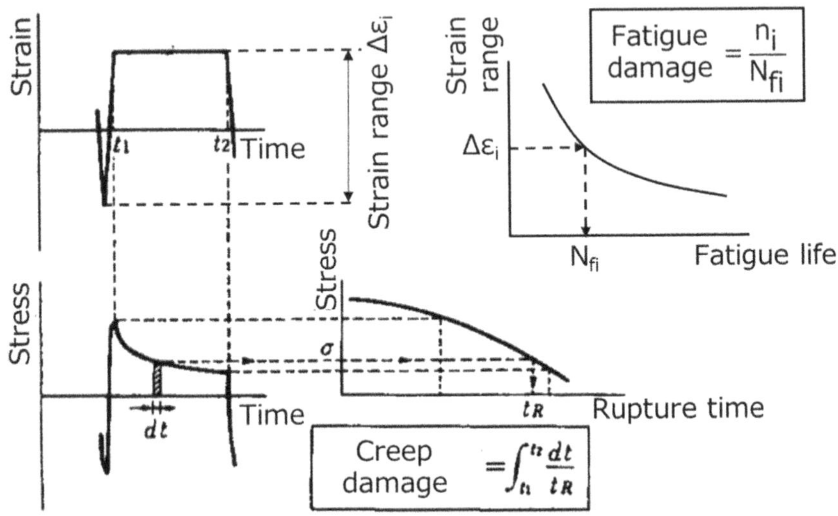

Figure 4.3. *Schematic diagram of creep and fatigue damage calculation*

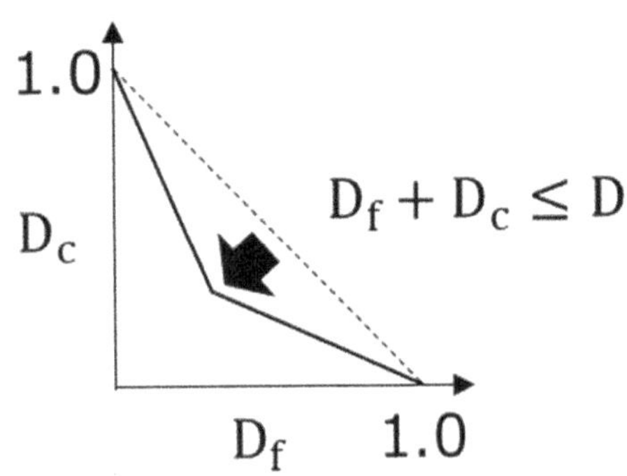

Figure 4.4. *Limit for creep-fatigue damage*

Another method to evaluate creep-fatigue damage is to use the ductility exhaustion method ($\varepsilon_{cj}/\delta_{cj}$) instead of the time fraction rule (t_j/t_{Rj}). Note that δ_{cj} is the creep rupture ductility (amount of creep strain at rupture) and ε_{cj} is the creep strain. The essential difference between time fraction and ductility exhaustion methods lies in whether primary creep is considered to contribute to creep damage. If only the secondary creep rate ($\acute{\varepsilon}_2$) contributes to creep damage, then the following equation holds, which means that the ductility exhaustion rule and the time fraction rule are equivalent.

$$\delta_{cj} = \dot{\varepsilon}_2 \, t_{Rj} \tag{4.4}$$

$$\varepsilon_{cj} = \dot{\varepsilon}_2 \, t_j \tag{4.5}$$

The number of thermal cycles postulated in the structural design of SFRs is not so large. Therefore, it is supposed that fatigue damage is small and creep damage is dominant. Creep damage is evaluated accounting for the three factors ①, ② and ③, as shown in Figure 4.3.

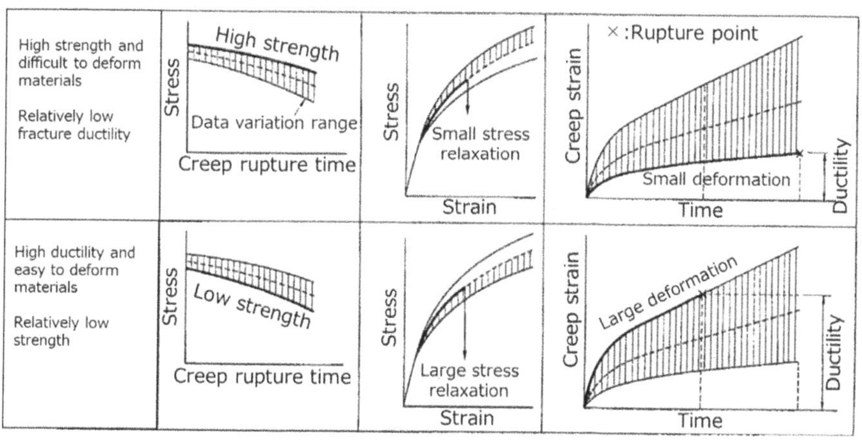

Figure 4.5. *Correlation between strength and ductility of materials*

From the view point of ①, materials with small yield ratio and high ductility are suitable. As for ②, easily crept materials (i.e. materials with high creep ductility) are suitable. With regard to ③, materials with high creep strength are suitable. Materials satisfying all of these requirements are ideal but a balance of them all is actually important. Statistical analyses based on the data of various materials were conducted to formulate cyclic stress–strain characteristics, creep strain-time (creep deformation) characteristics and stress-creep rupture time (creep strength) characteristics (Aoto et al. 1986; Wada et al. 1986; Yoshitake et al. 1986). It is necessary to determine structural material specifications considering the relationship between strength and ductility, taking data variabilities into account, as shown in Figure 4.5. Since components are not used under conditions where the creep effect is pronounced, generally it is better to think that creep ductility is more important than creep strength (see Table 4.2).

Creep effect	Material characteristics improving creep-fatigue property
Large scale creep deformation and significant degradation of long-term strength	Creep ductility should be more emphasized than creep strength, expecting stress relaxation.
Medium scale creep deformation and degradation of long-term strength	Balance of creep strength and creep ductility is important.
Small scale creep deformation and slight degradation of long-term strength	Creep strength should be emphasized, not expecting stress relaxation.

Table 4.2. *Degree of creep effect and material property factors that enhance creep fatigue*

4.3.2. *Improvement of creep-fatigue properties and materials*

In order to simultaneously improve the reliability and cost efficiency of plants, structural materials with improved creep-fatigue properties have been developed. There are many activities to develop new materials for SFR components. Among them, 316L(N), 316FR and modified 9Cr-1Mo steels are discussed in this section. For SFR components such as a reactor vessel, 316L(N) steel has been developed in Europe based on 316SS, controlling the chemical contents of carbon (C) and nitrogen (N). C is limited to the range of 0.04–0.08 weight% for conventional 316SS, but it is more strictly limited to the range of 0.02–0.03 weight% in 316L(N) steel. In addition, N is controlled to be in the range of 0.06–0.08 weight%. In Japan, significant efforts were made to improve creep-fatigue property of materials for SFR components. To achieve better high-temperature strength and ductility than conventional stainless steels such as 304SS and 316SS, in SFR condition, 316FR steel was developed based on 316SS optimizing the chemical compositions. 316FR steel developed in Japan has a C content lower than 0.02 weight% and a N content of 0.06–0.12 weight%. Furthermore, phosphorus (P) is optimized in 316FR steel targeting from 0.020 to 0.045 weight% within the range of conventional 316SS (i.e. 0.045 weight% or less). As a result, creep strength superior to conventional materials, 316SS and 304SS were achieved, as shown in Figures 4.6 and 4.7.

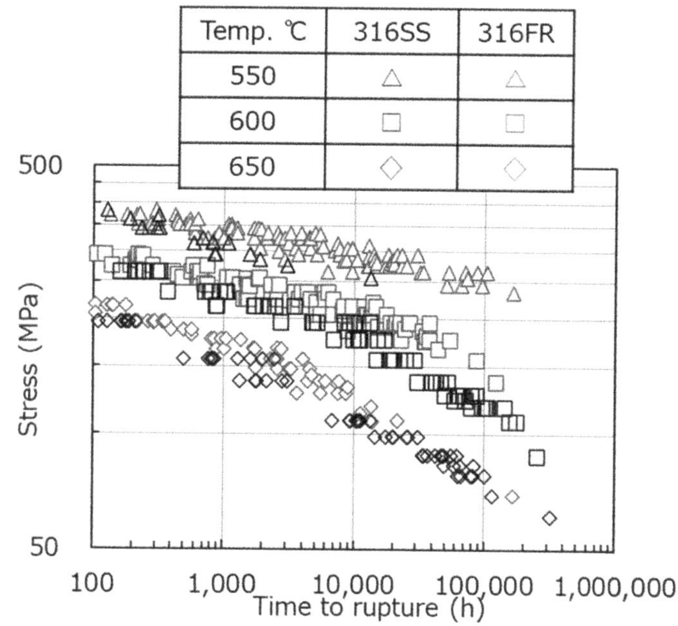

Figure 4.6. *Comparison of creep strength between 316SS and 316FR steel (NIMS 2005). For a color version of this figure, see www.iste.co.uk/balbaud/nuclearenergytoday.zip*

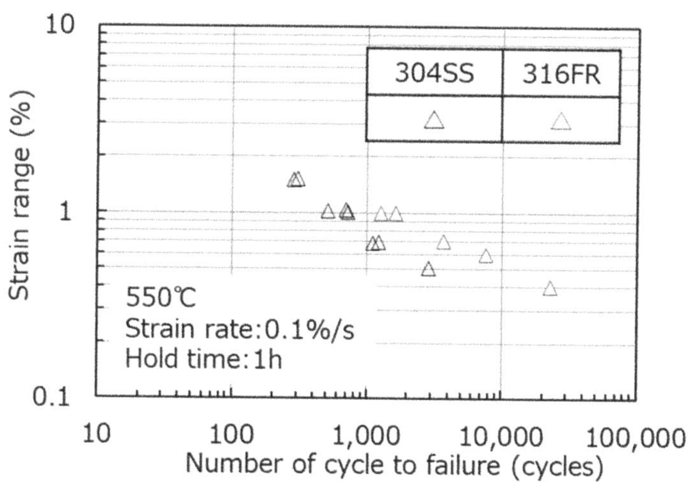

Figure 4.7. *Comparison of creep-fatigue strength between 304SS and 316FR steel. For a color version of this figure, see www.iste.co.uk/balbaud/nuclearenergytoday.zip*

Structural materials for SGs must have excellent SCC resistance as well as corrosion resistance at elevated temperatures. In addition, heat transfer efficiency also has to be taken into account when selecting a material. As for heat conduction coefficient, ferritic steels are superior to austenitic steels, but since the high-temperature strength of conventional 2.25Cr-1Mo steel is inferior to that of austenitic steels, ferritic steels with better high-temperature strength were strongly desired for advanced SFRs.

As summarized in Table 4.1, a separate type SG system was employed in Japanese prototype SFR "Monju". For the super heater, 321SS was chosen taking advantage of high-temperature strength. For the evaporator, 2.25Cr-1Mo steel was adopted considering SCC resistance.

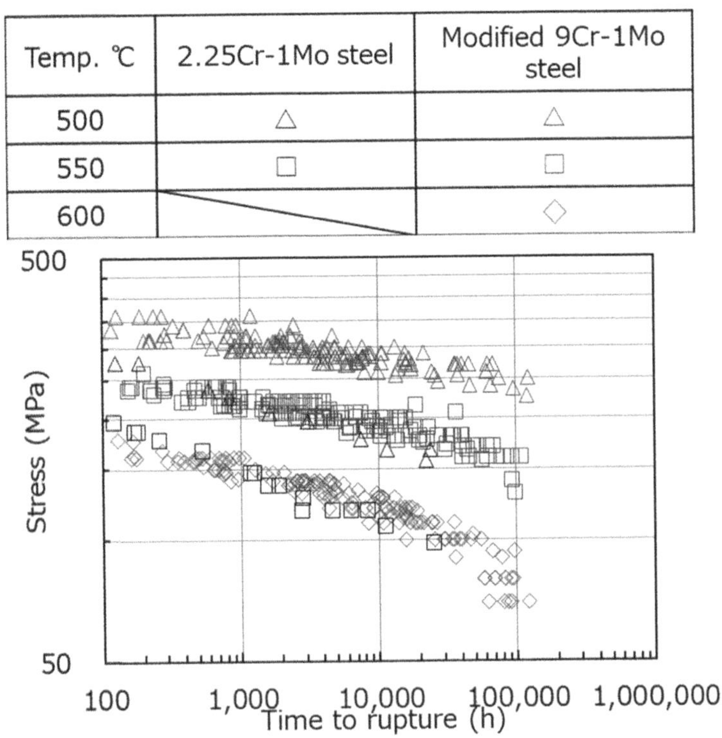

Figure 4.8. *Comparison of creep strength between modified 9Cr-1Mo steel and 2.25Cr-1Mo steel (NIMS 2014). For a color version of this figure, see www.iste.co.uk/balbaud/nuclearenergytoday.zip*

However, because such a separated SG system leads to higher costs, the concept of integrated SG has been investigated for practical use of SFRs. From this viewpoint, modified 9Cr-1Mo steel, developed in the United States, are planned to be applied to SGs of an SFR. The steel has superior creep and creep-fatigue strength to 2.25Cr-1Mo steel, as shown in Figures 4.8 and 4.9. Modified 9Cr-1Mo steel has already been used in ultra-supercritical power plants in Japan and are expected to be a promising candidate structural material for the future SFRs.

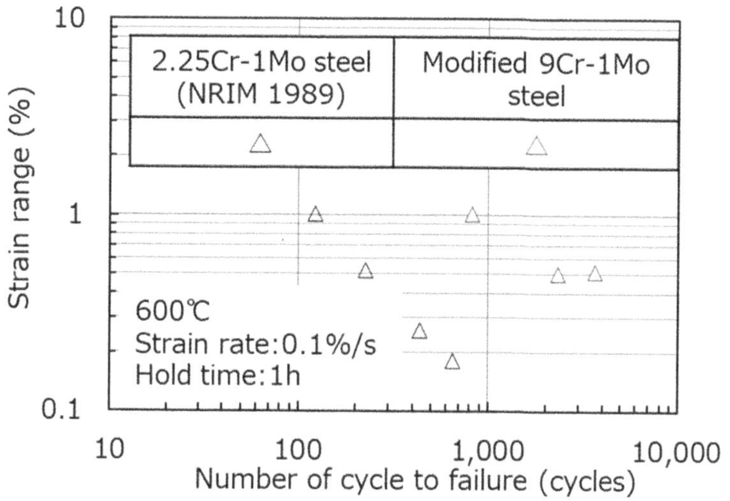

Figure 4.9. *Comparison of creep-fatigue strength between modified 9Cr-1Mo steel and 2.25Cr-1Mo steel. For a color version of this figure, see www.iste.co.uk/balbaud/nuclearenergytoday.zip*

4.3.3. *Creep-fatigue damage evaluation for 316FR steel and modified 9Cr-1Mo steel*

Figures 4.10 (Nagae et al. 2015) and 4.11 (Takaya et al. 2014), respectively, show the results of creep-fatigue damage evaluation for 316FR steel and modified 9Cr-1Mo steel using the linear damage summation rule as illustrated in section 4.3.1. Creep damage was evaluated by the time fraction rule. As shown in these figures, sufficiently accurate creep-fatigue evaluation is possible for the candidate structural materials for the future SFRs.

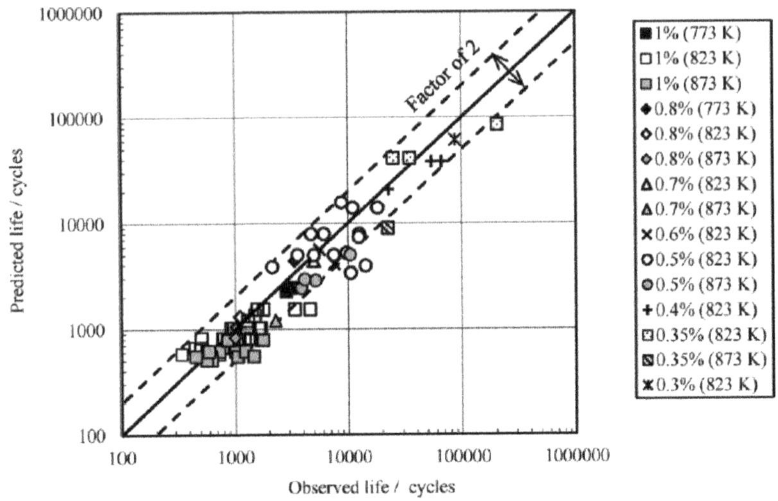

Figure 4.10. *Creep-fatigue damage evaluation of 316FR steel by linear damage summation rule (source: Nagae et al. 2015)*

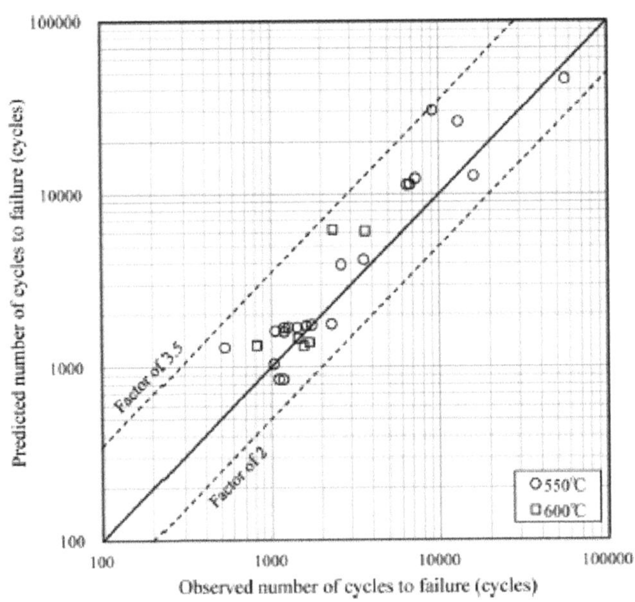

Figure 4.11. *Creep-fatigue damage evaluation of Mod.9Cr-1Mo steel by linear damage summation rule (source: Takaya et al. 2014)*

To enhance economic competitiveness and reduce radioactive wastes, the design period (expected service period) of SFRs tends to be 60 years. Correspondingly, time-dependent allowable stress values up to 500,000 h for 316FR steel and modified 9Cr-1Mo steel were published in the design and construction codes for fast reactors by the Japan Society of Mechanical Engineers (JSME). For the steel, long-term creep test data, longer than 150,000 h have been obtained. The region-splitting method (Kimura et al. 2004), which considers different failure mechanisms for short-term creep and long-term creep, was applied to obtained creep test data, to improve the accuracy of regression and determine the acceptable values (Onizawa and Hashidate 2019).

4.3.4. *Effect of operating environment on creep-fatigue properties*

4.3.4.1. *Effect of sodium environment on creep-fatigue properties*

When discussing the characteristics of structural materials, the effect of operating environment on material properties should be considered. Most material tests are conducted in air. Therefore, the separation of the effects of oxidation at elevated temperatures from intrinsic material properties is needed when assessing material properties in liquid sodium, which forms a reductive environment. Fatigue strength degradation caused by oxidation at elevated temperatures has been clarified by fatigue tests performed in a high vacuum where the effect of oxidation is limited. When the creep effect becomes dominant, the degradation in creep-fatigue strength caused by oxidation seems to become smaller (Tokimasa and Nitta 1986; Morishita et al. 1988; Asayama et al. 1989).

Tests in liquid sodium also show similar trends to those under vacuum. Less effect of sodium corrosion has been observed, unless temperature exceeds 600°C, because oxygen concentration can be controlled in sufficiently low levels in the test. In the case of tests in air, the formation of oxide scales on the surface of the specimen and its cracking induced crack initiation (Challenger et al. 1981; Nonaka et al. 1987) and many cracks were observed (Hirano et al. 1983; Hirano et al. 1984). The major cause of the degradation of fatigue strength in air results from the decrease in crack initiation life caused by oxidation. The effect of environment on crack growth rate is small (Kato et al. 1988; Maruyama et al. 1988).

Since a liquid sodium environment is a reductive one, there is general corrosion (e.g. thinning) and transformation of material surface caused by the elution of components from the steel. However, thinning is not a factor in crack initiation and few surface cracks are observed in the tests in liquid sodium. Since fatigue strength in a sodium environment is superior to that in air, the allowable stress values for structural design using the data obtained in air gives a conservative evaluation. However, as shown in Figure 4.12 for 304SS, cracks initiate at grain boundaries at temperatures higher than 650°C and the influence of test environment on fatigue strength becomes negligible.

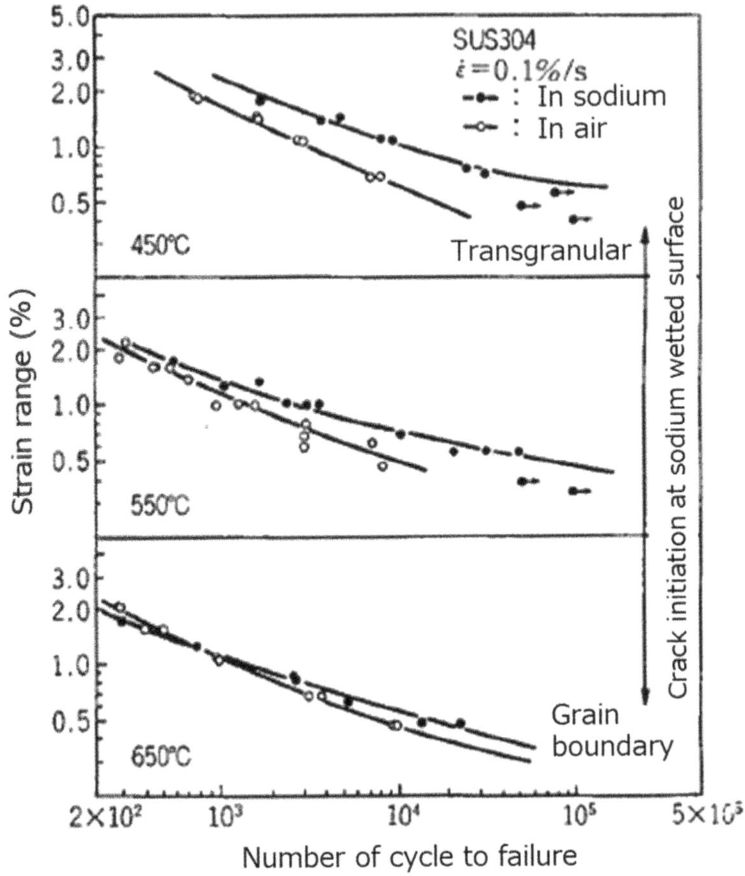

Figure 4.12. *Comparison of high temperature fatigue properties in air and in sodium*

In long-term creep-fatigue tests, where creep is dominant, cavity formation at grain boundaries becomes a major cause of failure. Therefore, the volumetric effect of cavity formation was more dominant than the surface effect caused by oxidation. As a result, significant difference between test results in air and in sodium was not observed.

In creep tests in air, many surface cracks were observed near the fracture zone, while cracks scattered over the entire surface in tests in sodium. As for the depth of surface cracks near the fracture zone, in the case of 304SS, cracks comparable to approximately one or two crystal grain sizes were often observed in the tests in air, while cracks larger than two crystal grain sizes were often observed in tests in sodium (Huthmann and Borgstedt 1988).

However, it has been confirmed that creep strength in sodium is comparable to that in air and the degradation in creep ductility is not significant. Therefore, it is relevant to assume that the interaction of grain boundary corrosion and cavity formation on the surface is not significant. In fact, it has been reported that there is no remarkable difference between creep-fatigue lives in air and in sodium (Wood and Baldw 1983; Wada et al. 1989).

Surface decarburization of ferritic steels in liquid sodium results in a slight degradation of creep strength, but extra creep damage accumulation is not significant because the initial stress becomes small and stresses become easier to relax due to softening.

As for surface carburization of austenitic stainless steels in sodium, it is necessary to consider the degradation of ductility. According to the results of creep tests in sodium for some carburized materials, creep strength seems to degrade slightly with carburization. Since strain hardening due to dynamic strain aging becomes more pronounced in tests with large strain ranges and initial stresses, the creep damage increases. As shown in Figure 4.13, when a noticeable carburization occurs, creep-fatigue life degrades in the carburized region (actually, very limited region contacted with liquid sodium) of the material. However, if the strain range is small, creep-fatigue life degradation hardly occurs even in a noticeably carburized material, because plastic deformation is negligible.

In addition, since the source of carbon in coolant sodium is usually decarburization of low chromium steel in contact to sodium, carbon activity in sodium is not so large. Carburization was observed merely up to about one grain depth from the surface. These results suggest that since the influence of carburization on material characteristics is limited in one or more grain depth from the surface, there is little influence on creep-fatigue damage, which is less affected by the surface. In fact, the results of long-term creep-fatigue tests in extremely carburizing conditions (e.g. the carbon concentration on the surface reaches up to 1%) showed that the surface prematurely cracked and a slight reduction of creep-fatigue life was observed at 1% total strain range. At about 0.4% strain range, which is closer to the actual designing conditions, crack initiated internally and creep-fatigue life was not reduced (Asayama et al. 1991).

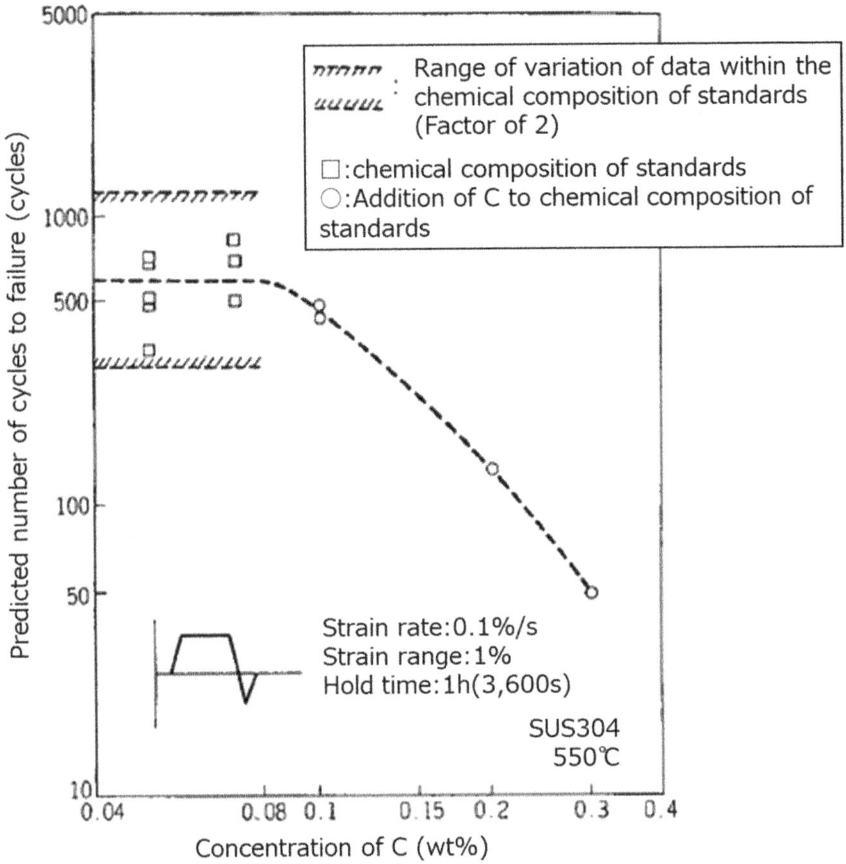

Figure 4.13. *Relationship between carbon content in SUS304 steel and creep-fatigue strength*

Figure 4.14 shows the results of evaluating the change in creep strength due to carburization for 316FR steel by simulating the amount of carburization during the creep test using diffusion equations (Watashi and Wakai 1995). ΔC_{mx} is the amount of carbon on the steel surface that is in equilibrium with the amount of carbon in sodium. ΔC_{gb} is the amount of carbon that diffuses to the grain boundary and was estimated to be between 0.10 and 0.30% because experimental values were not available. No degradation in creep strength caused by carburization is observed at elevated temperatures for long-term regions, thus no need to consider the degradation of creep strength caused by carburization under actual plant operating conditions.

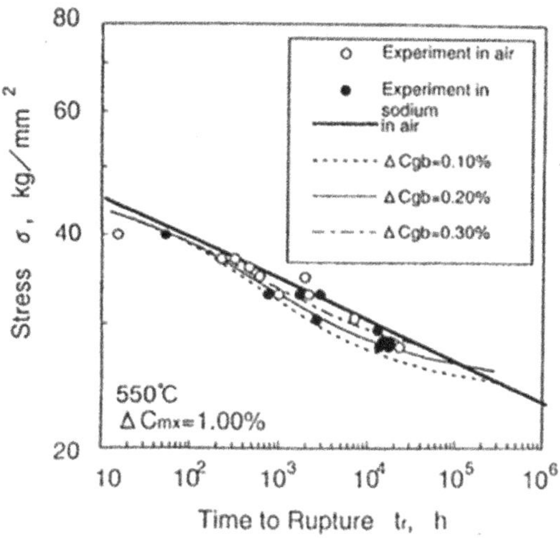

Figure 4.14. *Comparison of creep test results and simulation results of 316FR steel in sodium and air as a function of C at grain boundary*

4.3.4.2. *Effect of neutron irradiation on creep-fatigue properties*

Figures 4.15 and 4.16, respectively, compare the creep strength and creep rupture elongation of 304SS and 316FR steels, which were neutron irradiated in JOYO and JMTR to nonirradiated materials (Miyaji et al. 1997). In a neutron irradiation environment, boron ^{10}B in steel forms He. Since ^{10}B tends to precipitate at grain boundaries, He accelerates vacancy accumulation along interfaces of the grain boundary precipitates, leading to a decrease in creep strength and ductility (Yuzuhara et al. 1986). Fast neutron irradiation hardens material and results in large stresses at grain boundaries. As a result, creep-fatigue strength may be degraded by irradiation. Therefore, the properties of the steel can be improved by limiting the ^{10}B content in the steel. By limiting the total amount of fast neutron irradiation to below the level that ensures a tensile elongation of 10%, the hardening of material is indirectly suppressed. As for hardening with irradiation, surveillance tests showed that the properties required in the design were maintained.

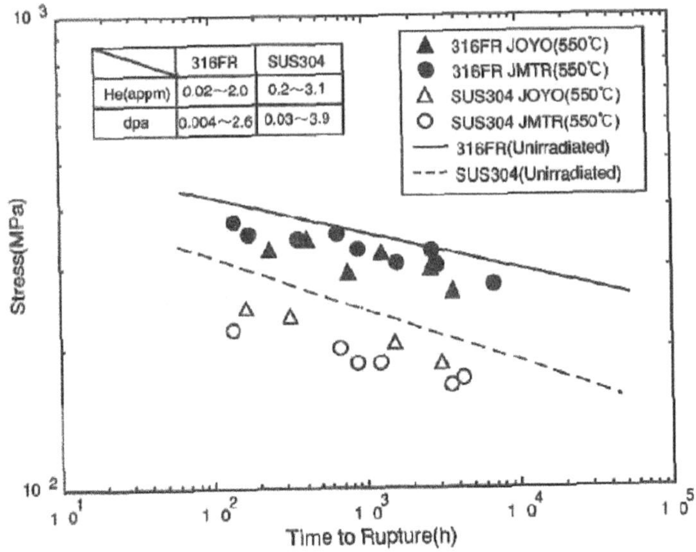

Figure 4.15. *Comparison of creep strength between irradiated and nonirradiated materials*

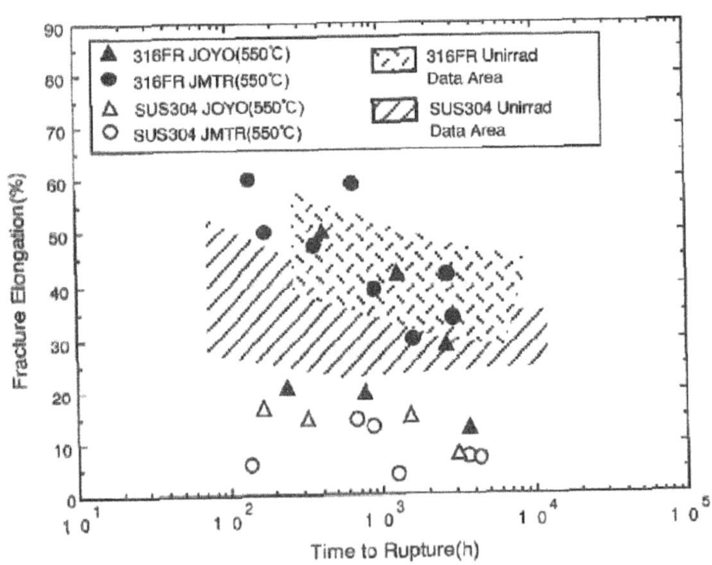

Figure 4.16. *Comparison of creep fracture elongation between irradiated and nonirradiated materials*

Figure 4.17 schematically illustrates microscopic interaction between environmental effects and creep damage. Since the details of failure mechanisms have not yet been clarified, a conservative extrapolation method to evaluate long-term region was applied to the test data and an appropriate safety factor was proposed. For such uncertainties, a realistic approach was employed by considering a sufficient safety margin.

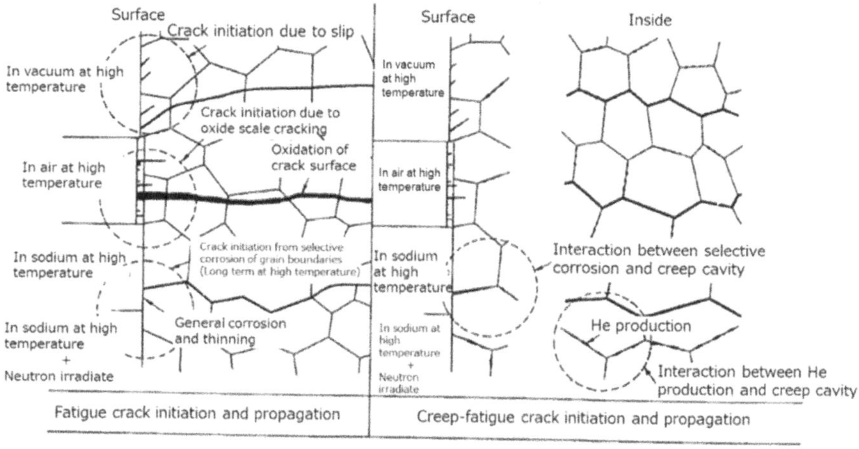

Figure 4.17. *Conceptual diagram of the interaction between environmental effects and creep cavity production*

4.4. Ongoing activities

In structural design of SFR components, creep-fatigue failure is the most important failure mode to be prevented. The Japanese prototype SFR "MONJU" was designed so that welded joints do not locate at high stress portions, but it would be necessary to locate them at high stress portions to enlarge design windows for future SFRs. Therefore, rational creep-fatigue evaluation methods have to be established.

Stress/strain distribution in the welded joints is caused by differences in material properties of various microscopic structures that consist of a welded joint. For the creep-fatigue evaluation of welded joints, it is necessary to consider possible enhancement of fatigue and creep damages caused by stress and strain redistribution within a welded joint.

4.5. Conclusion

SFRs are one of the most promising reactor systems to be deployed across the world to ensure reliable and sustainable energy supply. Development of structural materials suitable to the loading conditions and the environment of SFRs plays a key role in the improvement of their safety and economy.

Structural materials for SFRs are subjected to cyclic thermal stresses at high temperatures and require excellent creep-fatigue properties. Structural materials for SFRs including more recently developed low carbon nitrogen added austenitic stainless steels and high chromium ferritic steels exhibit better creep-fatigue properties compared with conventional ones and allow for designs with higher coolant temperatures and longer lives. The applicability of a widely accepted creep-fatigue evaluation method, the linear damage summation rule, for those materials was demonstrated with experimental data. The effects of exposure to sodium and neutron irradiation on creep-fatigue properties were investigated and it was shown that they can be treated in basically the same manner as conventional types of steel. General corrosion, decarburization of ferritic steels and carburization of austenitic steels could occur in sodium as was the case for conventional steels. However, these phenomena were not considered to have significant effects in the operating environment of actual plant. Neutron irradiation could have adverse effects on creep strength and ductility of austenitic stainless steels but material integrity would be maintained by limiting the total amount of fast neutron irradiation during operation to the same levels as conventional materials.

4.6. References

Aoto, K., Koakutsu, T., Wada Y., Hirano, M. (1986). The prediction of the stress relaxation behaviour of the high-temperature structural materials by creep strain equations. In *Proceedings of the International Conference on Creep*. JSME, Tokyo.

Asayama, T. and Asada, Y. (1989). Application of the overstress concept of creep-fatigue damage evaluation of 2.25 Cr-1Mo steel. *Transactions of the 10th International Conference on Structural Mechanics in Reactor Technology*. IASMiRT, Los Angeles.

Asayama, T., Kagawa, H., Komine, R., Wada, Y. (1991). Creep fatigue behaviour of SUS304 Stainless Steel tested in carburised sodium at 550°C. In *Int. Conf. on Mechanical Behavior of Materials*. Pergamon Press, Oxford.

Challenger, K.D., Miller, A.K., Brinkman, C.R. (1981). An explanation for the effects of hold periods on the elevated temperature fatigue behavior of 2¼Cr-1Mo steel. *Journal of Engineering Materials and Technology*, 103, 7–14.

GIF (2002). A technology roadmap for generation IV nuclear energy systems. U.S. DOE Nuclear Energy Research Advisory Committee and Generation IV International Forum.

Hirano, M., Komine, R., Kitao, K., Nihei, I., Yoshitoshi, A. (1983). Low-cycle fatigue properties of SUS 304 stainless steel in high temperature sodium. In *IWGFR-49*. International Atomic Energy Agency (IAEA), Vienna.

Hirano, M., Kitao, K., Nihei, I., Yoshitoshi, A. (1984). Effects of sodium on the low-cycle fatigue behavior of 2.25Cr-1Mo steel in flowing sodium at high temperature. In *Intern. Conf. on Liquid Metal Engineering and Technology*. British Nuclear Energy Society, London.

Huthmann, H. and Borgstedt, H.U. (1988). Influence of sodium on the creep-rupture behavior of Type304 stainless steel. In *Proc. of IV Int. Conf. on Liquid Metal Engineering and Technology*. SFEN, Paris.

IAEA (1992). LMFBR plant parameter 1991. IWGFR/80. Report, International Atomic Energy Agency, Vienna.

Imazu, A. (1986). Elevated temperature structural design technology in fast breeder reactor plants. *Journal of the Society of Mechanical Engineers*, 89, 542–548.

Kato, S., Komine, R., Hirano, M., Wada, Y. (1988). Low-cycle fatigue properties of SUS 304 stainless steel in elevated temperature sodium fluid. *Journal of the Society of Materials Science*, 37, 328–333.

Kimura, K., Sawada, K., Kubo, K., Kushima, H. (2004). Influence of stress on degradation and life prediction of high strength ferritic steels. In *ASME/JSME 2004 Pressure Vessels and Piping Conference*. ASME, New York.

Mannan, S.L., Chetal, S.C., Raj, B., Bhoje, S.B. (2003). Selection of materials for prototype fast breeder reactor. *Transactions of the Indian Institute of Metals*, 56(2), 155–178.

Maruyama, A., Kato, S., Komine, R., Hirano, M., Wada, Y., Kano, S., Nihei, I. (1988). Low-cycle fatigue properties of SUS304 stainless steel in elevated temperature fluid sodium. In *Int. Conf. on Liquid Metal Engineering and Technology*. SFEN, Paris.

Miyaji, N., Abe, Y., Asayama, T., Aoto, K., Ukai, S. (1997). Effects of neutron irradiation on tensile and creep properties of stainless steels. *Journal of the Society of Materials Science*, 46, 500–505.

Morishita, M., Tagichi, K., Asayama, T., Ishikawa, A., Asada, Y. (1988). Application of the overstress concept for creep-fatigue evaluation. In *Low Cycle Fatigue, STP 942*, Solomon, H.D., Halford, G.R., Kaisand, L.R., Leis, B.N. (eds). ASTM, New York.

Nagae, Y., Takaya, S., Asayama, T. (2015). Development of creep–fatigue evaluation method for 316FR stainless steel. *Journal of Pressure Vessel Technology*, 137(4), 041407-1–041407-5.

NIMS (2005). NIMS creep data sheet No. 45A. Report, National Institute for Material Section, Ibaraki.

NIMS (2014). NIMS creep data sheet No. 43A. Report, National Institute for Material Section, Ibaraki.

Nonaka, I., Kitagawa, M., Ohotomo, A., Wada, Y. (1987). Effect of compressive strain holding on the low cycle fatigue strength of 2¼Cr-1Mo steel at elevated temperatures. *Tetsu-to-Hagane*, 73, 2267–2274.

NRIM (1989). Fatigue Data Sheet No. 62. Report, National Research Institute for Metals, Tokyo.

Onizawa, T. and Hashidate, R. (2019). Development of creep property equations of 316FR stainless steel and Mod.9Cr-1Mo steel for sodium-cooled fast reactor to achieve 60-year design life. *Mechanical Engineering Journal*, 6, 1–15.

Ozaki, T. and Sasaki, Y. (1989). LWR and FBR. Their similarities and differences. *Nuclear Engineering*, 35, 72–80.

Takaya, S., Nagae, Y., Asayama, T. (2014). Development of creep-fatigue evaluation method for modified 9Cr-1Mo steel. *Journal of Pressure Vessel Technology*, 136, 031404-1–031404-8.

Tokimasa, K. and Nitta, I. (1986). Effect of air environment on creep-fatigue properties of 2¼Cr-1Mo steel. *Journal of the Society of Materials Science*, 35, 267–273.

Wada, Y. (1986). Quality improvement of structural materials for fast reactor. *The Piping Engineering*, 28, 105–112.

Wada, Y., Iwata, K., Aoto, K., Kawakami, Y. (1986). A new Ludwik type cyclic stress-strain equation for type 304 austenitic stainless steel is presented to represent the cyclic stress-strain hysteresis loops. In *Computational Mechanics '86*. Springer, Tokyo.

Wada, Y., Seki, S., Hasebe, S., Nihei, I. (1989). Mechanical properties in-air and in-sodium of bellows material. *ASME PVP*, 168, 107–112.

Watashi, K. and Wakai, T. (1995). Simulation of sodium effect on creep fracture of 316FR. *Pre-SMIRT 13 Seminar on Fracture, Fatigue and Life Prediction*, 145–167.

Wood, D.S. and Baldw, A.B. (1983). Creep I fatigue behaviour of type 316 steel in sodium. In *Int. Conf. on Liquid Metal Engineering and Technology*. BNES, London.

Yoshitake, T., Wada Y., Hirano, M. (1986). A statistical study of creep rupture and stress strain behavior of structural materials under elevated temperature conditions. In *Proceedings of the International Conference on Creep*. JSME, Tokyo.

Yuhara, S., Kikuchi, T., Oka, Y., Tani, S., Nakazawa, T. (1986). Neutron irradiation effects on creep-rupture properties of SUS304 stainless steel. In *Proceedings of the International Conference on Creep*. JSME, Tokyo.

Yvon, P. (2017). *Structural Materials for Generation IV Nuclear Reactors*. Woodhead Publishing, Cambridge.

5

Materials and Processes for Heavy Liquid Metal Cooled Reactors

Serguei GAVRILOV
SCK CEN Belgian Nuclear Research Centre, Mol, Belgium

5.1. Introduction

Heavy liquid metals such as liquid lead and lead bismuth-eutectic (LBE) are very prominent coolants for fast neutron spectrum nuclear systems such as LFR with its associated closed fuel cycle and for accelerator-driven subcritical systems developed for the efficient technological transmutation of high-level nuclear waste. The following advantageous features makes heavy liquid metal coolant very attractive for application in the aforesaid nuclear systems (Alemberti et al. 2020b).

– Low neutron absorption and moderation of lead and lead-bismuth coolants readily enable sustaining a hard neutron energy spectrum, providing good neutron economy and performance flexibility. It leads to very high efficiency in fuel utilization, the burning of long-lived actinide components of high-level waste and the utilization of the thorium fuel cycle.

– The very high boiling temperature of the lead coolant (1,749°C), which virtually eliminates coolant boiling and the related safety challenges present for other liquid coolants, enables further simplification of design and safety demonstration.

– The low partial vapor pressure of lead and LBE, which allows primary system operation at near atmospheric pressure, avoiding thus the need to maintain complex and expensive structures to provide pressure boundaries in LFR systems.

Materials and Processes for Nuclear Energy Today and in the Future,
coordinated by Fanny BALBAUD-CÉLÉRIER and Céline CABET.

– There are no rapid exothermic reactions between lead or LBE and air and water which supports simplified designs that do not require complex and expensive intermediate systems to isolate the primary coolant from its secondary coolant, normally water.

– The high operating temperature of LFRs, which enables high power conversion efficiencies.

– Low neutron-moderating capabilities of lead and LBE allow us to have wider fuel assembly subchannels to provide efficient coolant circulation at different modes.

– For accelerator driven system (ADS) systems heavy liquid metal serves also as the high efficiency target for the proton beam to produce neutrons by spallation reactions.

Despite providing the aforementioned advantages, the implementation of heavy a liquid metal coolants technology faces number of challenges:

– The operational experience of nuclear systems with HLM coolant is limited and not so well reported in peer reviewed sources.

– The compatibility with lead-based alloys of the previously qualified for application in nuclear systems steels is limited and those limits are still to be defined and to be appropriately justified.

– Lead has high melting point (327°C) that requires dedicated efforts to keep it in a liquid state and also it imposes additional requirements for the system design to perform necessary operations under shut-down conditions.

– High operating temperatures necessary for higher thermal efficiency impose needs for qualification of high thermal strength materials for which corrosion resistance at those temperatures is an additional challenge.

– Lead-based alloys are very dense and seismic resistance for the reactor blocks requires dedicated assessments.

– Liquid metal is opaque and requires the development of special systems for manipulation of the components under the free surface.

The portfolio of nuclear systems with heavy liquid metal coolant is very diverse: submarine propulsion reactors, experimental ADS facility, SMR, micro-reactors and full-scale NPP. The systems described in this chapter (see Table 5.1) show the variety of scales, designs and applications and are not an exhaustive list of all of the heavy liquid metal cooled nuclear systems under development.

Project	Developer	Coolant	Power, MW_{th}/ MW_e	Core inlet/outlet temperature(°C)	Fuel	Status
BREST-OD-300	NIKIET (Russia)	Pb	700/300	420/535	PuN–UN	Under construction
SVBR-100	AKME Engineering (Russia)	LBE	280/101	340/490	UO_2	Detailed design
MYRRHA ADS	SCK CEN (Belgium)	LBE	64/NA	220/400	MOX	Base design
ALFRED	FALCON (Ansaldo Nucleare, ENEA, RATEN-ICN)	Pb	300/125	390/430 (stage 1) 400/480 (stage 2) 400/520 (stage 3)	MOX	Base design
CLEAR-I ADS	INEST (China)	LBE	10/–	260/390	UO_2	–
SEALER	LeadCold (Sweden)	Pb	8/3	390/432	UO_2	Conceptual design
W-LFR	Westinghouse	Pb	950/450	390/530 (Phase I) 390/650 (Phase II)	UO_2 (Phase I) UN (Phase II)	Conceptual design

Table 5.1. *Selected projects of nuclear systems with heavy liquid metal coolants (state of April 2022)*

Overall, including replacement reactors, 15 LBE cooled reactors have been built for eight nuclear submarines and two ground-based facilities (Gromov et al. 1998). A total operational experience of about 80 reactor years has been accumulated. All of the built reactors were for military purposes leading to a rather scarce amount of available information in open literature. For the last two decades, Russia has actively conducted a research program on the development of reactors with heavy liquid metal coolants for civil applications (SVBR-100 and BREST-OD-300).

The BREST-OD-300 reactor has a pool-type design and an integral lead circuit accommodated in one central and four peripheral cavities of the concrete steel-lined vessel. The central cavity houses the core barrel together with the side reflector, the control rods, the spent fuel assemblies storage and the shell that partitions the hot and the cold lead flows. Four peripheral cavities (according to the loop number) accommodate the steam generator and reactor coolant pump units, heat exchangers of the emergency and normal cool down systems, filters and auxiliary components. The cavities have hydraulic interconnection. BREST-OD-300 uses a two-circuit heat transport system to deliver heat to a steam turbine. The fuel is uranium-plutonium mononitride (PuN-UN). The fuel assemblies are shroud less to maintain coolant flow around potential blockages. According to calculations, this allows us to maintain the maximum temperature below the established limit of 800°C even if the inlet flow of seven fuel assemblies is completely blocked. It is planned to accomplish the construction of the reactor system by 2026.

Figure 5.1. *Overall view of BREST-OD-300. Adapted from Lemehov et al. (2022). For a color version of this figure, see www.iste.co.uk/balbaud/nuclearenergytoday.zip*

Multi-purpose HYbrid Research Reactor for High-tech Applications (MYRRHA) is a multipurpose fast spectrum research facility that was developed in 1998 at Belgian Nuclear Research Centre (SCK CEN). MYRRHA is a hybrid reactor based on the ADS concept where a proton accelerator, a spallation target and a subcritical reactor are coupled (Aït Abderrahim et al. 2021). It can operate in both a subcritical mode when coupled with an accelerator and in a critical mode. The pool-type reactor is located in the reactor pit, which features a liner able to serve as a secondary confinement in case of a reactor vessel leakage or break (De Bruyn Aït Abderrahim 2022). The reactor cover closes the reactor vessel and supports all of the components. The reactor core consists of 211 positions for fuel assemblies, control and scram rods, the spallation target assembly, in-pile sections, reflector assemblies, instrumentation and surveillance capsules. From these core positions, 55 are accessible from the cover. Because of the structures above the core, the fuel and reflector assemblies have to be loaded from underneath by in-vessel fuel handling machine. Thanks to the buoyancy, the core support is from above and the core restrain system fixes the radial position of the fuel assembly to avoid small displacement during operation.

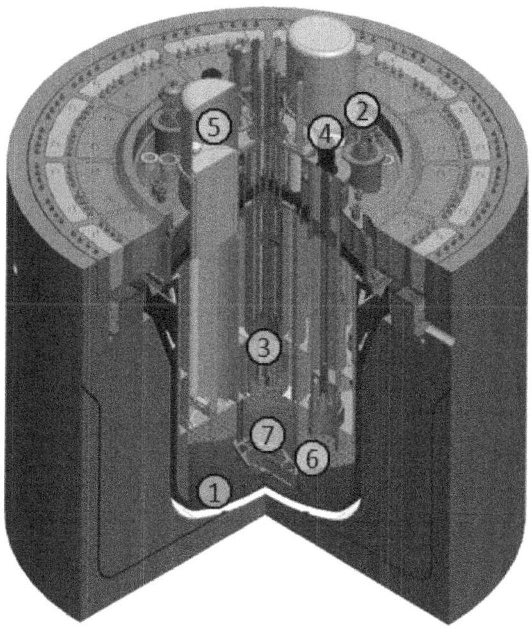

Figure 5.2. *Notional representation of MYRRHA reactor. Adapted from De Bruyn Aït Abderrahim (2022). For a color version of this figure, see www.iste.co.uk/balbaud/nuclearenergytoday.zip*

The development of the Advanced Lead-cooled Fast Reactor European Demonstrator (ALFRED) is led by Ansaldo Nucleare, ENEA and RATEN-ICN through the FALCON consortium (Alemberti et al. 2020a). ALFRED is a pool configuration design with lead as the primary coolant and supercritical water in the secondary circuits. Conceptually, the layout is very similar to that of an SFR. The reactor core, using MOX fuel in this case, is placed low in the reactor vessel to aid decay-heat removal by natural convection. The core is cooled by forced convection using mechanical coolant pumps. A safety vessel surrounding the reactor vessel retains lead coolant in the unlikely event of reactor vessel failure. To progressively gain the operating experience and to validate stepwise corrosion mitigation techniques, the staged approach has been put forward (Frignani et al. 2019). The staged approach involves initial use of the reactor at low power, low temperature and wide safety margins. The first stage will be based on materials, coolant chemistry conditions and protective measures proven through long-term tests in experimental facilities. In the consequent stages, the thermal power and core outlet temperature should be increased and will be justified on the basis of the accumulated operational experience and using the experimentally validated materials and corrosion protective measures (Alemberti et al. 2020a).

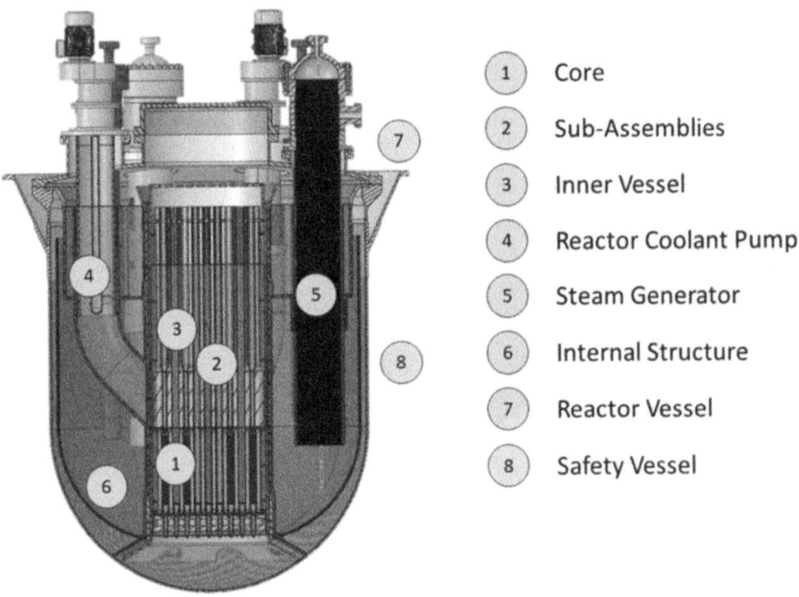

Figure 5.3. *ALFRED reactor coolant system sketch (Alemberti et al. 2020a). For a color version of this figure, see www.iste.co.uk/balbaud/nuclearenergytoday.zip*

SEALER (Swedish Advanced Lead Reactor) is a very small lead-cooled reactor designed for commercial power production in off-grid applications (Wallenius et al. 2018). To minimize the costs related to the fuel management, the fuel is never replaced during operation. The fuel is 2.4 tons of 19.75% enriched uranium oxide, and the rate of electricity production may vary between 3 and 10 MW, leading to a core-life between 10 and 30 years (at 90% availability).

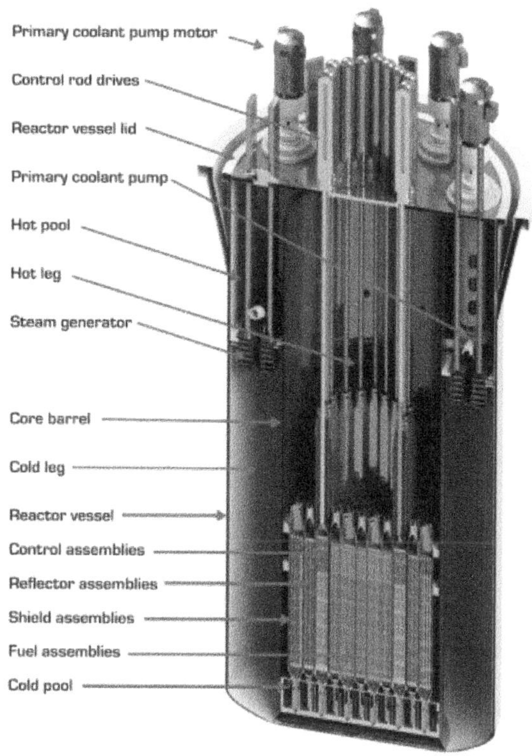

Figure 5.4. *SEALER reactor coolant system sketch (Wallenius et al. 2018). For a color version of this figure, see www.iste.co.uk/balbaud/nuclearenergytoday.zip*

Westinghouse is developing its next-generation high capacity nuclear power plant based on LFR technology. The Westinghouse LFR is envisioned to be developed through a staged approach with a lower power, near term demonstration LFR as the Phase I. This plant should operate with reduced duty conditions (coolant temperature, burnup of uranium oxide fuel) to maximize use of higher readiness materials and is intended to demonstrate LFR technology's overall safety and performance characteristics (Ferroni et al. 2022). For Phase II, the large plan with more demanding conditions for coolant temperature and burnup should be developed. Westinghouse LFR is pool configuration design with lead as the primary coolant and supercritical water in the secondary circuits. The uranium oxide fuel should be used in Phase I and uranium nitride for Phase II. The maximum lead coolant temperature is up to 530°C for Phase I and should be extended to 650°C for Phase II. The demonstration of key systems, components and materials should have started in 2022.

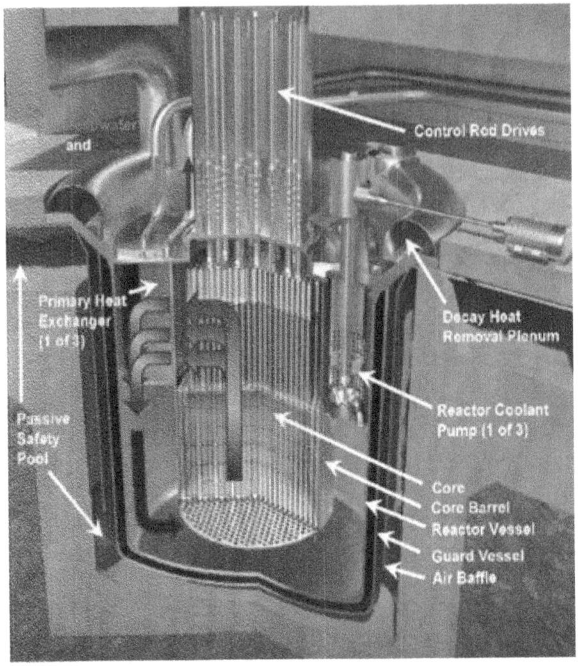

Figure 5.5. *Notional representation of the Westinghouse LFR primary system (Ferroni et al. 2022). For a color version of this figure, see www.iste.co.uk/balbaud/nuclearenergytoday.zip*

It should be mentioned also that liquid lead-lithium alloy is considered as tritium breeding material for fusion reactors. Heavy liquid metals are also considered as the working fluid in thermal energy storage for concentrated solar power systems (Rinaldi et al. 2021) and for the liquid electrode in high-capacity electric energy storages (Peplow 2014).

5.2. Requirements for materials

The ambitious goals of innovative nuclear systems with heavy liquid metal coolants impose the compatibility of the materials of the primary circuits with the hostile environment at elevated temperatures.

The fuel cladding is the first safety barrier to retain the fission products, including raising the pressure of fission gases during the fuel cycle. The following specific requirements apply for the candidate materials for cladding:

– corrosion resistance at the high temperature range under neutron irradiation during fuel cycle including back end of it;

– sufficient radiation resistance for swelling, radiation creep and radiation embrittlement;

– compatibility with the selected fuel composition (UO_2, MOX or UN-PuN);

– for the reactors with the closed fuel cycle additional requirements on cladding may be imposed by specific reprocessing technology.

Near core structural components as the core support structures and the core barrel are also subject to neutron irradiation and will accumulate the corresponding damage. This damage may be significant especially for small reactors with compact cores, where a substantial shielding on the periphery of the core is difficult to incorporate. Therefore, the selected materials should have sufficient radiation resistance and the effect of irradiation should be incorporated in the mechanical design of the corresponding components.

The candidate materials for the heat exchangers and for the steam generators should provide sufficient corrosion resistance in contact with both primary and secondary coolants. The design of the heat exchangers is also demanding for joining technologies of the selected candidate materials to be sufficiently qualified.

Specific requirements on the materials depend on the selected design approach for structural components and for materials of fuel assembly. For reactor coolant pumps, components as impeller and bearings must have the ability to resist erosion/corrosion under high liquid metal flow velocities. For specific design as ALFRED, it should be combined with corrosion resistance at elevated temperatures. For the reactor coolant pump, harsh conditions of high flow velocities as 10 m/s and operating temperatures above 430°C (Alemberti et al. 2020a) require the selection of corrosion-erosion resistant materials and a corrosion mitigation approach to provide the anticipated lifetime of said component.

Various materials properties as chemical composition, physical properties, corrosion resistance and mechanical properties of the selected candidate materials are required at the early stages of nuclear-systems development to perform safety assessments, neutronics calculations and mechanical design of structural and functional components. It makes it extremely challenging to rely, in the design, on materials which are not completely qualified for the specific application. The mechanical design requires availability of the design material properties based on targeted lab tests performed with few industrial heats of the material. Availability of the design properties for the selected candidate materials is one of the most essential requirements for a successful deployment of innovative nuclear systems.

The auxiliary systems for coolant conditioning and chemistry control to provide the expected performance of the candidate materials are also required. The most evident example for heavy liquid metal coolant technology is the requirements on the coolant chemistry control system to maintain dissolved oxygen concentration in liquid metal for corrosion mitigation purposes (Schroer 2011).

5.3. Material choices and basic properties

Depending on the availability of the previous experience with liquid metal coolant, different approaches for materials selection were adapted by the developers. For the reactors with heavy liquid metal coolants developed by the Russian Federation, the historical choice of the Si-doped ferritic/martensitic and austenitic steels is based on the previous experience of the construction and the operation of the submarine propulsion reactors with LBE coolant. For nuclear systems developed in other countries, the choice of the materials mainly relies on the experience gained with the sodium cooled fast reactors. Indeed, qualification of new materials for the structural components and for the fuel sub-assembly can take few decays of testing and characterization for which extensive testing capacities including irradiation facilities with fast neutron spectrum are required. Therefore, the choice for the near term realization projects is mainly based on the materials that are qualified by previous operation experience of nuclear reactors and for which compatibility with heavy liquid metal coolant is still to be demonstrated.

Steel 10X15H9C3B1(EP-302) was developed by the USSR for application in heavy liquid metals with a temperature limit of 600°C. It is an austenitic stainless steel with a controlled delta ferrite content of 2–5% and has acceptable welding properties. The recommended heat treatment is solution annealing at 1,050–1,100°C with cooling in air afterwards. To improve the thermal strength, it is recommended that an additional heat treatment at 750–800°C for 8 h is performed. The following issues with this steel have been identified:

– cracking in heat affected zone at operational temperatures above 480°C;

– phase instability and Ni_3Si precipitates after irradiation of 12–24 dpa at 460–575°C;

– deformation induced martensitic transformation at room temperature.

This steel is recommended for in-vessel structural components of BREST-OD-300 and SVBR-100 reactors (Karzov et al. 2009).

Steel 12X13M2C2(EP-852) – 12% chromium steel doped with Si for corrosion resistance – was developed at VNIINM for in-core components of submarine reactors with LBE coolant. It is ferritic/martensitic steel and the recommended heat treatment is normalization at 1,050°C and tempering at 700°C for 1 h. The recommended temperature limit is 600–620°C (Tarasikov et al. 2020).

EP-823 (16X12MVSFBR) is 12% chromium ferritic-martensitic steel alloyed with Mo, W, Nb and V for microstructure stability and doped with Si for corrosion resistance. This steel was developed to replace EP-852 for in-core component of fast reactors and reactors with heavy liquid metal coolant in particular. The heat treatment is normalization at 1,040–1,080°C for 1 h and tempering at 680–730°C for 2–3 h. In comparison with previously developed by the USSR 12% Cr steels, EP-823 has higher thermal strength and higher resistance to low-temperature irradiation embrittlement. The properties of this steel can be optimized by variation of tempering time and temperature. This steel is also recommended for the hexagonal wrapper of fuel assemblies for SFR, which are developed by the Russian Federation. More than 20 fuel assemblies with the wrapper fabricated from EP-823 were used in the core of BN-600 for three cycles and no issues were reported (Tarasikov et al. 2020). This steel is recommended for in-core components of BREST-OD-300 and SVBR-100 reactors.

T91 is ferritic-martensitic steel (also designated as modified 9Cr-1Mo steel or 9Cr-1Mo-V). Although the T91 material was originally developed for use in the Liquid-Metal Fast Breeder Reactor Program, its major application has been in fossil-fired power plants beginning in the early 1980s. The industrial heat treatment consists of normalizing at 1,040–1,080°C, which provides dissolution of carbides without significant grain growth and tempering at 750–780°C for homogeneous precipitation of carbides within the martensitic structure. T91 is also approved for nuclear service through the ASME and RCC-MRx nuclear design codes. This steel is selected as the candidate material for the proton beam window of MYRRHA ADS. At earlier stages of MYRRHA ADS and ALFRED development, T91 steel was considered as the candidate material for fuel assembly, the heat exchanger and the core support plate. However, for various reasons, austenitic steels were preferred to T91.

AISI 316L (also designated as X2CrNiMo17-12-2, DIN 1.404) is a low carbon grade of molybdenum-alloyed austenitic stainless steel. Because the material will not undergo a phase transformation under cooling after annealing, the heat treatment can be completed in one single treatment. This treatment, called solution annealing, is performed at sufficiently high temperature (1,050–1,100°C) to achieve good diffusion properties of the alloying elements in the steel. Typically, it is used in the solution annealed condition. 316L also has good overall experience feedback from Phénix SFR where it was used for the reactor vessel, the primary cooling systems, the core cover plug and other components (Guidez 2013). This steel is selected as the primary candidate material for the reactor pressure vessel and internals of MYRRHA ADS, ALFRED and W-LFR.

Steel DIN 1.4970 (also designated as X10CrNiMoTiB 15 15 and as 15-15Ti) is Ti-stabilized austenitic stainless steel developed for fuel cladding of SFR (Bergmann et al. 2003). In a cold-worked state, it provides high resistance to irradiation-induced swelling and creep. This approximately 20% cold-worked steel is considered as the candidate material for fuel cladding of W-LFR and MYRRHA ADS system. For ALFRED project, AIM1, which is a swelling-optimized variant of 15-15Ti, is selected for the cladding and for the hexagonal wrapper of fuel assembly (Alemberti et al. 2020a).

	EP-302	EP-852	EP-823	DIN 1.4970	T91	AISI 316L
C	0.08–0.12	0.1–0.15	0.14–0.18	0.08–0.12	0.06–0.15	< 0.03
Cr	14–16	12–14	10–12	14.5–15.5	7.9–9.6	16–18
Ni	8–10	< 0.5	0.5–0.8	14.5–15.5	< 0.43	10–14
Mn	0.4–0.8	< 0.6	0.5–0.8	1–2	0.25–0.66	2
Mo	–	1.2–2	0.6–0.9	1–1.4	0.8–1.1	2–3
Si	2.2–3	1.4–2.1	1–1.3	0.3–0.7	0.18–0.56	< 0.75
W	–	–	0.5–0.8	–	–	–
Nb	0.7–1	–	0.2–0.4	–	–	–
Ti	–	–	–	0.33-0.55	–	–
V	–	–	0.2-0.4	< 0.05	0.16-0.27	–
P	≤ 0.02	≤ 0.2	≤ 0.015	< 0.015	< 0.025	< 0.045
S	≤ 0.015	≤ 0.15	≤ 0.01	< 0.015	< 0.012	< 0.03

Table 5.2. *Chemical composition of the steels candidate for the components of primary systems of reactors with heavy liquid metal coolants (wt%, balance Fe)*

Iron chromium aluminum (FeCrAl) alloys and alumina-forming austenitic (AFA) stainless steel alloys show superior oxidation resistance at elevated temperatures due to the formation of a protective alumina (Al_2O_3) scale rather than the Cr_2O_3 scale that forms on conventional stainless steel alloys. Those alloys are considered to be prospective materials for the structural components and for fuel assembly for the late stages of the project development. Also, the weld overlay of these alloys on conventional steels is considered as a possible option for corrosion mitigation.

The surface treatments techniques including surface alloying and application of coatings are under development to provide the required corrosion(-erosion) resistance of the conventional steels.

5.4. Fabrication and joining

Fabrication of the fuel assembly for the reactors with HLM coolant is similar to those of SFR but specific surface treatments of the components to mitigate corrosion should be embedded in the process. The total surface area of the cladding tubes to be protected against corrosion for an average reactor core can be several hundred square meters that require higher performance for the surface treatment application technology. The quality control of the surface treatment is an essential part of the fabrication technology. For fuel assemblies with wrapper tubes, the internal surface should be also treated.

We should admit the complex shape of the cladding tubes used in the submarine rectors with LBE coolant, which is also intended for SVBR-100 reactor. This cladding is made up of ferritic/martensitic steels in the form of a solid-rolled tubular element with spiraled ribs located on the outer surface of the fuel rod (Derunov et al. 2020).

The processes most widely used for the welding of structural components of nuclear reactor is tungsten inert gas (TIG) welding process. TIG welding of thick-section items such as the reactor pressure vessel is an expensive and time-consuming practice involving extensive prework, including fixtures, tooling, preheating components and multiple weld passes. The latest developments in electron beam welding technology offer the opportunity to weld “thick-section” components in a single pass and eliminate the need for NDE, resulting in significant savings in time and cost. Furthermore, the preheating step can be eliminated, since the EB process is carried out in a locally created vacuum (Baufeld and Dutilleul 2019).

The experience feedback from previous systems is essential for the development and construction of any complex engineering installation and nuclear reactor in particular. Nuclear design codes are based on the experience feedback and provide powerful means for the development of new reactor components and systems. RCC-MRx code is quite distinguishable for the design of reactors with heavy liquid metal coolant since it incorporates feedback on experience from Phénix and Superphénix reactors in the form of various provisions for the procurement, fabrication, welding procedures and design properties of 316L and T91 steels, which are incorporated in the code.

5.5. Main material limitations

The future of innovative nuclear systems depends on the capability of the selected materials to withstand expected operating conditions and to maintain the required structural integrity at hypothetical accident conditions. The most constraining issues for reactor systems with HLM coolants arise from materials degradation phenomena as liquid metal corrosion (LMC), mechanically assisted degradation effects and environmentally induced cracking (EIC), among which liquid metal embrittlement (LME) is the most explicit. The effects of HLM environment on thermal creep of various steels were also observed in the lab tests and their assessment should be provided for the mechanical design. The phenomenology and experimental observations of the listed phenomena are discussed further in this chapter.

5.5.1. *Liquid metal corrosion*

LMC is currently considered as the major challenge for the heavy liquid metal coolant technology. LMC of conventional steels in liquid lead alloys is manifested in form of *dissolution* with the elements of *metallurgically influenced corrosion* and in form of *oxidation* due to interaction of dissolved in liquid metal oxygen with the steel compounds. Mechanically assisted degradation observed for materials in contact with lead alloys includes *erosion corrosion, fretting* and *corrosion fatigue.* For corrosion mitigation, a protective oxide film, which delays both mass transport of alloying elements to liquid metal and transport of oxygen from liquid metal to substrate bulk metal, is normally required. For effective corrosion mitigation, the protective oxide film should sustain various types of load, straining, temperature gradients, high flow velocities, wear and combination of those conditions.

The *dissolution* corrosion is governed by solubility of alloying elements of steels in liquid metal. The release flux of alloying elements is proportional to the difference between solubility and the actual concentration in liquid metal. Corrosion mitigation by saturation of liquid metals with the steel alloying elements may work for simple isothermal systems. For the systems with a temperature gradient, liquid metal saturated at high temperature will be oversaturated in cold leg which may cause deposition on the walls and for a worst-case scenario plug the flow paths.

Some alloy-forming elements as nickel and manganese in austenitic stainless steels have high solubility in liquid lead alloys. Chemical reactions of dissolved alloying elements with impurities in liquid metal as iron with oxygen can retain the concentration of metals with relatively low solubility below the solubility limits and therefore maintain the driving force for corrosion (Schroer 2021). Depending on the specific chemistry of liquid metal, dissolution corrosion appears as selective leaching or uniform dissolution. Dealloying of steel due to selective leaching can result in formation of voids and cause phase transformations since nickel serves as the stabilizer of fcc lattice in austenitic steels. *Metallurgically influenced corrosion* manifests in form of heavy liquid metal penetration in the corroded layer or further in the bulk along specific paths as twin or grain boundaries. The development of dissolution corrosion is enhanced in presence of the deformation defects, probably due to enhanced diffusion of heavy metal atoms along these defects.

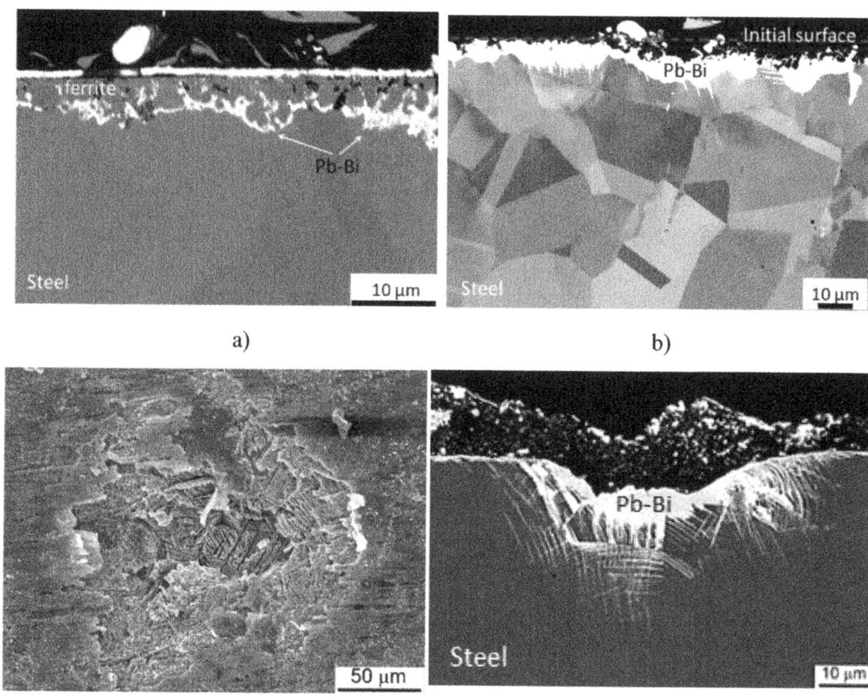

a) b) c) d)

Figure 5.6. *Morphological features of dissolution corrosion of 316L in LBE: (a) selective leaching low oxygen–500°C–2.500 h; (b) uniform dissolution low oxygen–500°C–5.000 h; (c) top view of pit-type damage; (d) cross-section of pit type damage flowing conditions 2 m/s–2 × 10^{-7}% oxygen–10.000 h (Gavrilov et al. 2021)*

The *oxidation* form of LMC may have beneficial effects if the protective oxide film forms, which prevents excessive mass transport of alloying elements to liquid metal. However, at elevated temperatures above 400°C, the oxide films formed on top of austenitic steels of submicrometer thickness or the thicker oxide scale of few micrometers formed on ferritic-martensitic steels may not efficiently delay the mass transport of alloying elements and/or oxygen. Local degradation of the oxide films may cause localized corrosion, forming pitting type corrosion damages on both austenitic and ferritic-martensitic steels. Thicker oxide layers formed on ferritic-martensitic steels can lose adhesiveness probably due to accumulation of the point defects and exfoliate from the surface in flowing liquid metal. The surface oxidation requires sufficient concentration of dissolved oxygen and therefore mass transport of oxygen to the surface might be beneficial for corrosion resistance if protective oxide film forms. In case of excessive oxidation and also in case of chemical reaction with corrosion products surplus of dissolved oxygen may have a negative effect by enhancing corrosion damages (Klok et al. 2018).

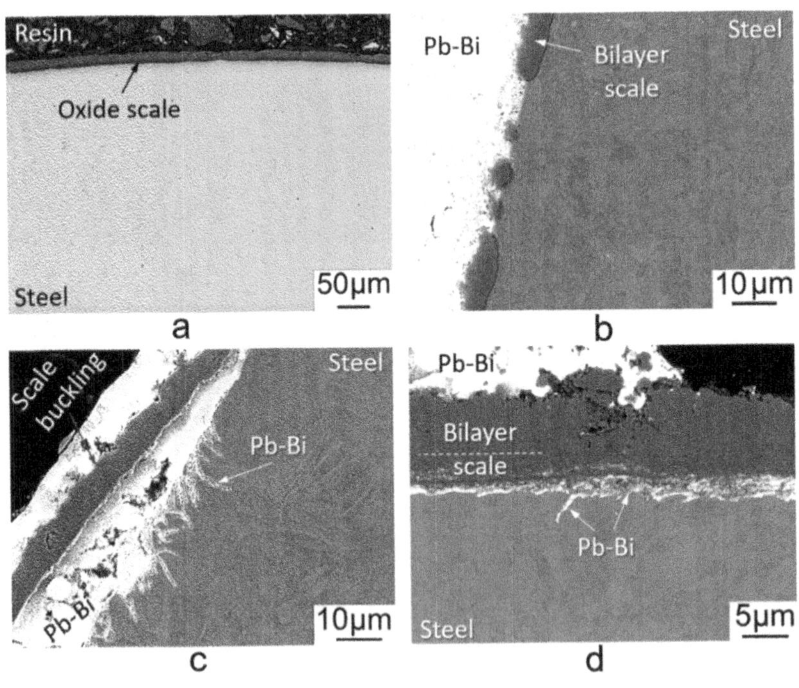

Figure 5.7. *Morphological features of formation of oxide scale on the surface of T91 steel at 400°C in flowing Pb-Bi eutectic (2 m/sec) with 10^{-7} mass% dissolved oxygen for 20,000 h. (a) major oxidation; (b) not completed oxidation; (c) oxide scale buckling along scale/steel interface and (d) penetration of Pb-Bi between scale and steel interface. Adapted from Tsisar et al. (2020)*

The corrosion products released from the steel surfaces may affect the coolant circulation and obstruct the flow in the most severe cases. The released iron and chromium can react with the dissolved oxygen to form oxides, which can deposit on the surface or agglomerate as small dispersed particles. These particles can be further redistributed in the circuits by the flow and accumulate on the liquid metal-gas interfaces. The formation of these oxides and their presence in the coolant affect the oxygen balance and can retain the concentration of iron in the liquid metal below its solubility limit. The small oxide particles in the boundary layer of the liquid metal flowing along the surface may also affect the flow. Dissolved nickel in LBE at high concentration can also form intermetallic phases with bismuth and in the worst case scenario plug the cold legs of the circulation loops (Schroer et al. 2021).

The effect of flow velocity on LMC is still not fully characterized to identify specific limits on flow velocities. The limit of 2 m/s was "inherited" from the submarine reactors program and was communicated verbally without any reference in open literature. The recent works (Choi et al. 2021) on the effect of flow on development of dissolution corrosion indeed do show that dissolution corrosion switch to another regime for flow velocities above 2 m/s, but the effect of flow is already quite significant at that velocity. The tests on the effect of flow at relevant operational conditions temperatures, exposure times and oxygen concentrations should still be done to confirm the validity of the 2 m/s limit.

The analysis of the available experimental data shows a significant increase in the corrosion rates for steels exposed to HLM above 400°C (Gong et al. 2022b) and therefore this temperature is considered to be the limit for unprotected conventional austenitic steel as 316L, DIN 1.4970 in LBE coolant. For liquid lead, the experimental data are much scarcer, however the existing results of long exposure (Kosek et al. 2022) demonstrate reduced corrosion rates and the ability to form more protective oxide film in lead in comparison to LBE environment at comparable oxygen concentrations. Therefore, the limits in liquid lead for unprotected conventional austenitic and ferritic-martensitic steels can probably be justified for temperatures above 400°C. These temperature limitations are also conditional and require, for their justification, accomplishing the ongoing qualification programs. The application of those limits also requires the coolant chemistry control in place and corresponding surveillance programs to confirm the corrosion assessment used to establish the limitations. The same temperature limit of 400°C, as mentioned above for unprotected austenitic stainless steels in contact with heavy liquid coolant, was established in earlier works dedicated to the submarine reactors (Yachmenyov et al. 1998).

5.5.2. *Liquid metal embrittlement*

The natural fear for any system containing liquid metal is the susceptibility of the materials to liquid metal embrittlement (LME). The actual effect of lead-based alloys on susceptible steels is far less spectacular than the disintegration of aluminum by liquid gallium, which is the most striking example. LME is a special case of environmental-induced cracking and can be defined as reduction of stress to initiate and/or to propagate crack in a susceptible material in contact with a liquid metal.

The tests performed in the second half of the last century in many laboratories revealed the occurrence of LME in lead and LBE for pure iron, for several carbon steels, for low-alloyed steels and for ferritic 12% Cr steels with and without tempering heat-treatment (Gong et al. 2022a). During the last two decays, many tests were performed for the qualification of two candidate steels: 316L and T91, for application in contact with HLM coolants. One of the necessary conditions for the appearance of LME is intimate contact between liquid and solid metals, which in some cases can be prevented by native oxide film or by lack of surface wetting. Since intimate contact was not provided during the whole duration of the mechanical test, the test may be not so conclusive on appearance and severity of LME effect. There is sufficient experimental evidence that ferritic martensitic steels and T91 in particular are susceptible to LME in contact with liquid lead-based alloys. Quite some experimental data on LME were obtained for the T91/LBE couple (Gong et al. 2022b), which allow us to draw some conclusions. In tensile tests, LME was manifested by reducing fracture strain, surface cracks in the necking region (Figure 5.7) and the specific pattern of fracture quasi-cleavage. The LME effect takes place in a temperature range, which borders varies with the test strain rate. Therefore, outside of the temperature range called the ductility trough, the effect is not observed. It also means that there is a certain range of the strain rate within which LME takes place at a specific test temperature. It was also found that irradiation (Stergar et al. 2016), Si doping (Van den Bosch et al. 2012) and hardening by tempering heat treatment at reduced temperatures (Legris et al. 2000) significantly enhance susceptibility of T91 to LME.

Contrary to T91, the tensile tests of 316L in LBE do not show evidence of LME. Also in an irradiation hardened state (Stergar et al. 2016), in cold worked conditions and when the intimate contact with liquid metal was provided by pre-exposure, no indication of LME was found on 316L.

The tests performed in the conditions when intimate contact of LBE at the crack tip was provided (Ersoy et al. 2016) did show a significant reduction of T91 fracture toughness in LBE and effect increased when lowering the displacement rates (Ersoy et al. 2021).

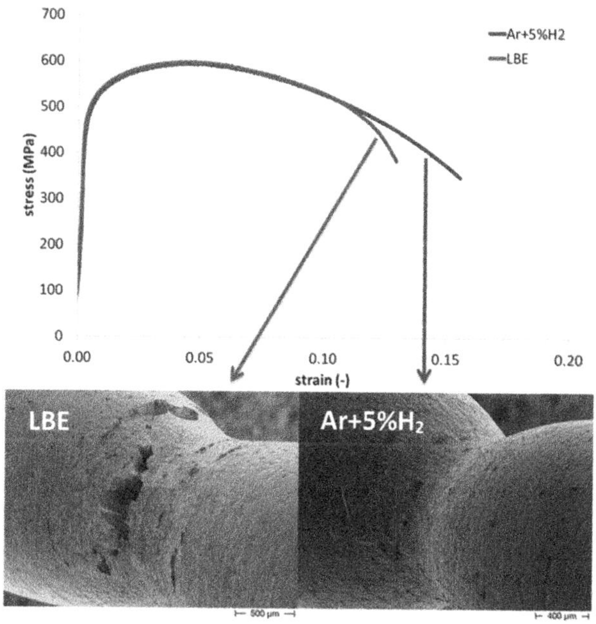

Figure 5.8. *Engineering stress–strain curves of tensile test of T91 in Ar-5%H2 gas and in LBE, below are the necking observed on the samples tested in both environments. For a color version of this figure, see www.iste.co.uk/balbaud/nuclearenergytoday.zip*

5.5.3. *Corrosion fatigue and corrosion creep*

The effect of the testing environment on fatigue resistance of steels is well known and the reduction of fatigue life was observed for steels tested in plain air, in high temperature water and in liquid metals when compared to those in vacuum. The mechanisms of how environment enhances nucleation and propagation of fatigue cracks are complex and not necessarily related to the material embrittlement. The "fatigue endurance trough" observed in low-cycle fatigue tests of T91 in LBE (Gong et al. 2016) has the same temperature range as the "ductility trough" derived from tensile tests and this is an argument to explain the reduction of fatigue life in LBE by LME effect. In case of 316L steel, weak or no reduction of fatigue life was observed for tests in LBE in comparison with air and the environmental effect was not linked to LME (Gong et al. 2022b).

The data on the effect of HLM on creep properties of steels are contradictory. For T91 steel tested in HLM, the literature data report a negligible effect in liquid lead (Yurechko 2011) and significantly enhanced creep rate and reduced time-to-failure in LBE (Jianu et al. 2009). For 15-15Ti steel, severe effects on creep rate and time-to-failure were found (Gong et al. 2020b). Since the tests are performed in the temperature range where severe corrosion of unprotected surface of steels takes place, the distinction between material loss due to corrosion and enhancement of the creep rate is yet to be clearly made. To overcome the limitations for materials application in the high temperature range, the issue of corrosion creep of steels in HLM should be thoroughly investigated and the effect on creep properties should also be assessed.

5.5.4. *Limitations for material qualification*

High nickel solubility as the main driving force of the dissolution process makes it very challenging, if at all feasible, to use nickel base alloys for the components in contact with heavy liquid metal. High thermal strength and enhanced creep resistance of nickel based alloys make them one of the most promising direction for further enhancement of temperature in advanced nuclear systems with sodium, gas and supercritical water coolants. Unfortunately, for the heavy liquid metal coolant, those materials are not applicable without development of the appropriate corrosion mitigation approach. The hard surfacing alloys for friction surfaces as Colmonoy are Ni-based and therefore their application is problematic due to LMC.

The failure modes of various functional and structural components are often synergetic effects of different degradation phenomena. It is difficult to predict potential failures on the only basis of dedicated laboratory experiments. However, the operational experience of various installations used for investigation of various aspects of HLM technology is a very valuable source of information to identify potential failure modes and to develop corresponding mitigation strategy. So far, very limited information has been published on materials performance operational feedback from nuclear and nonnuclear installations with heavy liquid metal coolant.

The main materials limitation is therefore the availability of materials with necessary corrosion resistance in the selected coolant temperature range from one hand and the qualification for application in structural components or as materials for fuel assembly.

5.6. Main material challenges and solutions under evaluation

The corrosion mitigation approach selected for SVBR-100 and for BREST-OD-300 systems (Table 5.1) is based on Si-doped steels and oxygen control in the coolant to maintain the protective oxide scales on the surfaces in contact with HLM coolant. No specific material challenges for materials qualification were reported in open literature for these two nuclear systems. We could assume that the main challenge is the detection of unforeseen materials degradation phenomena, which can cause components or systems failure.

The main materials challenges for the other near term projects with HLM coolants are to define and to justify the corrosion mitigation approaches for the austenitic stainless steels selected as the candidate materials in the majority of designs for both the structural components and for the fuel assembly. The solutions for LMC mitigation are normally based on two major principles, which are the modification of heavy liquid metal environment and/or the modification of steel. In most of these approaches, the main purpose is to form and maintain an appropriate barrier for the mass transport at the interface between liquid metal and steel. Once corrosion is initiated, the purpose is to delay its development. Environmental parameters that can be used to mitigate or reduce corrosion include temperature, dissolved oxygen concentration and liquid metal flow velocity.

Chemistry control in general and active oxygen control in particular are basic requirements for nuclear systems with HLM coolants (OECD/NEA 2015). The process of oxygen control in HLM is essentially of three different kinds:

– gas-phase control (OECD/NEA 2015);

– solid-phase control (OECD/NEA 2015);

– electrochemical oxygen pumping (Lim et al. 2014).

The purpose of the oxygen control system is to maintain what is dissolved in HLM within the preferred operational range. On the one hand, the dissolved oxygen concentration in LBE should be maintained in the range where the formation of the protective oxide film is possible. On the other hand, this concentration should not exceed the limit after which formation of lead oxide starts. The latter can jeopardize coolant circulation due to plugging of flow paths. Also maintaining sufficient level of oxygen concentration may be challenging in stagnant zones where convection mass transport is limited. Therefore, systems solely reliant on oxygen concentration would require design solutions to avoid stagnant zones. High consumption of oxygen on metal surfaces may also create low oxygen zones downstream of narrow channels as in the fuel assembly. Therefore, oxygen concentration control is a sophisticated technology, which is required but may not always be sufficient for the required corrosion mitigation. Availability of the corrosion loops with the oxygen control to perform long-term corrosion tests in the representative conditions of the reactor circuits conditions is the most critical issue for the on-going programs on corrosion mitigation.

The corrosion products release rate is one of the essential parameters to control for the development of appropriate coolant purification technology. However, the electrochemical methods used to measure corrosion rates in aqueous solutions cannot be applied due to the electronic conductivity of liquid metals. Moreover, weight loss methods are also not easily applicable due to the difficulties in cleaning corroded surfaces from liquid metal residuals, as well as complex morphology of LMC. Methodologies used to measure what the corrosion products release in liquid metal environments are therefore required; such as indirect measurements based on coulometric titration of oxygen (Gregori et al. 2022).

To protect conventional steels against corrosion phenomena in HLM, surface alloying to form in situ protective alumina scales is under development (Weisenburger 2011). The process consists of applying a dedicated coating and remelting a few tens micrometer thick layer by a pulsed electron beam in a vacuum chamber. The composition of the initial coating corresponds to the substrate steel composition with the addition of Al and with minor doping with reactive elements like Y or Zr. The deposit of heat by the electron beam to melt the surface layer is so local that the substrate material is not affected and retains its original properties. This method with the utilization of GESA (Gepulste Elektronenstrahl Anlage) facility was developed at KIT (Mueller et al. 2005).

Coating by means of different depositing techniques are indeed very promising mitigation options. The coating could be applied by a number of processes like PVD, air plasma spraying (APS) or vacuum plasma spraying (VPS) and low-pressure plasma spraying (LPPS). The oxide-based coating may have the advantage that there is no need of in situ formation of the oxide. This would relax the requirement on the oxygen control in the liquid metal. The pulsed laser deposition (PLD) is a promising technique to create nanoscale alumina coatings (Vanazzi et al. 2019). Thin alumina coatings applied by PLD show good radiation resistance under ion irradiation up to 450 dpa. Thanks to fine microstructure, the alumina coating can have a plastic response under stress and therefore maintain integrity and adhesiveness to the substrate while strained.

Ferritic alumina forming (FeCrAl) alloys show superior corrosion resistance in many extreme and high-temperature environments and were evaluated for application in contact with HLM coolant. In the recent study, low-alloyed FeCrAl steel specifically aimed for HLM applications showed promising oxidation properties and ductile behavior in liquid lead up to 750°C (Dömstedt et al. 2019). However, FeCrAl steels are prone to LME (Gong et al. 2020a) and there are no sufficient data at the moment for qualification of this steel as fuel cladding material. Using FeCrAl as weld overlay on top of conventional steels may help to overcome these issues, and this solution was recently successfully applied on thin wall 316L tubes by low power pulse laser welding technology (Yang et al. 2021).

Austenitic steels alloyed with aluminum have the potential to improve the corrosion resistance in HLM due to the formation of chemically stable alumina-based oxide films and are named as alumina-forming austenitic (AFA) steels (Ejenstam and Szakalos 2015). As for other steels with austenitic structure, AFA steels would probably not be susceptible to LME in HLM. However, assessments of the effect of HLM on mechanical properties and corresponding testing are yet to be performed.

For the candidate material of the pump impeller with sufficient corrosion–erosion resistance, the MAX phase Ti_3SiC_2 was evaluated. It was selected because of its easier machining and forming in comparison with other composite materials.

The additional challenge for austenitic stainless steels is to define the effect of liquid metal on the (thermo)mechanical properties as fatigue and creep and to incorporate these effects in the design approach.

The main challenge for the corrosion-resistant materials under development is to be qualified for the mechanical design. It also includes codification of the materials by integration in the nuclear manufacturing codes. The candidate materials for fuel assembly and for in/near core component should provide sufficient neutron radiation resistance. Due to limited availability of fast neutron irradiation capacities and very high cost of the materials irradiation campaigns, the advanced materials characterization as small specimen test technology is of critical importance for screening and qualification of the candidate materials.

To handle LME effect in the development of the reactor components, the following three options were identified:

– to use materials susceptible to LME with reduced requirements on the specific mechanical properties to accommodate LME effect;

– to use materials with negligible effect of LME on the design mechanical properties;

– to use the LME mitigation approach.

The first option to accommodate LME is to reduce the requirements on the mechanical properties to widen the range of candidate materials. The main challenge with this option is to provide conservative assessment of the required reduction of the mechanical properties. These properties include fracture toughness, fatigue life, fatigue crack growth rates and possibly creep rates for high-temperature applications. The phenomenology of LME and its parametric dependence is not sufficiently understood to establish the testing parameters to produce conservative test results. In the study of the effect of the displacement rate on fracture toughness (Ersoy et al. 2021), it was found that for the samples tested in LBE, the fracture toughness was significantly reduced with the decrease in the displacement rate. The value obtained at the lowest rate was already low enough to question the suitability of the material as a structural material. Another challenge is the level of maturity of the methodologies used to assess the environmental effects on the mechanical properties.

The second option is to use materials, which do not show susceptibility to LME. This reduces the range of candidate materials to austenitic stainless steels only. However, to demonstrate no indication of LME, screening by SSRT tests may not be sufficient for qualification. It may still be necessary to demonstrate that there is no effect of heavy liquid metal environment on the mechanical properties used in the components' design as fracture toughness, fatigue life, fatigue crack growth rate and creep properties. The corresponding materials-testing program also requires methodologies to perform the tests. In particular, intimate contact between steel and liquid metal should be provided during testing. In addition, tests should be performed for a wide range of parameters such as temperature, oxygen concentration in LBE and displacement rates.

The third option is to define appropriate LME mitigation approach for materials susceptible to LME. One of the approaches under discussion is to maintain integrity of the oxide layers on the surface of the components in order to prevent contact of liquid metal with the steel. However, actual failures observed in operation are often due to synergy between different degradations phenomena. In practice, initiation of the crack on the surface may occur as a result of thermal fatigue, allowing the liquid metal to penetrate in the crack and to provide contact at the crack tip. Another suggested mitigation approach is to establish operational temperatures above upper limit of the ductility trough. However, temperature transients occurring during reactor start-up may lead to high thermal stresses of components, while it still may be partially at a low temperature. The increase in the shutdown temperatures above the ductility trough range would therefore impose high demands on corrosion mitigation. Therefore, the third option to define the appropriate LME mitigation approach seems to be very attractive but its feasibility is yet to be demonstrated.

Surveillance programs to confirm that the developed mitigation techniques are sufficiently preventing undesirable materials degradation are yet to be developed and to put in place for the appropriate aging management (Chaouadi et al. 2011).

5.7. References

Aït Abderrahim, H., De Bruyn, D., Van Den Eynde, G., Fernandez, R. (2021). Accelerator driven subcritical systems. In *Encyclopedia of Nuclear Energy*, Greenspan, E. (ed.). Elsevier, Oxford.

Alemberti, A., Caramello, M., Frignani, M., Grasso, G., Merli, F., Morresi, G., Tarantino, M. (2020a). ALFRED reactor coolant system design. *Nuclear Engineering and Design*, 370, 110884.

Alemberti, A., Tucek, K., Takahashi, M., Obara, T., Kondo, M., Moiseev, A., Tocheny, L., Smith, C.F., Hwang, I.S., Wu, Y. et al. (2020b). Lead-cooled fast reactor (LFR) system safety assessment. Report, GEN IV International Forum.

Baufeld, B. and Dutilleul, T. (2019). Electron beam welding of large components for the nuclear industry. *MATEC Web Conf.*, 269, 02009.

Bergmann, H.J., Dietz, W., Ehrlich, K., Muhling, G., Schirra, M. (2003). Entwicklung des Werkstoffs X10CrNiMoTiB 15 15 als Strukturmaterial für Brennelemente. Report, FzK.

Chaouadi, R., Gavrilov, S., Fernandez, R., Lambrinou, K., Scibetta, M. (2011). Surveillance program assessment for critical components of next generation nuclear reactors. In *2011 – ICAPP – International Congress on Advances in Nuclear Power Plants*, Paris.

Choi, J., Hwang, I., Lee, Y. (2021). Flow accelerated corrosion of stainless steel 316L by a rotating disk in lead-bismuth eutectic melt. *JOM*, 73, 4030–4040.

De Bruyn, D. and Aït Abderrahim, H. (2022). MYRRHA, the Belgian prototype that fascinates the world. In *International Conference on Fast Reactors and Related Fuel Cycles FR22: Sustainable Clean Energy for the Future (CN-291)*. IAEA, Vienna.

Derunov, V.V., Mayorov, V.M., Pomeschikov, P.A., Rusanov, A.E., Smirnov, A.A., Shulepin, S.V., Sharikpulov, S.M. (2020). Fuel rod cladding, fuel rod and fuel assembly. Patent, US patent application 15/021,698.

Dömstedt, P., Lundberg, M., Szakalos, P. (2019). Corrosion studies of low-alloyed FeCrAl steels in liquid lead at 750°C. *Oxidation of Metals*, 91(3), 511–524.

Ejenstam, J. and Szakalos, P. (2015). Long term corrosion resistance of alumina forming austenitic stainless steels in liquid lead. *Journal of Nuclear Materials*, 461, 164–170.

Ersoy, F., Gavrilov, S., Verbeken, K. (2016). Investigating liquid-metal embrittlement of T91 steel by fracture toughness tests. *Journal of Nuclear Materials*, 472, 171–177.

Ersoy, F., Verbeken, K., Gavrilov, S. (2021). Influence of displacement rate and temperature on the severity of liquid metal embrittlement of T91 steel in LBE. *Materials Science and Engineering: A*, 800, 140259.

Ferroni, P., Liao, J., Stansbury, C., Wise, D., Wright, R., Lee, S.J., Ickes, M., Banyay, G., Franceschini, F., Metzger, K. et al. (2022). The Westinghouse lead fast reactor: Overview and progress in development. In *International Conference on Fast Reactors and Related Fuel Cycles FR22: Sustainable Clean Energy for the Future*. IAEA, Vienna.

Frignani, M., Alemberti, A., Tarantino, M. (2019). ALFRED: A revised concept to improve pool related thermal-hydraulics. *Nuclear Engineering and Design*, 355, 110359.

Gavrilov, S., Klok, O., Tsisar, V. (2021). Assessment of liquid metal corrosion for candidate materials and welds of MYRRHA primary circuit. Report, H2020-GEMMA Project No. 755269.

Gong, X., Chen, J., Hu, F., Xiang, C., Yu, Z., Xiao, J., Wang, H., Gong, H., Wang, H., Liu, C. et al. (2020a). Liquid metal embrittlement of an $Fe_{10}Cr_4Al$ ferritic alloy exposed to oxygen-depleted and-saturated lead-bismuth eutectic at 350°C. *Corrosion Science*, 165, 108364.

Gong, X., Yang, Z., Deng, Y., Xiao, J., Wang, H., Yu, Z., Yin, Y. (2020b). Creep failure of a solution-annealed 15-15Ti steel exposed to stagnant lead-bismuth eutectic at 550 and 600°C. *Materials Science and Engineering: A*, 798, 140230.

Gong, X., Chen, H., Zhang, F., Zhu, W., Ma, H., Pang, B., Yin, Y. (2022a). Degradation of tensile mechanical properties of two AlxCoCrFeNi (x = 0.3 and 0.4) high-entropy alloys exposed to liquid lead-bismuth eutectic at 350 and 500°C. *Fifth International Workshop on Spallation Materials Technology*, 558, 153364.

Gong, X., Short, M.P., Auger, T., Charalampopoulou, E., Lambrinou, K. (2022b). Environmental degradation of structural materials in liquid lead- and lead-bismuth eutectic-cooled reactors. *Progress in Materials Science*, 126, 100920.

Gregori, G., Aerts, A., Gladinez, K., Rosseel, K., Doneux, T., Lim, J. (2022). Electrochemical measurement of solubility product of metal oxides in liquid metals by coulometric titration of oxygen. *Electrochimica Acta*, 432, 141202.

Gromov, B.F., Toshinskii, G.I., Chekunov, V.V., Orlov, Y.I., Belomitcev, Y.S., Gorelov, I.N., Karabash, A.G., Leonchuk, M.P., Pankratov, D.V., Pashkin, Y.G. (1998). Designing the reactor installation with lead-bismuth coolant for nuclear submarines. The brief history. Summarized operation results. In *Heavy Liquid Metal Coolants in Nuclear Technology (HLMC-98)*, Gromov, B.F. (ed.). Russian Federation, IPPE, Obninsk.

Guidez, J. (2013). *Phénix: The Experience Feedback*. EDP Sciences, France.

Jianu, A., Muller, G., Weisenburger, A., Heinzel, A., Fazio, C., Markov, V.G., Kashtanov, A.D. (2009). Creep-to-rupture tests of T91 steel in flowing Pb-Bi eutectic melt at 550°C. *Journal of Nuclear Materials*, 394(1), 102–108.

Karzov, G.P., Kashtanov, A.D., Markov, V.G. (2009). Corrosion-mechanical strength of steels with γ- and α-lattice in contact with lead-based liquid-metal coolant. *Vopr. Materialoved.*, 59(3), 315–329.

Klok, O., Lambrinou, K., Gavrilov, S., Lim, J., De Graeve, I. (2018). Effect of lead-bismuth eutectic oxygen concentration on the onset of dissolution corrosion in 316 L austenitic stainless steel at 450 °C. *Journal of Nuclear Engineering and Radiation Science*, 4(3), 031019.

Kosek, L., Rozumova, L., Hojna, A., Aparicio, C., Vronka, M., Vit, J. (2022). Mechanism of localized corrosion issues of austenitic steels exposed to flowing lead with 10–7 wt.% oxygen at 480°C up to 16000 h. *Journal of Nuclear Materials*, 572, 154045.

Legris, A., Nicaise, G., Vogt, J.-B., Foct, J., Gorse, D., Vançon, D. (2000). Embrittlement of a martensitic steel by liquid lead. *Scripta Materialia*, 43, 997–1001.

Lemehov, V.V., Adamov, E.O., Orlov, V.V., Kaplienko, A.V., Moiseev, A.V., Smirnov, V.S., Yarmolenko, O.A., Sarkulov, M.K., Bazhanov, A.A., Chekov, M.E. et al. (2022). Pilot demonstrational fast reactor with lead coolant BREST-OD-300. In *International Conference on Fast Reactors and Related Fuel Cycles FR22: Sustainable Clean Energy for the Future (CN-291)*, Vienna.

Lim, J., Manfredi, G., Gavrilov, S., Rosseel, K., Aerts, A., Van Den Bosch, J. (2014). Control of dissolved oxygen in liquid LBE by electrochemical oxygen pumping. *Sensors and Actuators B: Chemical*, 204, 388–392.

Mueller, G., Engelko, V., Weisenburger, A., Heinzel, A. (2005). Surface alloying by pulsed intense electron beams. *Vacuum*, 77(4), 469–474.

OECD/NEA (2015). *Handbook on Lead-bismuth Eutectic Alloy and Lead Properties, Materials Compatibility, Thermalhydraulics and Technologies*. OECD/NEA, Paris.

Peplow, M. (2014). Liquid-metal batteries get boost from molten lead. *Nature*. doi: 10.1038/nature.2014.15967.

Rinaldi, A., Barbieri, G., Kosykh, E., Szakalos, P., Testani, C. (2021). Materials for high temperature liquid lead storage for concentrated solar power (CSP) air tower systems. *Materials*, 14(12), 3261.

Schroer, C. (2021). Fundamental interactions of steels and nickel-based alloys with lead-based liquid alloys or liquid tin. In *TMS 2021 150th Annual Meeting & Exhibition Supplemental Proceedings*. Springer, Cham.

Schroer, C., Wedemeyer, O., Konys, J. (2011). Aspects of minimizing steel corrosion in liquid lead-alloys by addition of oxygen. *Nuclear Engineering and Design*, 241(12), 4913–4923.

Schroer, C. Tsisar, V., Skrypnik, A., Wedemeyer, O. (2021). Effect of the solution of steel elements in flowing oxygen-containing lead–bismuth on the plugging of flow paths. In *H2020 GEMMA Deliverable D4.1 Assessment of Liquid Metal Corrosion for Candidate Materials and Welds of MYRRHA Primary Circuit*. SCK•CEN.

Stergar, E., Eremin, S.G., Gavrilov, S., Lambrecht, M., Makarov, O., Iakovlev, V. (2016). Influence of LBE long term exposure and simultaneous fast neutron irradiation on the mechanical properties of T91 and 316L. *Journal of Nuclear Materials*, 473, 28–34.

Tarasikov, V.P., Soloviev, V.A., Birgevoi, G.A., Porollo, S.I., Shulepin, S.V., Ivanov, A.A., Leont'eva-Smirnova, M.V., Mitrofanova, N.M., Naumenko, I.A. Moganov, E.M. (2020). Effect of neutron irradation on physico-mechanical properties of steel and alloys of domestic nuclear reactors, Fizmalit, Moscow (in Russian).

Tsisar, V., Gavrilov, S., Schroer, C., Stergar, E. (2020). Long-term corrosion performance of T91 ferritic/martensitic steel at 400°C in flowing Pb-Bi eutectic with 2 × 10–7 mass% dissolved oxygen. *Corrosion Science*, 174, 108852.

Van Den Bosch, J., Coen, G., Hosemann, P., Maloy, S.A. (2012). On the LME susceptibility of Si enriched steels. *Journal of Nuclear Materials*, 429(1–3), 105–112.

Vanazzi, M., Paladino, B., Iadicicco, D., Ferré, F.G., Di Fonzo, F., Bassini, S., Utili, M., Tarantino, M., Agostini, P., Beghi, M.G. (2019). Alumina nanoceramic coatings: An enabling technology for heavy liquid metal-cooled fast reactors. In *IAEA Technical Meeting on Structural Materials for Heavy Liquid Metal Cooled Fast Reactors*. IAEA, Vienna.

Wallenius, J., Qvist, S., Mickus, I., Bortot, S., Szakalos, P., Ejenstam, J. (2018). Design of SEALER, a very small lead-cooled reactor for commercial power production in off-grid applications. *Nuclear Engineering and Design*, 338, 23–33.

Weisenburger, A., Müller, G., Heinzel, A., Jianu, A., Muscher, H., Kieser, M. (2011). Corrosion, Al containing corrosion barriers and mechanical properties of steels foreseen as structural materials in liquid lead alloy cooled nuclear systems. *Nuclear Engineering and Design*, 241(5), 1329–1334.

Yachmenyov, G.S., Rusanov, A.Y., Gromov, B.F., Belomytcev, Y.S., Skvortsov, N.S., Demishonkov, A.P. (1998). Problems of structural materials corrosion in lead-bismuth coolant. In *Heavy Liquid Metal Coolants in Nuclear Technology (HLMC-98)*, Gromov, B.F. (ed.). Russian Federation, IPPE, Obninsk.

Yang, S., Xi, B., Zhang, Y., Wang, D., Liu, B., Szakálos, P., Ejenstam, J., Wallenius, J., He, G., Zhang, W. (2021). Development of a welding process to overlay FeCrAl alloy on a thin wall austenitic stainless steel tube. *Nuclear Materials and Energy*, 27, 100958.

Yurechko, M., Schroer, C., Wedemeyer, O., Skrypnik, A., Konys, J. (2011). Creep-to-rupture of 9%Cr steel T91 in air and oxygen-controlled lead at 650 degC. *Journal of Nuclear Materials*, 419(1–3) 320–328.

6

Materials and Processes for Molten Salts Reactors

Céline CABET[1], Yann DE CARLAN[2] and Laure MARTINELLI[2]

[1] *CEA, Service de Recherche en Matériaux et Procédés Avancés, Université Paris-Saclay, Gif-sur-Yvette, France*

[2] *CEA, Université Paris-Saclay, Gif-sur-Yvette, France*

6.1. Introduction with general description of the systems

A molten salt reactor (MSR) is a nuclear system in which halides of fissile and/or fertile elements, such as uranium, plutonium and/or thorium are combined with carrier salts to form fluids. MSRs primarily refer to homogeneous core designs to distinguish them from concepts in which molten salts without dissolved fissile element are used to cool solid fuel assemblies. Although they share many characteristics stemming from the molten salt thermo-physical properties, the development path for these two types of concepts is different, particularly for the issues of structural materials and corrosion control, which are the subjects of this chapter.

MSRs have a number of advantages over conventional reactor designs. They run at low pressures – hydrostatic typically <0.4 MPa – and so do not need the large pressure vessels required in light water reactors (LWR). The MSR containment vessel does not need the huge volume and ability to deal with steam pressure buildup, as is the case for LWRs. Unlike other reactors, MSRs have no water in the primary fluid so there is no possibility of hydrogen being produced within the containment. MSRs can run on a variety of fuels and can even burn transuranic waste produced by other reactors (LeBlanc 2010). These designs have very strong

Materials and Processes for Nuclear Energy Today and in the Future, coordinated by Fanny BALBAUD-CÉLÉRIER and Céline CABET.

negative temperature and void coefficients, which have a short response time, supporting safety and allowing automatic load, following operation. As the fuel is in a molten state, meltdown of the core is not a concern. In the event of an emergency, the fuel salt may automatically be drained to passively cooled, critically safe, drain tanks. Generation of minor actinides is another major issue with nuclear reactors; most fission products in MSRs are from stable halides that will stay within the salt during any leak or accident.

MSRs can operate with thermal, epithermal or fast neutron spectrums, and with a variety of fuels. Therefore, there is a range of MSR systems. However, materials selection and qualification in the harsh operating environment are common tall orders. Also, availability of materials compatible with molten salts over long-term operation is one of the technical viability issues on the development path of MSR systems as a whole. Innovative material solutions will most certainly have to be developed to meet the high-level requirements issuing from operation at high temperature, in a highly corrosive medium and possibly under neutron irradiation. Fabrication and joining of such complex material architectures, as well as control of the melt chemistry to mitigate corrosion, are other challenges on the path to designing such reactor concepts.

6.1.1. *Background*

R&D activities of nuclear applications of molten salts started in the United States in the late 1940s as part of a program to develop a nuclear-powered airplane (Serp et al. 2014). In 1954, a 2.5 MWth MSR, called aircraft reactor experiment (ARE), was built and fueled with UF_4 dissolved in a mixture of zirconium and sodium fluorides, moderated with beryllium oxide. It operated successfully for nine days at steady-state outlet salt temperatures approaching 900°C. The reactor was found to be stable and self-regulating.

Construction of a test reactor called molten salt research experiment (MSRE) began in 1962. For simplicity, it was designed with a single fluid and a tank of graphite with flow channels. The initial fuel salt composition was ^{7}LiF-BeF_2-ZrF_4- $^{235}UF_4$ (65-29.1-5.0-0.9 mole%). The secondary salt was LiF-BeF_2 (66-34 mole%). The maximum power was approximately 8 MWth and the outlet temperature was about 650°C. All metallic parts of the system in contact with the salt were made from the nickel-based alloy INOR-8 (later called Hastelloy-N), which was specially developed to overcome corrosion problems encountered with

Inconel 600 in the ARE program. It first operated with uranium 235 enriched at 33%; then, the uranium was replaced with ^{233}U, the fissile nuclide of the thorium cycle.

Figure 6.1. shows the global layout of a graphite-moderated MSR. The MSRE provided facilities for testing fuel salt, graphite and structural alloy under reactor operating conditions and had a successful run for almost five years between 1965 and 1969. One important result that came out of the MSRE program is the conclusion that MSR is a practical concept (MacPherson 1985).

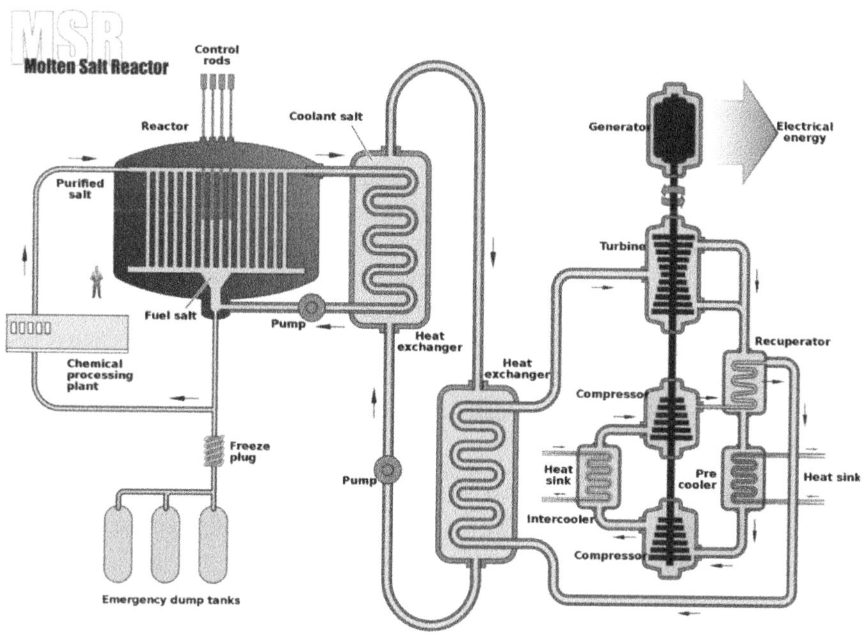

Figure 6.1. *Example of an MSR scheme. For a color version of this figure, see www.iste.co.uk/balbaud/nuclearenergytoday.zip*

The next step was for Oak Ridge National Laboratory to design a large-scale reactor, the molten salt breeder reactor (MSBR). A 2000 MWth one-fluid, two-region MSR using $LiF-BeF_2-ThF_4-UF_4$ (72-16-12-0.4 mole%) with graphite moderator was investigated. Its peak operating temperature was to be 705°C with eutectic NaF-$NaBF_4$ as secondary coolant. The U.S. MSR program closed down in the early 1970s, stopping the development of the MSBR. Ever since, the MSBR design and ARE/MSRE operational experience have served as baseline for many new MSR concepts.

6.1.2. *MSR research emphasis*

There is now renewed interest in the MSR concept, which is one of the six Gen IV reactors selected for further international development. The GIF Roadmap (GIF 2002) listed "materials compatibility with molten salt fuel" as one of the technology gaps to be resolved for the MSR viability assessment. Since the Gen IV selection process in the early 2000s, deployment objectives of MSRs have significantly changed. Some designs seek simpler, less ambitious MSRs that do not breed new fuel, do not call for a 60-year design life, do not propose online fuel reprocessing or which use the well-established enriched uranium fuel cycle. Also, since the mid-2010s, a worldwide revival of the molten salt technology has been expressed in a growing number of initiatives. Open and closed cycles have been envisioned, fueled with thorium, uranium (low enriched or enriched), plutonium and/or minor actinides. Small, intermediate and large power reactors have been proposed. The number of concepts makes it impossible to establish a comprehensive and up-to-date list. They may be categorized into at least four main families (see Figure 6.2):

– *concepts with a solid core or fixed-fuel and a molten-salt coolant*, mostly fueled with triso particles and using a graphite prismatic assembly or graphite pebble bed. All rely on fluoride-based molten salts;

– *liquid fuel MSRs in thermal spectrum*, fueled with a fluoride salt, frequently based on FLiBe. Most concepts derive from the MSRE/MSBR and propose to use a graphite moderator. Different reprocessing strategies are proposed (no batch or on-line);

– *liquid fuel MSRs with two fluids* (breeders), mainly differ from the previous category by their fuel cycle (breeder). Fuel is uranium-233 bred from thorium in FLiBe blanket salt. They operate with a graphite moderator;

– *liquid fuel MSRs in fast spectrum*, which allow the achieving of a fully closed cycle and offer the capability to burn spend fuel from other types of reactors. The first systems were based on fluoride salts but the more recent concepts tend to use chloride fuels, which favor the plutonium solubility.

Table 6.1 shows main reactor technology and operating parameters for a short list of concepts. More concepts are analyzed in IAEA (2020).

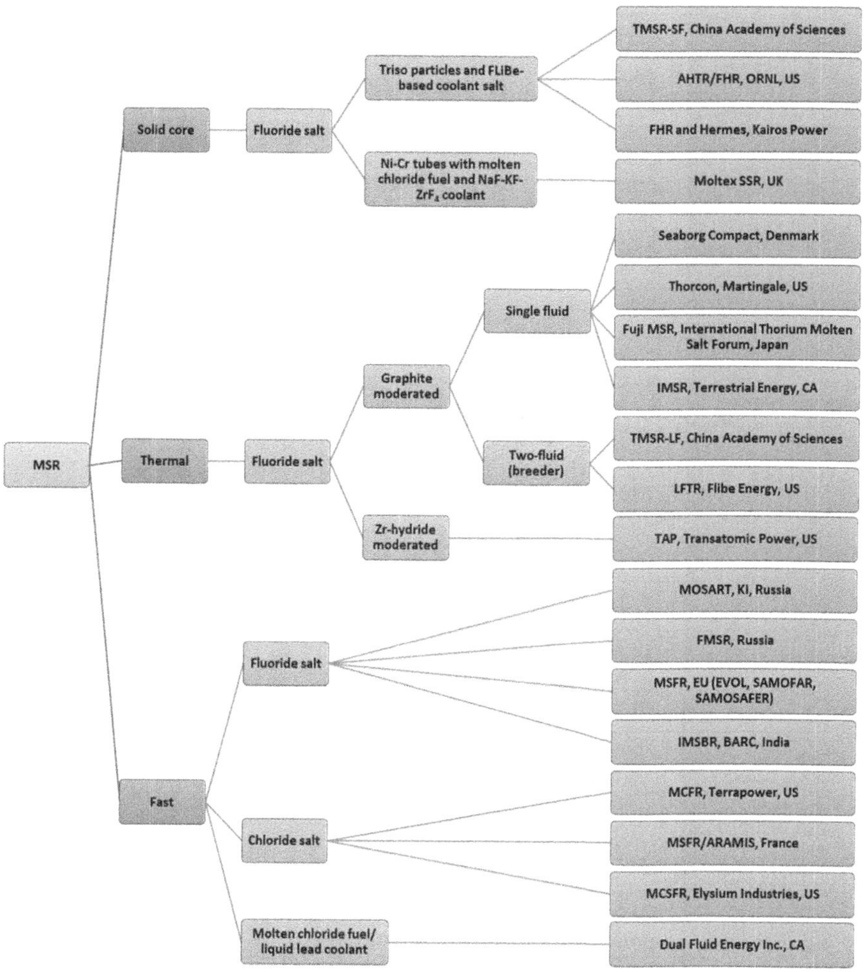

Figure 6.2. *Variety of MSR concepts (tree view). For a color version of this figure, see www. iste.co.uk/balbaud/nuclearenergytoday.zip*

	FUJI	KAIROS	IMSR	MCFR	MSFR
Thermal/ electrical capacity MW(t)/MW(e)	450/200	3,000	440/195	125/50 to 3,000/1,200	3,000
Fuel type	Molten fluoride salt	TRISO particles in graphite matrix	Molten fluoride salt	Molten chloride salt	Molten chloride salt
Coolant	Fluoride fuel salt	Li_2BeF_4	Fluoride fuel salt	Chloride fuel salt	$NaCl/MgCl_2/PuCl_3$
Moderator	Graphite	Graphite	Graphite	None	None
Secondary coolant	$NaBF_4$-NaF	Nitrate salt	ZrF_4-KF		
Core inlet/outlet coolant temp. (°C)	565/704	675/775	620/700	650/750	550–650/700
Fuel type	Molten fluoride salt	TRISO particles in graphite matrix	Molten fluoride salt	Molten chloride salt	Molten chloride salt
Fuel enrichment (%)	2.0 (0.24%U233 + 12.0%Th) or Pu or LEU	19.75	<5%low enriched U	10%Pu fissile/U+Pu or 15% enriched HALEU	
Design life (years)	30	20 (vessel), 80 (plant)	56	Unlimited, 15–40 for components	
RPV height/ diameter (m)	5.4/5.34		10.0/3.7	9/4	
Distinguishing features	No online reprocessing but spent fuel salt is reprocessed	Solid fuel fluoride salt cooled	Core unit is completely replaced every 7 years	Fast spectrum, no in-core structure	Very negative feedback coefficient

Table 6.1. *Reactor major technical parameters*

6.2. Requirements for materials

This chapter will more specifically address the issues related to materials and processes for homogeneous core designs, both in thermal and the fast neutron spectrum. Core of *thermal spectrum MSRs* is largely graphite with fuel salt channels. Use of interior moderation/shielding should minimize neutron embrittlement of a reactor vessel. *Fast spectrum MSRs* will have little (or no) in-core structural materials. Consequently, in-core control elements are unlikely. A key issue is protecting a reactor vessel from a radiation environment by employing reflectors (plus a fertile salt region if applicable), whose properties will be crucial in the core design. With such a variety of MSR systems under consideration (see section 6.1.2), some of which are still at a conceptual stage, suitable specifications for structural materials are difficult to determine. Metallic materials are required and will have to sustained in harsh environments where long-term corrosion resistance is a deciding factor.

Although these requirements will not be further developed in this chapter, recyclability/reusability of structural materials and minimization of wastes will have to be considered in the materials selection process and in the system design to guarantee public acceptance of MSR concepts and compliance to environmental regulations.

6.2.1. *Graphite performance (thermal spectrum)*

The reactor core, moderator and reflectors of a thermal spectrum advanced MSR constitutes large quantities of graphite. For MSBR, the core was envisioned to be built of about 20 tons of graphite. Radiation fluence (>1 MeV) is a key design issue for graphite damage. However, graphite is a well-known material in the nuclear industry due to its extensive use in gas-cooled reactors: AGR, Magnox, etc., that have been operational since the 1960s. Its performances have been extensively tested in terms of thermal and radiation behavior, dimensional stability and stresses generated during irradiation. Sufficient data exists for selecting nuclear-grade graphite, component design and supporting safety cases related to the use of graphite in new nuclear systems. However, in the MSR core specific environment, salt and gas permeation of graphite can affect microstructural properties and radiation behavior but also facilitate diffusion, deposition and retention of fission products and tritium. Due to the significant void volume of nuclear graphite, fission products and tritium retention necessitates seal coatings or pore impregnation to reduce open porosity. Alternatively, very fine-grained graphite grades with low Xe permeability are being developed. The reader is referred to Lee et al. (2020) for a review of the current technologies for mitigating salt and gas transport into graphite. Graphite will not be further discussed in this chapter.

6.2.2. *Neutron reflector (fast spectrum)*

Most fast chloride fueled reactors are still in an early project phase and limited information is available on the reflector design and material. Based on the inferred environmental conditions, we may propose a list of material requirements:

– high temperature mechanical strength under moderate load (<10 MPa);

– long-term high temperature microstructure stability;

– resistance to corrosion by molten chlorides;

– resistance to irradiation under fast neutron spectrum and at high dose (5–20 dpa/year).

6.2.3. *Structural materials*

However promising, the development of molten salt based nuclear systems relies on the availability of structural materials that can withstand long-term operation for the construction of vessels, pipes, pumps or heat exchangers that will operate in contact with the molten salt. Using the operating parameters listed in Table 6.1, the liquid fuel salt enters the reactor vessel at 520–565°C and exits at around 700°C for fluoride-cooled MSRs, and possibly a slightly higher temperature for chloride-cooled MSRs. For economic viability, year-long to decade-long design life is targeted with component replacement strategy significantly imposing materials duty. These settings impose severe requirements on the structural materials, including:

– metallurgical stability at *high temperature*;

– mechanical strength at high and low temperature (mechanical loads including cyclic loads of parts and components being largely undetermined at this stage);

– resistance to moderate *irradiation* (with neutron spectrum depending on application, thermal or fast reactors);

– resistance to high-temperature oxidation (outer side of the pipes);

– resistance to long-term thermal and environmental exposures setting-up highly corrosive conditions.

Materials readiness for MSR components will also involve status of standardization and codification (production of materials data and terms of procurement for nuclear construction codes such as ASME, section II, division 5 or RCC-MRx) and will require that procedures exist for fabrication, joining and control.

For molten salt systems, corrosion is a key feasibility issue in the development path. Fundamental analysis of the phenomena, experimental characterization of processes and their kinetics, modeling and lifetime prediction methods and, ultimately, validation experiments are all required for credible design, safety analysis and licensing of reactor concepts.

Structural materials for the pipes of a secondary circuit operating a molten salt will have to meet an equivalent set of specifications, though less stringent (no irradiation, no evolution of the salt chemistry due to nuclear reactions, easiest chemical control, etc.).

6.3. Legacy materials and basic properties

Piping for salt containment of the ARE was made of Inconel 600. This nickel base alloy corroded too rapidly at the design temperature (to a depth of 0.25 mm in 1,000 h at 816°C) and it was not strong enough. Then, a material screening program was launched; a series of alloys were exposed to a molten fluoride salt at 650°C and only two alloys were found sufficiently resistant to corrosion in fluorides: Hastelloy-B (Ni-28% Mo-5% Fe) and Hastelloy-W (Ni-25% Mo-5% Cr-5.5% Fe). However, both aged at the service temperature and became quite brittle because of the formation of Ni-Mo inter-metallic compounds. A significant material development program was undertaken with a series of Ni-Mo alloys investigated. Based on the free energy of formation, Ni and Mo are among the less likely metals to form stable fluorides and Mo is a solid solution strengthener in Ni-based alloys. It was shown that 6–8% Cr is necessary to maintain oxidation resistance at high temperature (outside part of pipes). These observations led to the selection of an alloy initially called INOR-8, then Hastelloy-N after commercialization (Hastelloy is a registered trademark of Haynes International, Inc.). The alloy chemical composition is shown in Table 6.2. All of the structural components for the MSRE were made from Hastelloy N.

The general corrosion of Hastelloy N was well controlled over the MSRE operation with a global metal loss estimated to be less than 5 μm per year. However, examination of surveillance specimens identified two material related concerns requiring further attention (McCoy 1978):

– Hastelloy N, which is a nickel base alloy, appeared to be sensitive to *irradiation embrittlement*. Though the dose was quite low, some delay neutrons interacted with ^{58}Ni via (n,α) transmutation reactions. The produced helium accumulated in the form of intergranular bubbles and caused a drop in creep ductility. Addition of carbide forming elements was explored in a continuation of the ORNL alloy development program. Additions of Zr and Hf were abandoned because they induced poor weldability. Additions of Ti were found to be highly effective by forming titanium carbides that trap helium at their interface and thus prevent its migration toward grain boundaries. Post-irradiation creep tests of Ti-modified Hastelloy N specimens irradiated at different temperatures showed that Ti carbides are less beneficial above 700°C and lose their ability to mitigate He embrittlement at 760°C. The Ti-modified Hastelloy N was extensively developed.

– The inside surface of Hastelloy N piping and components exhibited a significant *intergranular cracking*. It was concluded that the cracks issued from interaction of metallic elements with one of the fission products supposedly forming brittle compounds at grain boundaries. The presence of tellurium in the grain boundaries was later proven by one of the very early applications of Auger spectroscopy (Keiser 1977a). Although the Te flux was fairly low in the MSRE, roughly estimated to 109 atoms Te/cm^2/s, cracks as long as 0.33 mm long were discovered. A series of new alloy compositions based on Hastelloy N with additions of niobium and titanium were examined using laboratory corrosion tests in Te-containing salts. These elements were chosen with the assumption that they may form stable compounds with Te. It was found that Nb additions were beneficial but Ti additions did not mitigate Te embrittlement. Furthermore, Ti, along with Nb, would counteract the beneficial effects of niobium. Based on these results, there was a focus on Nb additions for mitigation of Te embrittlement.

RRC-Kurchatov also dedicated significant work to the development of Ni-Mo-Cr alloys with improved radiation resistance and reduced propensity to tellurium-induced embrittlement (Ignatiev et al. 2008). This endeavor resulted in the HN80M alloy family (Ni-12Mo-7Cr), whose contents are various alloying elements such as Nb, Al, V, Ti, Re, Mn and Y. Al, Mn and Nb showed the largest effect on the intergranular corrosion by tellurium in LiF-NaF-BeF_2 melt containing Te at 700°C and Al and V were effective against radiation embrittlement. The notable addition of tungsten was also tested under the label HN80MTW. GH3535 is a Chinese low-Cr, Mo-containing, Ni-base alloy. Extensive characterization of the alloy was carried out by the Shanghai Institute of Applied Physics and the Australian Nuclear Science and Technology Organization. The Czech alloy MONICR appears to be very closely related to Hastelloy N chemically.

– The properties of Hastelloy N were investigated for the MSRE program. ASTM standards are available for a number of product forms (Wright and Sham 2018). Some specifications have also been incorporated in the ASME Boiler and Pressure Vessel Code for use up to 704°C: tensile properties, relatively short-time creep-rupture data and some creep curves. However, Hastelloy N, which might be the most advanced alloy for use in MSRs, is not currently qualified for construction of nuclear components at high temperature (ASME section III, division 5 or equivalent). Qualification for design and construction of high-temperature nuclear components would require a considerable testing effort, including long-time aging and creep rupture of base metal in the requested product forms and weldments, as well as creep-fatigue characterization.

	Nominal composition	Modified Hastelloy N
Mo	16	11-13
Cr	7	6-8
Fe	<4	0.1
Mn	<0.8	0.15–0.25
Si	<1	0.1
V	<0.5	
Cu	<0.5	
W	<0.5	
C	0.06	0.06
P	0.015	0.01
S	0.02	0.01
B	0.01	0.001
Ti	Ti + Al < 0.5	2
Nb		0–2

Table 6.2. *Chemical composition of Hastelloy-N in wt% (Rosenthal et al. 1972)*

6.4. Main material limitations

The first natural candidates for structural application of MSRs remain the nickel-based grades. They already benefit from a significant experience in the nuclear industry, although less extensive than austenitic stainless steels of type 300 or ferritic-martenstic alloys. They show good formability even for complex components and can be used from conventional metallurgy, powder metallurgy or even additive manufacturing. This section will further explore their determining properties: corrosion in molten fluoride or chloride salt and irradiation embrittlement.

6.4.1. *Corrosion in molten fluorides*

Corrosion in molten fluoride was extensively reviewed in Calderoni and Cabet (2012) and Guo et al. (2018). It is different from that in most oxidizing media including water and molten nitrates or carbonates, for instance. The product of oxidation of a metal by a fluoride melt tends to be completely soluble in the corroding media; hence alloys will not form a protective oxide scale, which might inhibit (or at least limit the rate of) corrosion and *passivation is precluded.* Corrosion in molten fluoride essentially originates from an electrochemical process, that is, an electronic exchange. The reduction-oxidation reaction involves one (or more) metallic element(s) of the alloy and dissolved oxidizing species. The oxidation electrode reaction for a metal M, first of all Cr, corresponds to:

$$M + nF^- = MF_n + ne^- \qquad [6.1]$$

MF_n being the corresponding metallic ion, whatever its exact speciation in the fluoride melt (with $0 < n < 6$).

The reduction electrode reaction considering the reference electrochemical species $F^-/F_2(g)$ is:

$$F_2(g) + 2e^- = 2F^- \qquad [6.2]$$

Hence, the generalized oxidation reaction is:

$$M + n/2F_2(g) = MF_n \qquad [6.3]$$

The theoretical oxidation potential for metal dissolution or selective metal dissolution from an alloy is obtained from the Nernst equation applied to reaction [6.3]. An example of calculated and experimentally measured potentials using sweep linear voltammetry is shown in Table 6.3 for a LiF-NaF melt at 900°C (Fabre et al. 2013). Note that linear sweep voltammetry can also be applied to complex alloys for which the chemical activity is not easily known to measure experimental oxidation potentials. The reduction potentials of major salt cations are far below the oxidation potentials of structural metals. So, *molten fluoride salts are not intrinsically corrosive* to the constituent elements in the containment alloys, because of this greater thermodynamic stability of the molten fluoride salt constituents as compared to the fluoride of transition metals. Guo et al. (2018) gives the following mole fractions of cations for pure metals in equilibrium with FLiBe at 700°C (assuming the activity coefficient of MF_n is 0.01 in the melt): 3×10^{-21} for Ni(II), 9×10^{-17} for Fe(II), 3×10^{-12} for Cr(II) and 3×10^{-6} for Al (II). As such, it provides an assessment of the chemical activity of a given element in a static medium and it constitutes a thermodynamic basis for the choice of structural materials in molten salt systems provided the potentials were known.

Despite the promising picture derived from equilibrium thermodynamic potentials, various phenomena can alter the corrosion process with a drastic change in the corrosion mechanism and dissolution rate. The most commonly encountered phenomena are discussed hereafter.

	Computed	Experimental
Na	−2.23	−1.86
Cr	−0.8	−0.68
Fe	−0.41	−0.42
Ni	0	0
W	+0.05	+0.44
Mo	+0.23	+0.19
Ag	+0.99	+0.52
Au	+2.02	+1.57
F^-	+2.46	

Table 6.3. *Oxidation potentials of metals referenced to Ni/NiF$_2$ computed from E°(M/MF$_n$) and experimentally determined from linear sweep voltammetry in LiF-NaF at 900°C (Fabre et al. 2013)*

6.4.1.1. *Impurity-driven corrosion*

Together with the salt constituents (LiF, NaF, BeF_2, etc.), the liquid fuel or coolant will unavoidably contain oxidizing impurities, such as HF or H_2O, and metal ions, arising from initial pollution of the salt, in-leakages, dissolution of residual oxide scales, etc. Addition of oxidants can have a strong effect on the fluorine potential because F_2, HF, metallic cations are strong oxidants and may oxidize a metal M, following reaction [6.3] or similar. Water is one of the main pollutants and dissolves with production of HF. A high HF pressure results in the increase of the melt potential and oxoacidity. Therefore, the constituents of the alloy will easily be oxidized to the metallic fluorides, causing alloy corrosion. Wang et al. (2015) provides experimental evidence of accelerated anodic metal dissolution rate due to water additions.

In addition, fission of uranium or plutonium causes changes in the molten salt redox (Toth et al. 1995). Transmutation of ^{7}Li and ^{6}Li produces tritium fluoride, analogous to HF and transmutation of beryllium produces free fluorine. Hence, during reactor service, redox potential of the salt continuously increases.

6.4.1.2. *Role of alloying elements and microstructure*

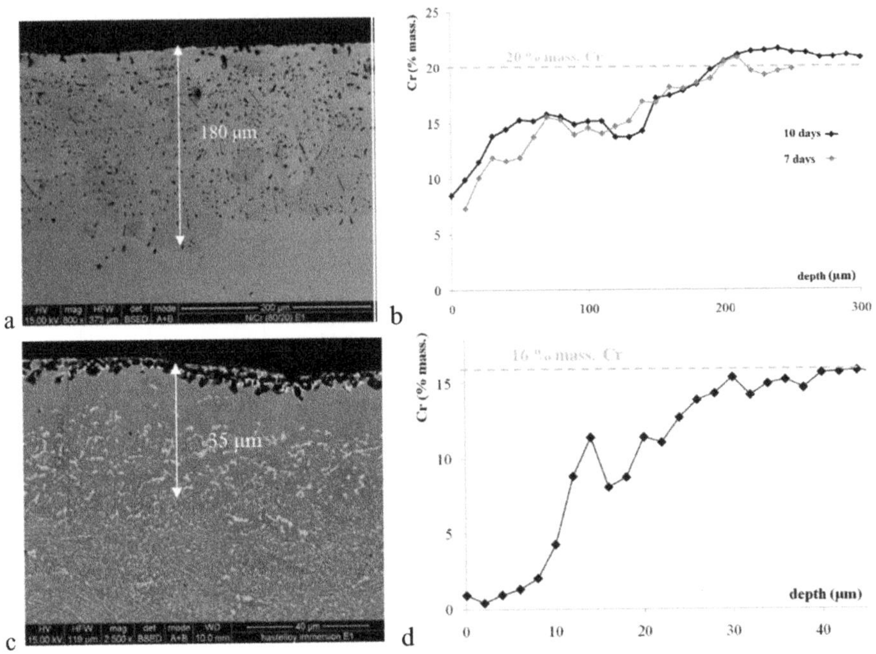

Figure 6.3. *Nickel base alloys immersed in LiF-NaF at 800°C (a and b) Ni-20Cr for 10 days; (c and d) Hastelloy C276 for 14 days; (a and c) cross-section image; (b and d) Cr profile from the outer surface by EDX (Fabre 2011). For a color version of this figure, see www.iste.co.uk/balbaud/nuclearenergytoday.zip*

As implied from the thermodynamic oxidation potentials shown in Table 6.3, lab tests show that chromium is the most reactive metallic elements, followed by iron and then nickel. Various experiments confirmed that nickel-base alloys in general are superior to other commercial steels for the containment of the fluoride salts, and alloys containing substantial quantities of chromium are more intensively attacked with selective dissolution of chromium (see Figures 6.3(b) and (d)). Chromium removal by oxidative dissolution causes injection of metallic vacancies at the alloy/molten salt interface. Depending on the alloy microstructure, the vacancies uniformly condensate in the bulk, producing discrete voids distributed throughout the alloy or condensate more specifically at grain boundaries or at some inclusions with formation of confined voids (this is especially visible on Figure 6.3(a)). In addition, the intensity of the corrosion is drastically reduced for commercial Hastelloy-C276 as compared to model Ni-20Cr alloy (Figure 6.3(a) and (c)). This may be due to both thermodynamic and kinetic effects with a likely lesser chromium activity in Hastelloy C276 than in Ni-20Cr, the alloying effect of Mo that should reduce the Cr diffusion within Hastelloy-C276 as well as different grain size and precipitation state. From the standpoint of general corrosion in molten fluorides, the most suitable metallic materials are nickel- molybdenum- tungsten-containing alloys with few or no chromium and iron. Iron containing alloys and steels may be used provided the salt chemistry is closely controlled (see section 6.5).

6.4.1.3. *Dissimilar material corrosion*

Small electrical forces can locally arise from the presence of dissimilar materials. Electric coupling, called galvanic coupling, between a nobler material – or a specific nobler area in a material – and a more reactive material dipped into the molten fluoride salt may produce sustained corrosion of the less stable metal.

The use of graphite crucibles for corrosion testing was shown to modify the test results with an acceleration of the alloy Cr-dealloying and deposition of Cr onto graphite. Such a difference in the Hastelloy N corrosion was reported for samples exposed to FLiBe salt for 1,000 h at 700°C in graphite or in nickel capsules (Zheng and Sridharan 2018) with chromium carbide particles up to ~1.5 mm on the surface of the specimens tested in a graphite crucible. The same phenomena were absent when nickel crucibles were used in place of graphite. A difference in welds versus base metal as well as non-homogeneity within a complex alloy such as grain boundaries or inclusions is also a potential source for galvanic coupling.

6.4.1.4. *Thermal gradient-driven corrosion*

Along the circuit of an industrial power conversion system, heat transfer to the secondary coolant induces a significant temperature difference that in turn causes convection and chemical gradients. Hence, oxidation products are conveyed by convection from hotter areas where the alloy dissolves to cooler spots where deposition may proceed. Therefore, a mass transfer occurs which continuously supports corrosion (Wang et al. 2018). Such a sustained corrosion was observed by ORNL using thermal- and forced-convection salt loops with material deposition on the cold leg (Keiser 1977b).

6.4.1.5. *Molten fluoride corrosion data*

Experimental data are available for the molten FLiNaK eutectic and the MSRE fuel salt. However, they remain rather limited for other molten salts and are insufficient for selecting alloys with appropriate long-term corrosion performance.

Raiman and Lee (2018) surveyed 11 corrosion studies in molten fluoride salt and analyzed the effect of salt purity, temperature, sample material, container material and experimental method (loop, capsule and crucible). The various works were inconsistent due to the lack of standardization among experimental test protocols with salt purity having the strongest correlation with corrosion rates.

6.4.2. *Corrosion in molten chlorides*

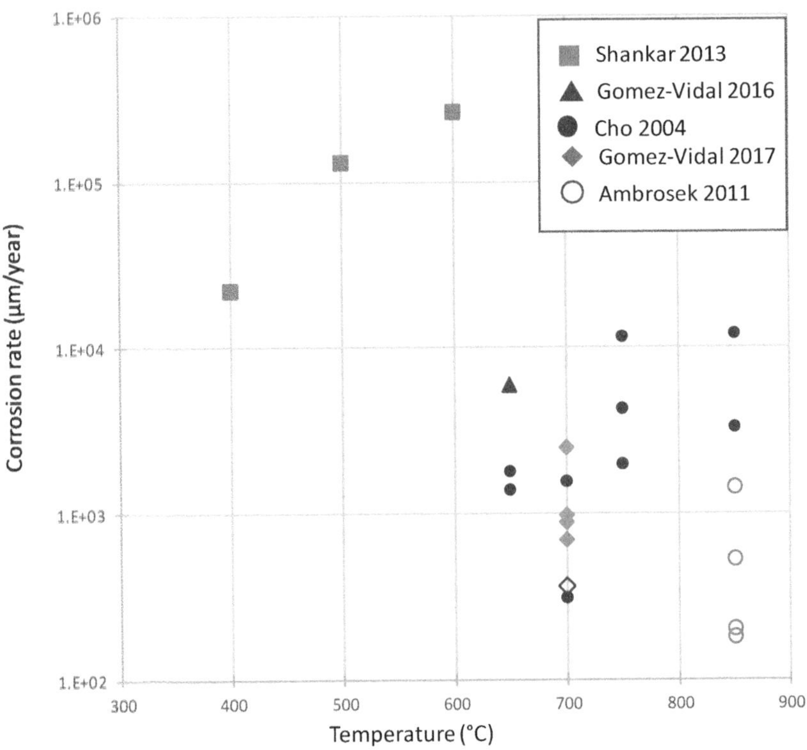

Figure 6.4. *Corrosion rates of Alloy 800 (Fe-30Ni-19Cr) at various temperatures in molten chlorides: KCl-MgCl*$_2$ *(Ambrosek 2011), NaCl-KCl (Gomez-Vidal 2017), NaCl-KCl-MgCl*$_2$ *(Ding et al. 2018), NaCl-LiCl (Gomez-Vidal and Morton 2016), LiCl-KCl (Shankar et al. 2013a) and LiCl-(Li*$_2$*O) (Cho et al. 2004). For a color version of this figure, see www.iste.co.uk/balbaud/nuclearenergytoday.zip*

As for fluoride melts, none of the likely chloride-based salts are aggressively corrosive to stainless steels or nickel-base alloys at likely reactor temperatures without impurities being present (Indacochea et al. 2001). However, ingress of impurities, first of which is water, cannot be prevented in a large-scale MSR and long-term operation will require a high degree of material compatibility. The literature on corrosion in chlorides is less abundant than that in fluorides. The results are scattered, as shown in Figure 6.4 for the example of Alloy 800 in various chlorine salts and various temperatures. The main reasons for these inconsistencies in the corrosion rate may be attributed to (1) different nature of the salts, (2) impurity level in salts, (3) materials in contact (crucible, gaseous atmospheres, etc.) and (4) experimental methods for assessing the corrosion rate.

6.4.2.1. *Impurities in molten chlorides*

It is generally considered that the presence of impurities in the chloride salts is the main cause of corrosion. These impurities are O_2, O^{2-}, H_2O (Grégoire et al. 2020). A corrosion test, performed under sweeping pure argon with a purified salt is generally less corrosive than without any purification (Liu et al. 2019). Deep purification can be obtained by thermal treatments, chlorination using CCl_4, HCl, carbochlorination with CO and Cl_2, $COCl_2$ or CCl_4 (Kurley et al. 2019), by addition of solid or liquid reducing metals (Mg, Li) or by salt electrolysis.

In order to avoid pollution, corrosion tests can be carried out in a glove box under inert gas or other gas-tight conditions. The test crucible, containing the salt and the alloy specimen, can also participate in polluting the salt during the corrosion test and altering the test results.

6.4.2.2. *Resistance of materials*

6.4.2.2.1. Steels and Ni base alloys

Most of the tested alloys are stainless steels or Cr containing Ni-based alloys and it is commonly accepted that steels have lower corrosion resistance than nickel-based alloys. Surface oxides that usually form at the alloy surface when exposed to high temperature and act as an efficient barrier against further corrosion are not protective in most molten chlorides. The activity of oxygen anions in the melt or oxoacidity defined as $pO^{2-} = -\log(a_{O^{2-}})$ is thus a determining factor in corrosion processes. Potential-oxoacidity diagrams in molten chloride present the stability of soluble and insoluble metallic species as a function of the potential and the pO^{2-}.

Metallic oxides can dissolve in melts at low pO^{2-} as the acidic dissolution of nickel oxide through:

$$NiO \rightarrow Ni + O^{2-}$$

On the other hand, metallic oxides can dissolve at high pO^{2-} as the basic dissolution of chromium oxide through:

$$2Cr_2O_3 + 4O^{2-} + 3O_2 \rightarrow 4CrO_4^{2-}$$

When in contact with chloride salts, the steels and alloys thus form salt-soluble (by acidic or basic dissolution) or unprotective oxides. Depending on the system, these oxides may be chromia (Zahs et al. 2000) or mixed oxides containing an element of the salt, such as $NaCrO_2$, Li_2CrO_4 (e.g. Cho et al. 2004), $ZnCr_2O_4$ (e.g. Liu et al. 2019; Zhuang et al. 2019) or $MgCr_2O_4$ (e.g. Ding et al. 2018) for Na-, Li-, Zn- or Mg-containing salts. Porous and poorly protective spinel-type oxides have also been reported: $NiFe_2O_4$, $NiCr_2O_4$, $(NiFeCr)_3O_4$ and $FeCr_2O_4$ (e.g. Shankar et al. 2013b; Ding et al. 2018; Liu et al. 2019; Zhuang et al. 2019). A preferential dissolution of Cr in comparison to Fe and Ni is also observed (Shankar et al. 2013b). Counterintuitively, pure Ni is less resistant to corrosion than Cr-containing Ni (Shankar et al. 2013b).

Intergranular corrosion, exacerbated by the presence of C and by temperature increase, is also frequently observed (Cho et al. 2004; Shankar et al. 2013b).

Alumina is less soluble than chromia, leading to a better corrosion resistance of alumina forming alloys than chromia forming alloys. Consequently, alumina forming Ni-based alloys are said to be the best compromise (Ding et al. 2019).

6.4.2.2.2. Alloying elements

Ti (e.g. Hofmeister et al. 2015), Zr, Hf, Ta, Mo and Nb (e.g. Grabke et al. 2004) are beneficial as they impede the precipitation of Cr carbides at grain boundaries.

Pure refractory metals such as tungsten, molybdenum and tantalum present a very good resistance to chlorine salt and can be beneficial as alloying elements in alloys (see Sun et al. 2018).

6.4.2.2.3. Ceramics

Many ceramics present good resistance to corrosion, as verified by Takeuchi et al. (2005) in NaCl-KCl at 750°C under Cl_2 bubbling.

6.4.3. *Tellurium embrittlement*

Structural materials for MSR homogeneous concepts may be damaged by reaction with fission products. As was discussed in section 6.3, during MSRE operation, the Hastelloy N suffered from embrittlement due to interaction with the fission product tellurium. Speciation of tellurium in the molten salt has not been clarified. In the lab, additions in molten fluoride salt of Te, Cr_3Te_4 and Cr_5Te_6 were responsible for intergranular cracking (McCoy 1978). Diffusion of tellurium is mainly intergranular, inducing the formation of nanovoids that may evolve into cracks and the reduction of diffusion of active elements (Jiang et al. 2022). Formation of intermetallic compounds such as Ni_xTe_y and Cr_xTe_y at grains boundaries certainly contributes to cracking (Delpech et al. 2010). Although fundamentally different, such a corrosion mode may also be curbed by controlling the melt redox state (see section 6.5). Microstructure tuning through addition of carbide forming elements also showed a beneficial effect on the propensity to Te-induced embrittlement (section 6.3).

6.4.4. *Irradiation embrittlement*

As they are less used in nuclear reactor cores than austenitic stainless steels of type 300 or ferritic-martenstic alloys, the irradiation behavior of nickel-based alloys has been less investigated. Their behavior as a function of chemical composition, irradiation conditions, etc., is therefore much less well known. Like most metal grades, they show a noticeable hardening under irradiation due to the formation of irradiation loops but the two main factors limiting their use under irradiation are undoubtedly their potential swelling and the formation of helium bubbles, which embrittles the material. Swelling (see Figure 6.5) is caused by the condensation of vacancies in the form of micro-cavities in the material. In the Nimonic PE16 grade (Fe 45Ni 17Cr 3Mo TiAlSi) irradiated with neutrons at 513°C at 58 dpa, many cavities are found, some of which are attached to titanium carbides (Figure 6.5, right-hand side).

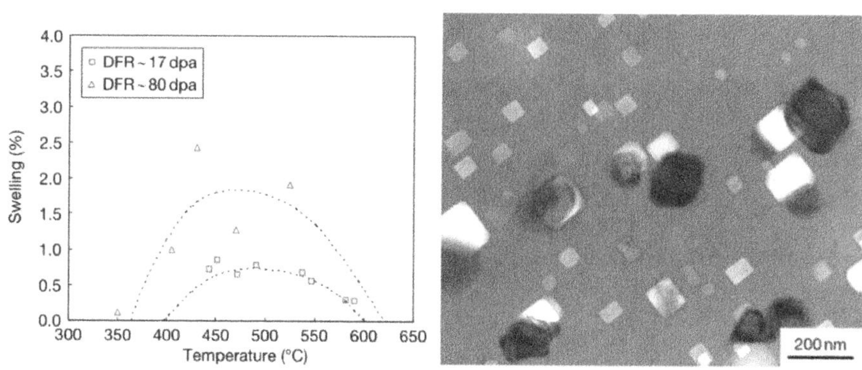

Figure 6.5. *Swelling of Nimonic PE16 alloy as a function of temperature for two different damage levels and micrograph of cavities after neutron irradiation at 513°C and 58 dpa (reproduced from Boothby 2012). For a color version of this figure, see www.iste.co.uk/balbaud/nuclearenergytoday.zip*

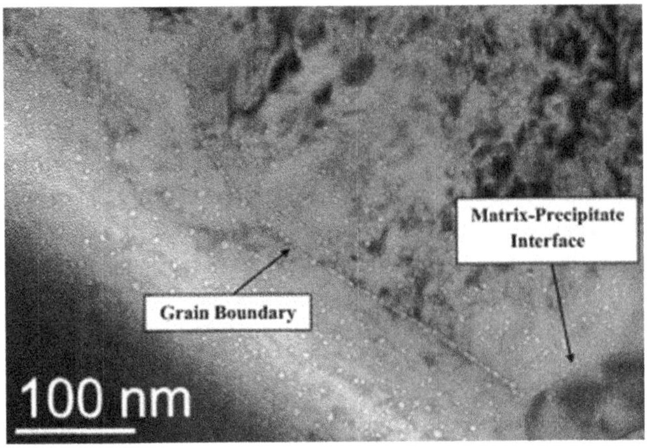

Figure 6.6. *Formation of helium bubbles (1,800 appm of He) in Inconel X-750 (Ni 15Cr 7Fe TiCo) irradiated at 330°C and 55 dpa (reproduced from Judge et al. 2015)*

Helium is easily formed during neutron irradiation of nickel-rich grades. It generally comes from the transmutation of nickel with its different isotopes and secondary of boron 10. The helium/dpa ratio strongly depends on the neutron spectrum of the irradiation. This is why the type of reactor, with epithermal or fast neutron spectrums, will strongly condition the production of helium in the material. Helium is feared in structural materials because it can migrate, condense at interfaces in the form of bubbles (see Figure 6.6) or at an even finer scale. This may greatly embrittle the alloy. To limit this phenomenon, ORNL started working on Hastelloy-N grades enriched with titanium and carbon so that the titanium carbides could trap helium and prevent it from weakening the grain boundaries (see section 6.3). Current knowledge on the behavior of nickel-based alloys is not sufficient to be able to specify the window of use for grades of interest that exhibit good corrosion resistance. Irradiations in fast spectrum will surely be necessary to specify the admissible temperatures and doses. Embrittlement under irradiation is a current problem and developments around nano-reinforced grades to limit this phenomenon are in progress.

6.5. Mitigation of corrosion by salt purification and chemical control

As for most cases of environmental degradation in nuclear systems, mitigating the corrosion in molten salt systems includes careful selection of materials and control of the melt chemistry.

6.5.1. *Molten fluorides*

As detailed in section 6.4, corrosion originates from electrochemical reactions with dissolved oxidizing agents. The propensity for corrosion of the metallic elements by the melt is often assessed by the "redox conditions". Olander (2002) discussed the fundamentals of this redox state in fluoride salts and associated methods to control it in FLiBe. He stated that the redox potential should properly be termed *the fluorine potential* based on the partial molar Gibbs free energy of the fluorine gas:

$$\Delta\overline{G}_{F_2} \equiv RT\ln p_{F_2} \qquad [6.4]$$

The "redox potential" controls the equilibrium concentrations of structural metals dissolved in the fluoride-based salts (see equations [6.3] and [6.4]). It is thus of major significance to extensively purify the initial salts and to continuously adjust the oxidation potential of the melt during operation. The fuel salt used for MSRE which contained UF_3 and UF_4 has been fairly well studied (Toth et al. 1995). In uranium-containing fluoride mixtures, the fluorine potential p_{F2} can be expressed as:

$$\ln p_{F2} = 2.\ln[UF_4]/[UF_3] + 2/RT.\Delta G° \qquad [6.5]$$

with $\Delta G°$ the standard free energy change of the reaction:

$$UF_3 + 1/2\ F_2(g) = UF_4 \qquad [6.6]$$

In the MSRE, the redox was adjusted by dipping beryllium – a strong reducing agent – in the Flibe-based salt with the ratio UF_4/UF_3 continuously maintained in the range 10 (the ratio had to stay above 10 to avoid disproportionation UF_3 into U and UF_4) to 100 (the ratio had to stay below about 100 for corrosion mitigation). This chemical control successfully limited dissolution of Hastelloy N to a metal loss less than 5 mm per year.

In addition, controlling the molten fuel chemistry may also allow suppressing or reducing of the tellurium-induced intergranular corrosion (Surenkov et al. 2020). This can be done by adjustment of the UF_3/UF_4 ratio below 80 at 700°C and 40 at 760°C. It is thought that tellurium activity is controlled by "complexing" reactions with chromium or other salt constituents, for example, as a telluride rather than as elemental tellurium by the reaction, where M is Cr or some other metal:

$$2UF_3 + MF_2 + xTe^0 \leftrightarrow MTe_x + 2UF_4 \qquad [6.7]$$

However, defining the redox state for fluoride salts in general – free of an element with two valence states – is not straightforward. Achieving equilibrium with a gas phase of controlled properties is a potential mean. As HF is a corrosive agent by itself, stripping with HF was shown to be unsuccessful to reduce alloy attack (Nagasaka et al. 2009). Use of HF/H_2 bubbling with a set ratio was experimentally proved to be more effective. However, a purification procedure with no introduction of HF is strongly recommended to prevent corrosion reactions. Solid NH_4HF_2 was successfully used in Afonichkin et al. (2010) for removal of oxygen-containing impurities from metallic fluorides and for conversion of uranium and thorium dioxides to anhydrous tetrafluorides.

6.5.2. *Molten chlorides*

In molten chlorides, salt purification and chemical control are still required as useful techniques for limiting structural alloy corrosion. However, redox control within a chloride salt melt would be more complex (and as yet unproven) than in a fluoride salt because of the more complex chemistry that arises from the larger number of ionization states. The use of $MgCl_2$ as a major constituent of the salt appears to be especially problematic as this salt is highly hydroscopic. In addition, redox control by adjusting the ratio $[U^{IV}]/[U^{III}]$, which was successfully used in the MSRE, cannot a priori fix a sufficiently low potential to mitigate corrosion in molten chloride salts.

6.6. Main material challenges and solutions under evaluation

As detailed in section 6.4, challenges on structural materials for MSR are related to resistance to corrosion and irradiation. Hastelloy-N has a maximum temperature for long-term operation of approximately 650°C and is susceptible to radiation embrittlement by the formation of helium, when exposed to irradiation. Material development first consists of incrementally changing its composition and microstructure. Introduction of fine titanium carbide or intermetallic could significantly increase the maximum permissible dose by allowing helium to be trapped in situ. The modifications mainly concern optimizations of Ti, Nb and Al contents. Advanced solutions (Li et al. 2022) are also being studied with the use of powder metallurgy and the manufacture of oxide dispersion strengthened (ODS) materials.

Other material families are under study for core materials (reflector). Firstly their corrosion resistance must globally be assessed. In NaCl-$MgCl_2$ at 700°C (immersion test of 500 h), refractory metals such as Ta-, W- and Mo-based alloy TZM showed lesser dissolution than conventional Ni alloys and steels (Raiman et al. 2019). Ceramics are also investigated. SiC or SiC_f/SiC materials have experienced significant developments in recent years and are worth being evaluated for fast MSRs (e.g. Lee et al. 2019). Note that managing the salt purity, especially in oxygen ions, remains very important to mitigate the corrosion resistance of SiC.

Experiments are being conducted on different coatings that could exhibit an excellent corrosion resistance in molten salts at high temperature: MAX phase such as Cr_2AlC, alumina, etc. Bimetallic structures could represent an interesting solution for MSRs. This involves making a thick deposit of an alloy with good resistance to salt corrosion (e.g. optimized nickel alloy, tungsten base alloy, Mo base alloy, etc.) onto a structural material, which could have the advantage of being code-qualified or at least well characterized with good mechanical behavior and irradiation resistance (e.g. stainless steel 316L or F/M steel T91). The deposition techniques can be very varied: cold spray, TIG welding, laser cladding (Zhou et al. 2021) with microstructures and adhesion properties that will need to be evaluated.

6.7. References

Afonichkin, V., Bovet, A., Shishkin, V. (2010). Salts purification and redox potential measurement for the molten LiF-ThF4-UF4 mixture. In *Proceedings of the First ACSEPT International Workshop*, Lisbon.

Ambrosek, J. (2011). Molten chloride salts for heat transfer in nuclear systems. Thesis, University of Wisconsin.

Boothby, R.M. (2012). Radiation effects in nickel-based alloys. In *Comprehensive Nuclear Materials*, Konings, R.J.M. (ed.). Elsevier, Oxford.

Calderoni, P. and Cabet, C. (2012). Corrosion issues in molten salt reactor (MSR) systems. In *Comprehensive Nuclear Materials*, Konings, R.J.M. (ed.). Elsevier, Oxford.

Cho, S.H., Zhang, J.S., Shin, Y.J., Park, S.W., Park, H.S. (2004). Corrosion behavior of Fe-Ni-Cr alloys in the molten salt of LiCl-Li_2O at high temperature. *J. Nucl. Mater.*, 325, 13–17.

Delpech, S., Cabet, C., Slim, C., Picard, G.S. (2010). Molten fluorides for nuclear applications. *Material Today*, 13(12), 34–41.

Ding, W., Shi, H., Xiu, Y., Bonk, A., Weisenburger, A., Jianu, A., Bauer, T. (2018). Hot corrosion behavior of commercial alloys in thermal energy storage material of molten $MgCl_2$/KCl/NaCl under inert atmosphere. *Sol. Energy Mater. Sol. Cells*, 184, 22–30.

Ding, W., Shi, H., Jianu, A., Xiu, Y., Bonk, A., Weisenburger, A., Bauer, T. (2019). Molten chloride salts for next generation concentrated solar power plants: Mitigation strategies against corrosion of structural materials. *Sol. Energy Mater. Sol. Cells*, 193, 298–313.

Fabre, S. (2011). Comportement de métaux et alliages en milieux fluorures fondus. Thesis, Université de Toulouse.

Fabre, S., Cabet, C., Cassayre, L., Chamelot, P., Delepech, S., Finne, J., Massot, L., Noel, D. (2013). Use of electrochemical techniques to study the corrosion of metals in model fluoride melts. *J. Nucl. Mater.*, 441(1–3), 583–591.

Generation IV International Forum (2002). A technology roadmap update for generation IV nuclear energy systems [Online]. Available at: https://www.gen-4.org/gif/jcms/c_9352/technology-roadmap [Accessed 16 August 2022].

Gomez-Vidal, J.C. (2017). Corrosion resistance of MCrAlX coatings in a molten chloride for thermal storage in concentrating solar power applications. *Npj Mater. Degrad.*, 1, 1–9. doi: 10.1038/s41529-017-0012-3.

Gomez-Vidal, J.C. and Morton, E. (2016) Castable cements to prevent corrosion of metals in molten salts. *Sol. Energy Mater. Sol. Cells*, 153, 44–51.

Grabke, H.J., Spiegel, M., Zahs, A. (2004). Role of alloying elements and carbides in the chlorine-induced corrosion of steels and alloys. *Mater. Res.*, 7, 89–95.

Grégoire, B., Oskay, C., Meissner, T.M., Galetz, M.C. (2020). Corrosion mechanisms of ferritic-martensitic P91 steel and Inconel 600 nickel-based alloy in molten chlorides. Part I: NaCl – KCl binary system. *Sol. Energy Mater. Sol. Cells*, 215, 110659.

Guo, S., Zhang, J., Wu, W., Zhou, W. (2018). Corrosion in the molten fluoride and chloride salts and materials development for nuclear applications. *Progress in Materials Science*, 97, 448–487.

Hofmeister, M., Klein, L., Miran, H., Rettig, R., Virtanen, S., Singer, R.F. (2015). Corrosion behaviour of stainless steels and a single crystal superalloy in a ternary LiCl-KCl-CsCl molten salt. *Corros. Sci.*, 90, 46–53.

IAEA (2020). Advances in small modular reactor technology developments. 2020 Edition. Report, IAEA, Vienna.

Ignatiev, V., Surenkov, A., Gnidoi, I., Fedulov, V., Uglov, V., Afonichkin, V., Bovet, A., Subbotin, V., Panov, A., Toropov, A. (2008). Compatibility of selected Ni-based alloys in molten Li,Na,Be/F salts with PuF3 and tellurium additions. *Nuclear Technology*, 164, 130–142.

Indacochea, J.E., Smith, J.L., Litko, K.R., Karell, E.J., Raraz, A.G. (2001). High-temperature, oxidation and corrosion of structural materials in molten chlorides. *Oxidation of Metals*, 55, 1–16.

Jiang, L., Wang, K., Leng, B., Ye, X., Chen, S., Liu, R., Liang, J., Li, C., Li Z. (2022). Tellurium segregation induced intergranular corrosion of GH3535 alloys in molten salt. *Corrosion Science*, 194, 109944.

Judge, C.D., Gauquelin, N., Walters, L., Wright, M., Cole, J.I., Madden, J., Botton, G.A., Griffiths, M. (2015). Intergranular fracture in irradiated Inconel X-750 containing very high concentrations of helium and hydrogen. *J. Nucl. Mater.*, 457, 165–172.

Keiser, J.R. (1977a). Status of the tellurium-hastelloy N studies in molten fluoride salts. Report, Oak Ridge National Laboratory ORNL/TM-6002, Oak Ridge, TN.

Keiser, J.R. (1977b). Compatibility studies of potential molten-salt breeder reactor materials in molten fluoride salts. Report, Oak Ridge National Laboratory ORNL/TM-5783, Oak Ridge, TN.

Kurley, J.M., Halstenberg, P.W., McAlister, A., Raiman, S.S., Dai, S., Mayes, R.T. (2019). Enabling chloride salts for thermal energy storage: Implications of salt purity. *RSC Adv.*, 9, 25602–25608.

LeBlanc, D. (2010). Molten salt reactors: A new beginning for an old idea. *Nuclear Engineering and Design*, 240, 1644–1656.

Lee, J.J., Raiman, S.S., Katoh, Y., Koyanagi, T., Contescu, C.I., Hu, W., Yang, Y. (2019). Chemical compatibility of silicon carbide in molten fluoride salts for the fluoride salt-cooled high temperature reactor. *J. Nucl. Mater.*, 524, 119–134.

Lee, J.J., Arregui-Mena, J.D., Contescu, C.I., Burchell, T.D., Katoh, Y., Loyalka, S.K. (2020). Protection of graphite from salt and gas permeation in molten salt reactors. *J. Nucl. Mater.*, 534, 152119.

Li, C., Lei, G., Liu, J., Liu, A., Ren, C.L., Huang, H. (2022). A potential candidate structural material for molten salt reactor: ODS nickel-based alloy. *J. Mater. Sci. Technol.*, 109, 129–139.

Liu, T., Xu, X., Liu, W., Zhuang, X. (2019). Corrosion of alloys in high temperature molten-salt heat transfer fluids with air as the cover gas. *Sol. Energy*, 191, 435–448.

MacPherson, H.G. (1985). The molten salt reactor adventure. *Nuclear Science and Engineering*, 90(4), 374–380.

McCoy, H.E. (1978). Status of materials development for molten salt reactors. Report, Oak Ridge National Laboratory ORNL/TM-5920, Oak Ridge, TN.

Nagasaka, T., Kondo, M., Muroga. T., Noda. N., Sagara. A., Motojima. O., Suzuki, A., Terai, T. (2009). Fluoridation and oxidation characteristics of JLF-1 and NIFS-HEAT-2 low-activation structural materials. *J. Nucl. Mater.*, 386–388, 716–719.

Olander, D. (2002). Redox condition in molten fluoride salts: Definition and control. *J. Nucl. Mater.*, 300, 270–272.

Raiman, S.S. and Lee, S. (2018). Aggregation and data analysis of corrosion studies in molten chloride and fluoride salts. *J. Nucl. Mater.*, 511, 523–535.

Raiman, S.S., Muralidharan, G., Mayes, R.T., Kurley, J.M. (2019). Compatibility studies of cladding candidates and advanced low-Cr nickel superalloys in molten NaCl-$MgCl_2$. Report, Oak Ridge National Laboratory ORNL/TM-2019/1132, Oak Ridge, TN.

Rosenthal, M.W. (1972). The development status of molten-salt breeder reactors. Report, Oak Ridge National Laboratory ORNL-4812, Oak Ridge, TN.

Serp, J., Allibert, M., Benes, O., Delpech, S., Feynberg, O., Ghetta, V., Heuer, D., Holcomb, D., Ignatiev, V., Kloosterman, J.L. et al. (2014). The molten salt reactor (MSR) in generation IV: Overview and perspectives. *Progress in Nuclear Energy*, 77, 308–319.

Shankar, A.R., Kanagasundar, A., Mudali, U.K. (2013a). Corrosion of nickel-containing alloys in molten LiCl-KCl medium. *Corrosion*, 69, 48–57.

Shankar, A.R., Thyagarajan, K., Mudali, U.K. (2013b). Corrosion behavior of candidate materials in molten LiCl-KCl salt under argon atmosphere. *Corrosion*, 69, 655–665.

Sun, H., Zhang, P., Wang, J. (2018). Effects of alloying elements on the corrosion behavior of Ni-based alloys in molten NaCl-KCl-$MgCl_2$ salt at different temperatures. *Corros. Sci.*, 143, 187–199.

Surenkov, A., Ignatiev, V., Presnyakov, M., Wang, J., Li, Z., Yang, X., Dai, Z. (2020). Effect of the [U(IV)]/[U(III)] ratio on selective chromium corrosion and tellurium intergranular cracking of Hastelloy N alloy in the fuel LiF-BeF2-UF4 salt. *EPJ Nuclear Sci. Technol.*, 6, 4.

Takeuchi, M., Kato, T., Hanada, K., Koizumi, T., Aose, S. (2005). Corrosion resistance of ceramic materials in pyrochemical reprocessing condition by using molten salt for spent nuclear oxide fuel. *J. Phys. Chem. Solids*, 66, 521–525.

Toth, L.M., Del Cul, G.D., Dai, S., Metcalf, D.H. (1995). Molten fluoride fuel salt chemistry. *AIP Conference Proceedings*, 346, 617.

Wang, Y., Liu, H., Yu, G., Hou, J., Zheng, C. (2015). Electrochemical study of the corrosion of a Ni-based alloy GH3535 in molten (Li,Na,K)F at 700°C. *Journal of Fluorine Chemistry*, 178, 14–22.

Wang, Y., Zeng, C., Li, W. (2018). The influence of temperature gradient on the corrosion of materials in molten fluorides. *Corrosion Science*, 136, 180–187.

Wright, R.N. and Sham T.L. (2018). Status of metallic structural materials for molten salt reactors. Report, Idaho National Laboratory, ART Program, Idaho Falls, ID.

Zahs, A., Spiegel, M., Grabke, H.J. (2000). Chloridation and oxidation of iron, chromium, nickel and their alloys in chloridizing and oxidizing atmospheres at 400-700°C. *Corros. Sci.*, 42, 1093–1122.

Zhang, D., Liu, L., Liu, M., Xu, R., Gong, C., Zhang, J., Wang, C., Qiu, S., Su, G. (2018). Review of conceptual design and fundamental research of molten salt reactors in China. *International Journal of Energy Research*, 42(5) 1834–1848.

Zheng, G. and Sridharan, K. (2018). Corrosion of structural alloys in high-temperature molten fluoride salts for applications in molten salt reactors. *JOM*, 70, 1535–1541.

Zhou, S., Xu, T., Hu, C., Wu, H., Hu, Liu, H., Ma, X. (2021). Effect of different topologies on microstructure and mechanical properties of multilayer coatings deposited by laser cladding with Inconel 625 wire. *Surface and Coatings Technology*, 421, 127299.

Zhuang, X., Liu, W., Xu, X. (2019). Hot corrosion of different alloys in chloride and carbonate molten-salt mixtures under argon atmosphere. *Sol. Energy*, 189, 254–267.

7

Materials and Processes for Fusion Reactors

Giacomo AIELLO
EUROfusion – Programme Management Unit, Garching, Germany

7.1. Introduction and general description of the systems

In a fusion reactor (Figure 7.1), fusion reactions take place within a hot plasma confined inside a vacuum chamber (vacuum vessel [VV]) by means of magnetic fields and heated by means of particles injection and electromagnetic radiation. The most extensively investigated confinement concept is the "tokamak", on which most of the existing fusion machines are based (e.g. JET, ASDEX, WEST in Europe, JT-60 in Japan, TFTR in the United States). It is also the configuration retained for ITER as well as the future European DEMOnstration reactor. DEMO, whose operation is foreseen to start around the middle of this century, should be the first fusion machine to produce electricity and connect to the grid.

With current technology, the reaction most readily accessible is between the nuclei of the two heavy isotopes of hydrogen – deuterium (D) and tritium (T):

$$^{2}D + ^{3}T \rightarrow ^{4}He\ (3.5\ MeV) + n\ (14.1\ MeV) \qquad [7.1]$$

Neutrons escape the plasma unaffected by the magnetic field and release their energy by elastic or inelastic collisions inside the volume of the plasma facing components (PFCs). Alpha particles instead radiate their energy by interacting with

Materials and Processes for Nuclear Energy Today and in the Future,
coordinated by Fanny BALBAUD-CÉLÉRIER and Céline CABET.

the magnetic field and other nuclei in the plasma, and eventually escape the plasma boundary (last magnetic closed surface). They are directed toward the bottom of the plasma where they are stopped at the surface of the plasma facing walls, giving rise to very high and localized heat loads. Due to the different locations and modalities in which the energy is transferred to the PFCs, the engineering – and materials – used in the power extraction components is very different. Two main systems are in charge of power extraction in a fusion reactor, the (Breeding) Blanket system and the Divertor system (Figure 7.1).

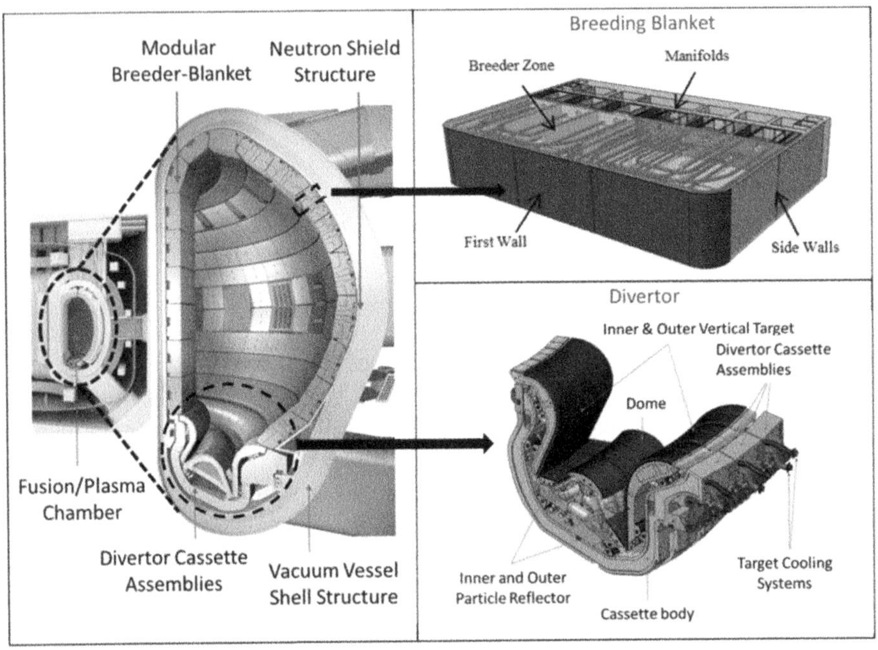

Figure 7.1. *Breeding Blanket and Divertor systems in a fusion reactor (EUROfusion 2020a). For a color version of this figure, see www.iste.co.uk/balbaud/nuclearenergytoday.zip*

The Breeding Blanket (BB) system is an in-vessel and plasma facing system consisting of large block-like modules surrounding the plasma. This system – which covers ~80% of the plasma surface – is designed to absorb the largest proportion of the plasma neutron and heat flux. Additionally, neutrons traversing the blanket are used to react with lithium in order to produce tritium, which is then extracted, processed and re-injected in the VV to fuel the reaction ("self sufficiency" requirement)[1]. A blanket module can be schematically represented as a box containing Li-compounds (liquid or solid). All the external walls of the box – and in particular the wall directly facing the plasma (First Wall [FW]) – are cooled by internal channels to extract the power deposited as the heat flux. Additional cooling elements – tubes or plates – are located inside the box, in the so-called Breeding Zone (BZ), to extract the power deposited by the neutrons. A system of manifolds is located in the rear part of the box to insure distribution of all fluids. The heat recovered in the BB provides the primary source for driving steam turbines used for electric power generation. Extraction of high-grade heat energy is therefore another requirement for the BB, whose coolant must operate at temperatures that allow for efficient power conversion cycles.

The Divertor (DIV) system is an in-vessel and plasma facing system in which all particles escaping the plasma (α-particles mainly, but also tritium and deuterium as well as plasma impurity particles) are diverted through an appropriate configuration of the magnetic field lines in order to collide with dedicated target structures (Figure 7.1). They are subsequently extracted from the VV chamber through magnetic channeling. The helium "ash" and the impurities are then completely removed, and the fuels are cycled back into the VV chamber. The DIV system consists of a series of cassette assemblies, each one composed of integrated components and systems. The key components in the DIV cassette assembly are the inner and outer vertical targets that are the primary plasma strike regions. The Dome and Reflector Plates protect the cassette body from direct contact with the plasma in situations where control of plasma is lost. The cassette body provides the structural support and insures coolant distribution to the other components. The cooling system in the DIV must mainly dissipate the heat energy coming from particles colliding with the targets. Although this is only a minor part of the fusion energy, the heat fluxes are very localized and lead to very high temperatures in the structures. The DIV system operates therefore with a (relatively) low temperature coolant. The overall heat energy dissipated by the DIV cooling system is a secondary source (pre-heating) for generating electric power.

---

1 Existing fusion machines are mainly devoted to the study of plasma physics and as such do not need to operate with a D-T "burning plasma". ITER features a burning plasma in the later phases, but uses existing tritium produced in current fission installations (mainly CANDU-type reactors). The blanket in these machines is usually referred to as "Shielding Blanket". DEMO should be the first reactor to demonstrate tritium production and features; therefore, a "Breeding Blanket".

7.2. Requirements for materials

The history of the development of fusion technology is intrinsically linked with the development of advanced materials. We can remember, for example, that it was the development of superconducting materials that allows present-day machines to reach the high magnetic fields and plasma densities necessary to achieve the fusion reaction. At the same time, the choice of materials for a given machine is always linked to its performance targets and programmatic goals.

This chapter will focus on the requirements for materials used in DEMO and possibly in the first generation of fusion power plants (FFPs). Indeed, DEMO will not be fully representative of an FPP, but must instead test technologies that could be extrapolated to early commercial FPPs (Eurofusion 2018). From the point of view of in vessel components (IVCs), this implies that DEMO will operate like a component test facility featuring a "driver" blanket with limited performance for most of the machine, while giving the possibility to test advanced blanket concepts at selected emplacements (Federici et al. 2019a). The driver blanket should be able to withstand a maximum integrated neutron fluence at end of life equivalent to ~20 dpa, after which it will be replaced. A second phase is foreseen to reach ~50 dpa (Stork et al. 2014a). "Baseline" materials for phase 1 have already been selected, as presented hereafter. Requirements and corresponding developments for advanced materials for phase 2 are described in section 7.5.

7.2.1. *Requirements for the structural material of the BB system*

The blanket is a key component in DEMO since it must insure the essential functions of high-grade heat extraction and tritium production. Blanket components are also exposed to the highest levels of neutron irradiation and their degradation affects the lifetime of the machine. The following requirements stem equally from programmatic and performance goals.

7.2.1.1. *Low neutron-induced activation*

The fusion reaction does not produce per se radioactive isotopes. However, the neutrons escaping the plasma will cause activation of the IVCs materials by means of transmutation reactions. Since the blanket represents the biggest system inside the VV, it is essential that the materials used present low (or reduced) activation characteristics. This is a necessary condition to minimize the waste burden[2] to future generations and realize the potential of fusion as a clean energy source.

7.2.1.2. *Stability under prolonged (fast) neutron irradiation*

Several critical phenomena are known to occur in steels when exposed to neutron irradiation. Besides hardening and embrittlement, irradiation induced-creep and swelling can also lead to dimensional instabilities. The latter is of particular concern in fusion reactors, since effects of the high energy fusion neutrons will be worse for the same fluence than in a fission reactor. Indeed, above a few MeV incident energy, the neutrons interact with the lattice atoms to produce transmutation products, in particular, helium and hydrogen. The helium atoms are capable of coalescing into microscopic helium bubbles located at grain boundaries, which would lead to further embrittlement and swelling of the material.

A specific fusion neutron source, which is under development, will be needed to experimentally investigate this phenomenon (Ibarra et al. 2019). Until then, a decision will have to be based on data collected in fission material test reactors complemented by modeling of irradiation-induced effects at high energies (Gilbert et al. 2021).

7.2.1.3. *Suitable physical and mechanical properties in the temperature range of operation*

The temperature range of operation of blanket structures is the result of the heat loads acting on the components and the operating parameters of the coolant. The choice of the final coolant for the DEMO driver blanket is still open; two options are being considered (Federici et al. 2019b): water or helium (Table 7.1). However, in order to limit the cost and complexity of developing different structural materials for the different blanket concepts, the same temperature range of 300–550°C is defined for the design of the blanket system. The structural materials shall present sufficient strength and ductility properties in this range of temperatures. Moreover, because of the strong thermal gradients in the FW, a lower coefficient of thermal expansion (CTE) and a higher coefficient of thermal conductivity are sought after properties for fusion structural steels in order to reduce the magnitude of (cyclic) thermal stresses.

---

2 Note that the main sources of radioactivity in a fusion reactor during operation is the tritium inventory. Tritium has a short decay time (12.3 years), but the management of tritiated waste is nonetheless an issue to be considered at decommissioning.

7.2.1.4. *Chemical compatibility with coolants, breeder and neutron multiplier materials*

In order to reach tritium self-sufficiency, lithium and a neutron multiplier material will have to be present inside the blanket structures. Considering the requirements for low activation and stability under neutron irradiation, the possible choices and combination of those are quite limited. Table 7.1 lists the materials considered for the two different concepts, HCPB (helium cooled pebble beds) and WCLL (water cooled lithium lead), selected for the DEMO driver blanket. The structural material will have to be compatible with the other materials used in the Blanket. Lithium, in particular, is highly reactive and corrosive and it is never used in its pure natural form. Water corrosion is another issue to consider.

Concept	Coolants	Breeders	Neutron multipliers
Helium cooled pebble bed (HCPB)	He 300–500°C, 8 MPa	Li_2TiO_3 Li_4SiO_4 (ceramic pebbles)	Be Pb
Water cooled lithium lead (HCLL)	H_2O 290–315°C, 16 MPa	Pb-15.7Li 300–500°C liquid eutectic	Pb (in PbLi)

Table 7.1. *Selection of functional materials for the design of the DEMO starter blanket*

7.2.2. *Requirements for the materials of the Divertor system*

The targets in the Divertor system are the most thermally charged components in a fusion machine, while at the same time being exposed to extremely high particle fluxes and neutron irradiation (although at lower levels than the Blanket). They are also a key component in insuring plasma stability. The Divertor targets use a plasma facing material with the function of shielding the target's structure ("Armor" material) and a structural material containing the coolant ("heat sink" material). The coolant selected for the DEMO Divertor system is water at relatively low temperature and pressure (~130°C, 4 MPa). The structural material for the Cassette body will instead be the same as the one used in the Blanket system. Requirements for Armor and heat sink materials are listed hereafter.

7.2.2.1. *Low erosion and compatibility with plasma operation*

The impact of diverted plasma particles on the target's surfaces leads to a progressive erosion of the plasma facing material. Erosion by the plasma is one of the phenomena that determines the life of the targets and the frequency at which they must be changed. More importantly, the eroded particles from the Armor material can enter the plasma, where they absorb and progressively release energy without contributing to the fusion reaction. High atomic number (Z) elements, which are transported to the "core" zone of the plasma, where most of the fusion reactions occur, are particularly detrimental. Low Z elements instead stay near the plasma edge, where their impact on plasma operation is lower. The chosen Armor material must therefore have low erosion rates and/or low Z[3].

7.2.2.2. *Limited tritium retention*

Tritium nuclei escaping the plasma are also diverted toward the Divertor's target, where they can be "trapped" inside the Armor material. Combined with the continuous erosion/re-deposition process, this leads to the formation of radioactive ashes that can circulate inside the VV and deposit on all IVCs' surfaces. This represents a safety concern for the operation of the reactor as well as during maintenance phases. The selected Armor material should therefore present low tritium retention.

7.2.2.3. *Suitable physical and mechanical properties to withstand strong thermal cycling*

The number and intensity of thermal cycles to which the Divertor targets are exposed is much stronger w.r.t. all other IVCs. The intensity of the heat loads on the targets is a direct consequence of plasma confinement physics and is estimated for DEMO to vary between 10 MW/m^2 and 20 MW/m^2 during normal plasma operation[4]. This leads inevitably to very high temperatures at the plasma facing side of the targets. Materials used in the Divertor will have to present good mechanical resistance to (thermal) fatigue and this over a very large range of temperatures. Moreover, high thermal conductivity and low CTE would help to reduce the intensity range of thermal stresses.

7.2.2.4. *Low activation and resistance to neutron irradiation*

Requirements of low activation and resistance to neutron irradiation also apply to the targets but are considered less important than for the BB structural materials. On the one hand, neutron fluxes are a factor of 2 lower at Divertor position and the

---

3 For plasma compatibility, the Blanket's First Wall is also coated with a thin layer (0–2 mm) of Armour material. It has, however, contrary to the Divertor, no protective or structural function w.r.t the blanket structure, since particle fluxes are much lower at those locations.

4 Thermal loads during accidental transients (disruptions) can be much higher than during normal operation and can indeed lead to failure of the Armour. By design, it is assumed that Divertor targets may have to be replaced following a disruption event.

component's masses are much smaller. On the other hand, targets already have very stringent requirements for plasma operation.

7.3. Material choices and basic properties

Over the last three decades, several alternatives have been considered for the materials of IVCs, all presenting advantages and drawbacks (Stork et al. 2014b; Federici et al. 2017; Tavassoli 2002). For the DEMO reactor, considering the time constraints and the necessary R&D efforts, consensus was found on the choice of the baseline materials, while in parallel, more advanced options are developed as presented in section 7.5.

7.3.1. *EUROFER97 as structural material for the BB*

Of the different class of structural materials considered for fusion applications, Reduced Activation Ferritic Martensitic (RAFM) steels have been selected for DEMO because of their excellent stability under neutron irradiation and their relatively advanced industrial development (Van der Schaaf et al. 2000). Indeed, metals and alloys with a "body-centered-cubic" (bcc) crystal lattice structure, show much lower swelling and irradiation-induced creep (Séran and Le Flem 2017), than metals with "face-centered-cubic" (fcc) lattices such as austenitic steels (Figure 7.2). For this reason, austenitic steels selected for the ITER blanket – whose maximum exposure is planned to reach 3 dpa – cannot be used in DEMO and even less in future FPPs (>100 dpa). Moreover, nickel – a basic element in austenitic steels – activates under neutron irradiation, giving rise to long-lived radioisotopes (Ni^{63} and Ni^{59}), which would lead to the necessity of long-term nuclear waste storage of IVCs in underground repositories[5].

RAFM steels were a highly successful development of the 1990s worldwide fusion programs and were developed from what can be termed "second-generation" Ferritic Martensitic (FM) steels containing 9%–12% wt. of chromium. EUROFER97 in particular is one such steel developed by the European Union (Van der Schaaf et al. 2003), derived from the conventional modified 9Cr–1Mo steel, where high activation elements and impurities (Mo, Cu, Nb, Ni) are replaced by their equivalent low activation elements (e.g. W, V and Ta) or kept to the lowest achievable level with current production methods (Figure 7.3). EUROFER has been produced (semi)industrially in four batches of increasing tonnage from 3.5 to 27 t. The

5 The current target for the structural materials of DEMO is to achieve low level waste classification after ~100 years.

production process was the same for all batches, that is, vacuum induction melting (VIM), followed by a remelting step by vacuum arc remelting (VAR). Plates up to 50 mm thick, bars and tubes have been already produced. Plates are normalized at a temperature between 940 and 980°C, hot-rolled and then tempered between 740 and 760°C. Chemical analysis has confirmed that impurity levels can be controlled by a careful selection of raw materials in the melt. EUROFER97 has been recently introduced in the RCC-MRx nuclear construction code, although it is still in the Probationary Phase Rules tome (Lebarbé et al. 2020).

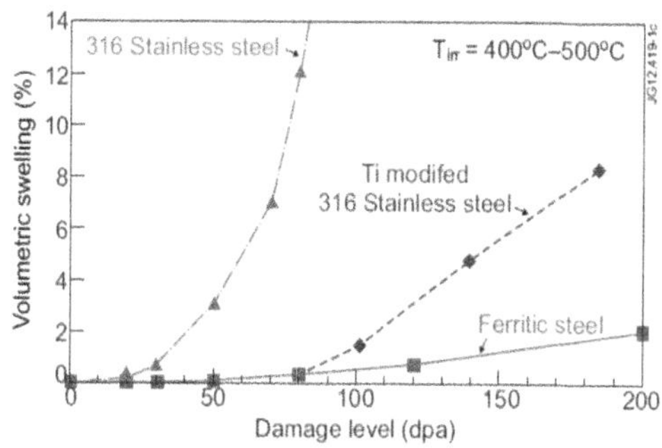

Figure 7.2. *Comparative swelling under fission neutron irradiation for ferritic and austenitic steels. For a color version of this figure, see www.iste.co.uk/balbaud/nuclearenergytoday.zip*

EUROFER 97 MPH - Chem. Comp. wt%			
	min	max	Comments
Cr%	8.5	9.5	Target 9
C%	0.09	0.12	Target 0.11
Mn%	0.2	0.6	Target 0.4
P%	-	0.005	
S%	-	0.005	
V%	0.15	0.25	
N%	0.015	0.045	Target 0.030
W%	1	1.2	Target 1.1
Ta%	0.1	0.14	Target 0.12
Ti%	-	0.02	
Nb%	-	0.005	ALAP
Mo%	-	0.005	ALAP
Ni%	-	0.01	ALAP
Cu%	-	0.01	ALAP
Al%	-	0.01	ALAP
Si%	-	0.05	
Co%	-	0.01	ALAP
B%	-	0.002	ALAP
As+Sn+Sb+Zr%	0	0.05	
Fe%-	Balance	Balance-	

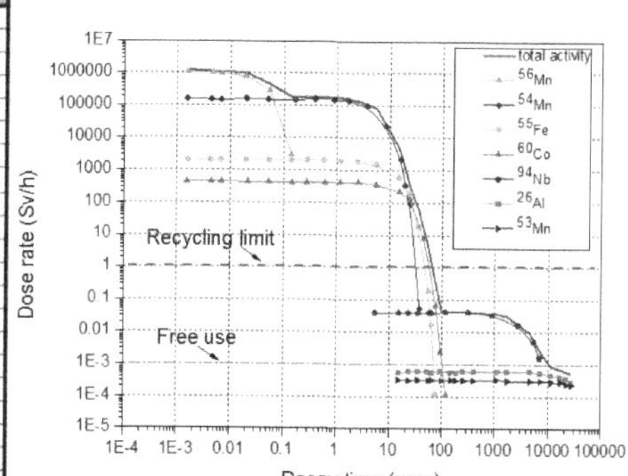

Figure 7.3. *Chemical composition of EUROFER97 in wt%, Fe balance. Substitution elements are highlighted in blue and impurities in red (left). Long decay time elements in EUROFER97 and the total activity in the function of years after neutron irradiation for a fluence 2.5 × 10^{20} n/m^2) (right). For a color version of this figure, see www.iste.co.uk/balbaud/nuclearenergytoday.zip*

7.3.1.1. *Basic physical properties*

Belonging to the class of Ferritic–Martensitic steels, EUROFER97 presents lower CTE and better thermal conductivity than austenitic steels (Figure 7.4). This leads to a thermal stress factor ($S_m \cdot E \cdot \lambda / [(1-\nu) \cdot \alpha]$ – where E is the Young Modulus, S_m is the allowable stress, ν is the Poisson's ratio, λ ισ the thermal conductivity and α is the thermal expansion coefficient) – roughly two times higher than for austenitic steels in the temperature range of interest and to proportionally lower thermal stresses.

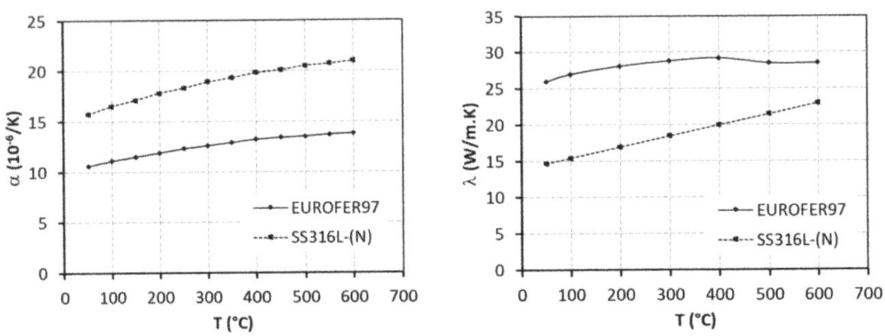

Figure 7.4. *Thermal expansion and thermal conductivity coefficients of EUROFER97 versus SS316L-(N) steel*

7.3.1.2. *Tensile and ductility properties before and after irradiation*

Figure 7.5 shows the evolution of the yield strength (*Sy*), ultimate tensile strength (Su) and allowable primary design stress (S_m), as well as the evolution of the total (ε_t) and uniform (ε_u) elongation of EUROFER97 compared to those of the 316L-(N) (ITER Grade) steel. In the whole range of temperatures relevant for IVC operation, EUROFER97 exhibits higher design stresses but much lower strain hardening capability.

Neutron irradiation below 350°C causes a progressive increase in strength and a corresponding decrease in ductility properties, as shown in Figure 7.6 (Gaganidze and Aktaa 2013). Change of properties stabilizes between 10 and 20 dpa, except for the uniform elongation that practically disappears after ~2 dpa. The effects of neutron irradiation are mitigated for irradiation temperatures above 350°C and seem to disappear above 400°C. The increase of strength properties is beneficial for design, but the loss of uniform elongation coupled with strong thermal stresses can lead to additional damage modes, most notably immediate plastic flow localization, which must be taken into account in the design (Aiello et al. 2011). On the other hand, contrary to austenitic steels, total elongation remains comparably high (~10%), which is beneficial from the point of view of safety (no catastrophic failure of the components).

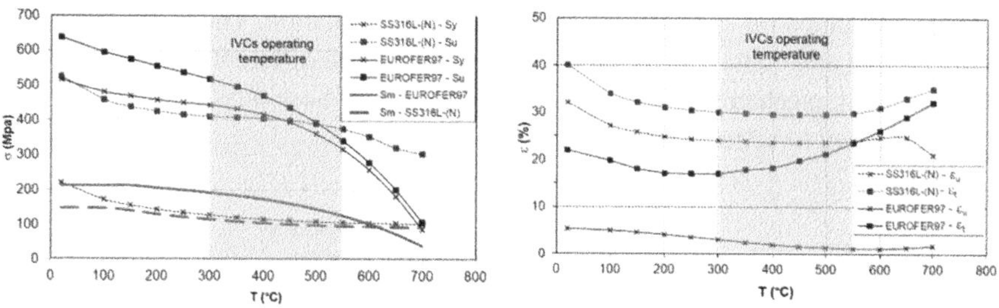

Figure 7.5. *Tensile (left side) and ductility (right side) properties of EUROFER97 versus SS316L-(N) steel. For a color version of this figure, see www.iste.co.uk/balbaud/nuclearenergytoday.zip*

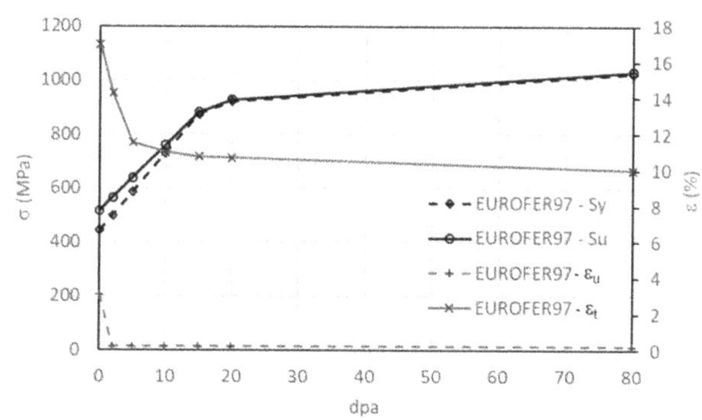

Figure 7.6. *Evolution of strength and tensile properties of EUROFER97 after irradiation at 300°C. For a color version of this figure, see www.iste.co.uk/balbaud/nuclearenergytoday.zip*

7.3.1.3. *Fatigue and creep–fatigue interaction properties*

EUROFER97, like all FM steels, exhibits very pronounced softening under low cycle fatigue (LCF) tests (Figure 7.7). Cycling softening has a detrimental effect on damage modes such as ratcheting and creep–fatigue interaction at long hold times, lowering the effective yield stress of EUROFER97 during operation. These issues are indeed common to all FM steels and will need to be taken into account in the design process. A considerable effort has recently been devoted to developing design rules specific to EUROFER97 under the operating conditions of the DEMO Breeding Blanket (Özkan and Aktaa 2015). Irradiation, on the other hand, is supposed to have a beneficial effect on fatigue and creep damage, although present results are obtained on specimens tested only after irradiation. The evolution of the microstructure and consequent material damage could be different under concurrent irradiation and mechanical loading. In pile testing, whose feasibility is limited, would be needed to better investigate these phenomena.

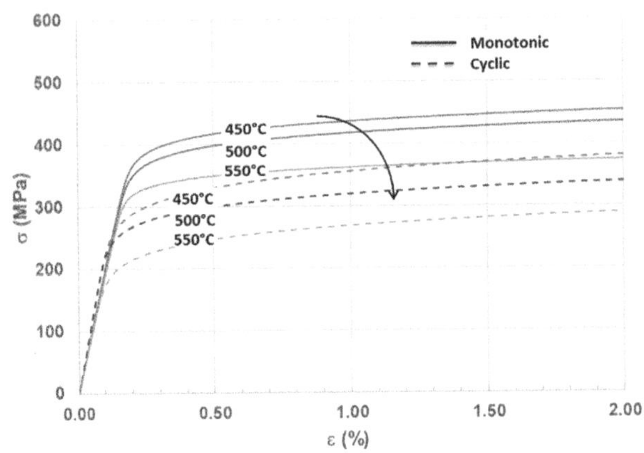

Figure 7.7. *Monotonic and reduced cyclic stress–strain curves for EUROFER97 at different temperatures. For a color version of this figure, see www.iste.co.uk/balbaud/nuclearenergytoday.zip*

7.3.1.4. *Ductile to brittle transition temperature (DBTT)*

As a bcc material, EUROFER presents a sharp transition from ductile to brittle behavior as the temperature decreases. The Charpy DBTT or TTKV temperature is obtained using either standard Charpy V or KLST specimens, whenever the criterion is below –50°C (Figure 7.8, left). The scatter is large, some data being as low as –130°C. The low ductile-brittle transition temperature means that any pressure retaining components made from EUROFER97 and operated at room temperature or above is safe against brittle fracture. Irradiation below 350°C causes, however, a shift of the DBTT toward higher temperatures. At high irradiation temperature (400–450°C or higher) limited data are available, but there is experimental evidence that the irradiation effect is strongly reduced (Figure 7.8, right).

Another cause of embrittlement is the high rates of transmutation He in a fusion neutron spectrum (~10 appm/dpa). Due to the absence of a specific fusion irradiation facility, the helium effects are presently studied in different simulation experiments mainly using the boron doping technique (Gaganidze and Aktaa 2013). These results should be regarded with care, since boron doping itself can affect the mechanical properties of the material as well as their evolution with neutron irradiation. Additional irradiations with alternative doping or enrichment techniques are planned.

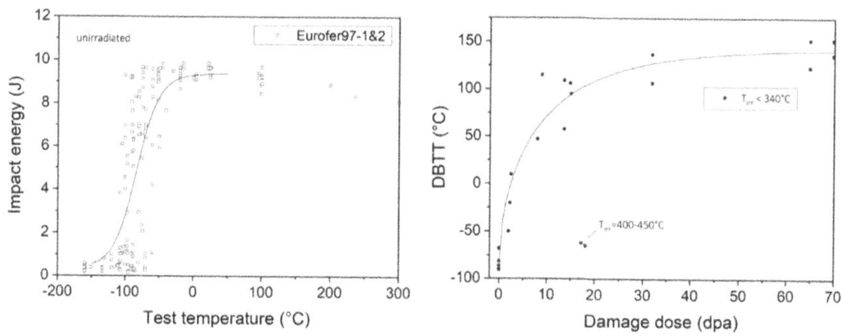

Figure 7.8. *DBTT of EUROFER97 (left) and corresponding irradiation shift versus dose for EUROFER97 for irradiation temperatures below 350°C (right) (EUROfusion 2022). For a color version of this figure, see www.iste.co.uk/balbaud/nuclearenergytoday.zip*

7.3.1.5. *Chemical compatibility with other materials*

The design of the blanket must take into account the compatibility between the different materials listed in Table 7.1. If compatibility with He and ceramic pebble beds does not seem to be an issue[6], corrosion effects by the PbLi eutectic and water have been extensively investigated.

In case of PbLi, the kinetics of the corrosion reactions with EUROFER lead to a progressive dissolution of the material in the eutectic without formation of a passivating layer. This effect is stronger with increasing velocity and temperature, although less important than in austenitic-type steels (Sannier et al. 1991). Moreover, the magnetic fields inside the VV can strongly modify the corrosion process because of magneto-hydro-dynamics (MHD) effects, affecting both the flow patterns at the steel/eutectic interface and the corrosion kinetics (Moreau et al. 2011; Carmona Gázquez et al. 2015). Indeed, present results indicate that corrosion barriers will be needed. Alumina-based coatings are therefore being developed using

6 Assuming a proper control of the coolant chemical composition.

different deposition techniques. Preliminary results indicate excellent stability of the protective layers, even under irradiation (Utili et al. 2021).

Water corrosion of steels is a well-known and extensively investigated phenomenon in fission reactors. However, comparatively few data exist on the corrosion of (RA)FM steels in flowing water at PWR conditions. Past studies hint at corrosion rates 10–40 times higher in FM steels with respect to stainless austenitic steels (Raquet and Sèjourné 1999), since the protective oxide layers in steels with low Cr content does not appear to be stable. One particular concern is the phenomenon of (irradiation assisted) stress corrosion cracking ((IA)SSC) (Maday 2002; Van Dyck and Bosch 2005). Another issue is the presence of activated corrosion products (ACPs) in the primary heat transfer system, which represent an additional hazard in safety analyses. Activities to extend the domain of application of ACP inventory calculation codes to materials and environmental conditions found in fusion blankets are on-going (Terranova and Di Pace 2021). If results should show that corrosion in water must be prevented, the alumina-based coatings developed for PbLi or other surface engineering techniques could be used.

7.3.2. *Tungsten and copper-chromium-zirconium as armor and heat sink materials for the Divertor targets*

As already explained, the armor (plasma facing) material of the Divertor targets must be compatible with the regimes of plasma operation. Combining this with the requirements of reduced activation properties leads to the exclusion of most elements, leaving essentially carbon (C) and beryllium (Be) among the low-Z elements, and tungsten (W) among the high-Z elements as the only viable choices (Richou et al. 2020).

Carbon was used in first generation tokamaks with non-burning plasmas, but tritium retention is too high to enable reliable plasma operation and tritiated dusts caused by erosion present a safety concern. Beryllium on the other hand presents tolerable tritium retention, but has a relatively low melting temperature (1,290°C vs. 3,420°C for carbon) and degradation of material properties is already important, starting from 650–750°C. Its use is not compatible with long-pulse/high-performance plasma operation, but it is selected as armor material for the ITER FW. Finally, tungsten has the highest melting temperature of all natural elements (3,420°C), presents limited erosion rates, low tritium retention (compared with carbon) and good mechanical properties up to 1,200–1,400°C. Under the characteristic neutron energy spectra in fusion components, activation is also limited (Taylor and Pampin 2006). Tungsten is therefore the material selected as armor for ITER as well as DEMO Divertor targets.

Tungsten is brittle at low temperatures and its machinability is limited because of weak grain boundaries. It cannot therefore be used for the structural parts (pressure containment function) of the targets, which are routinely exposed to cycling and thermal shocks. Another material must therefore be used for these structural parts that are usually referred to as "heat sink". Considering that the limiting factor in the design of the heat sink is the critical heat flux, materials with good ductility and high thermal conductivity leading to low wall temperatures are preferred. On the basis of these criteria, the ternary alloy copper–chromium–zirconium (CuCrZr) has been selected for the ITER and DEMO Divertor targets.

7.3.2.1. *Tungsten properties*

W and W alloys are commercially available from a number of producers in many forms and through different production routes. Sintered W is the most readily available and cheapest grade and the one finally selected for ITER and DEMO. The W powder is obtained by hydrogen reduction of ammonium paratungstate into tungsten oxides (WO_3, WO_{3-x}) and then metal at high temperature, and is controlled by temperature and water vapor partial pressure. Minimum tolerable purity of W is set to 99.94 wt%, and the limits to representative impurities (C, O, N, Fe, Ni, Si) are set to <0.01 wt%. Different W products such as sheets, plates, rods, round blanks are obtained via a powder metallurgical route including mixing of the alloying powders, isostatic pressing, sintering and different thermo-mechanical treatments (forming). Subsequently, the stress relieving heat treatments (HTs) are applied in order to reduce high dislocation densities introduced during the fabrication process. The microstructure of the tungsten products is very sensitive to the production route and applied thermo-mechanical treatment. Cross-rolled tungsten plates are characterized by the presence of pancake-shaped grains elongated into a rolling direction that confer the material anisotropic mechanical properties. W and W alloys tend to delaminate in the direction parallel to the direction of deformation. To avoid failure due to the delamination, the orientation of the texture has to be perpendicular to the surface of the W/Cu joint (i.e. parallel to the incident heat flux).

7.3.2.1.1. DBTT and recrystallization

Tungsten being a bcc material, it presents a sharp transition from ductile to brittle behavior. Indeed, the DBTT of the as-produced material is already considerably above room temperature (>300°C), and as the other properties, it can be different in different directions. The value of the DBTT depends also on the HT. For annealing temperatures above 1,200–1,300°C, which can in fact be reached during operation, the DBTT is significantly reduced. This is caused by recrystallization taking place (Figure 7.9), with the associated change in grain structure from highly deformed and dislocation-rich grains to more equi-axed grains (i.e. the removal of cold work). Recrystallization affects all mechanical properties of the material and should be avoided during operation.

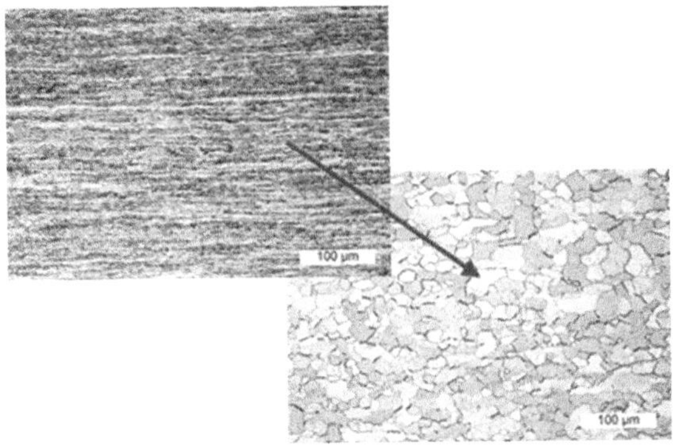

Figure 7.9. *Optical micrograph of a tungsten sheet, cold worked and stress relieved (above) and recrystallized (below) (Plansee 2022). For a color version of this figure, see www.iste.co.uk/balbaud/nuclearenergytoday.zip*

7.3.2.1.2. Basic mechanical properties

Tensile properties measured on stress relieved IGP W bars (ITER Grade Product) at two temperatures in as-received and recrystallized state are shown in Figure 7.10 (Wirtz et al. 2016). Although the strength properties are comparable for L and T directions, the elongation and ductility properties show significantly different behavior to be accounted for in design. For temperatures below 300°C, the material is brittle in the transversal direction. Recrystallization causes an increase in elongation properties but a considerable corresponding decrease in strength properties.

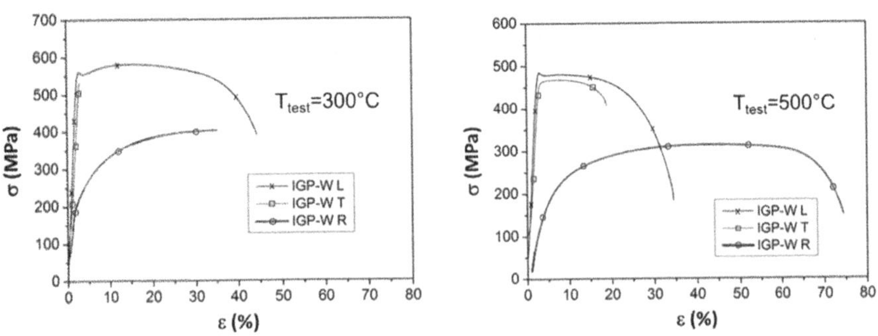

Figure 7.10. *Stress–strain curve for the longitudinal (IGP WL), transverse (IGP-WT) and recrystallized (IGP-WR) material. For a color version of this figure, see www.iste.co.uk/balbaud/nuclearenergytoday.zip*

The effects of recrystallization also cause a significant decrease in fatigue properties (Schmunk and Korth 1981), as shown in Figure 7.11.

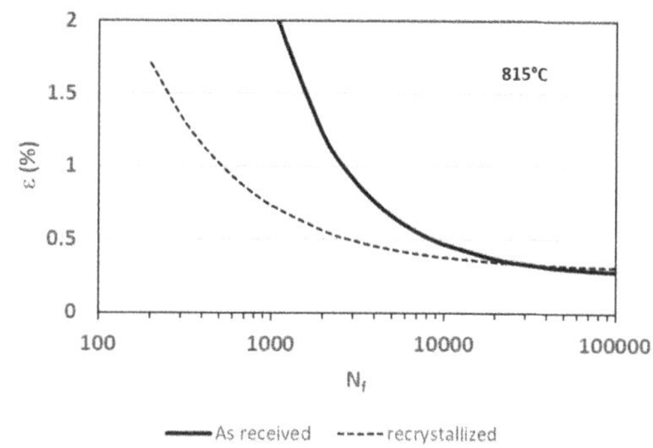

Figure 7.11. *Mean fatigue curve of as received and recrystallized W tested at 815°C*

Pure W and W alloys have low fracture toughness, even at temperatures above the DBTT. Figure 7.12 shows the fracture toughness data for stress-relieved tungsten. The limited available data show that the fracture toughness is significantly less than 100 MPa√m at all temperatures, with a typical value of ~30 MPa√m at 1,000°C.

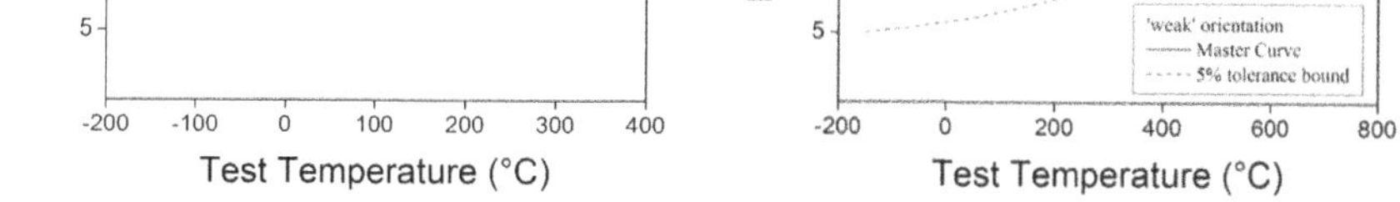

Figure 7.12. *Mean fracture toughness measured on the rolled rod and forged round blank polycrystalline tungsten products in their "strong" (left) and "weak" (right) orientations (EUROfusion 2020c). For a color version of this figure, see www.iste.co.uk/balbaud/nuclearenergytoday.zip*

7.3.2.1.3. Effects of neutron irradiation

Determination of the effects of neutron irradiation on W is complicated by the lack of a specific fusion neutron source. Irradiation in fission reactors to the same fast neutron fluence yields a much larger amount of metallic transmutation products (Hf, Ta, Re and Os) than in fusion systems (Sawan 2014). Even the small addition of Re or Hf can have a strong influence on physical and mechanical properties of W and must be avoided by appropriate shielding of irradiation rigs in MTRs. Irradiation has shown that even at very low doses, a significant shift in DBTT can be observed for samples irradiated at a temperature below 600°C (Gaganidze et al. 2021). More surprisingly, however, the shift was still important at high temperatures as shown in Table 7.2 (Pintsuk et al. 2022). The reasons for this lack of recovery at high temperatures needs still to be fully understood, but based on these limited data, we can suppose that tungsten tiles might become fully brittle at the expected level of irradiation in DEMO.

Condition	DBTT (°C)
Unirradiated	400
1 dpa/400°C	1,100
1 dpa/600°C	≥1,100
1 dpa/800°C	1,000–1,025
1 dpa/1,000°C	≥1,200

Table 7.2. *Preliminary trends for the DBTT of IGP tungsten*

7.3.2.2. *CuCrZr properties*

Several Cu alloys (DS Cu Glidcop Al25, CuCrZr, CuNiBe, etc.) were initially screened for application as heat sink for the ITER Divertor, and the CuCrZr alloy was finally selected due to its high fracture toughness, industrial availability and cost (Barabash et al. 2007). CuCrZr is commercially available in a wide range of products from a number of different producers and the material specifications are covered in several international standards (e.g. DIN EN 12167, UNS C18150, etc.).

Copper-chromium-zirconium is a precipitation-strengthened alloy. For DEMO application, alloying elements Cr (0.6–0.9 wt.%) and Zr (0.07–0.15 wt.%) are recommended in order to reduce the scatter of properties and to improve toughness and weldability (Appello and Fenici 1988). Cr content must stay within a narrow range, high enough above the solubility limit to account for Cr inhomogeneity and low enough to prevent coarse Cr precipitates, which may cause embrittlement. Zr is required to improve homogeneity of precipitates, ductility and fatigue resistance. It increases the solubility of Cr in the Cu-Cr-Zr system. It must also be approximately equal to or below the solubility limit at the solution annealing temperature to prevent the formation of brittle intermetallic phases. Phosphorous impurities should be kept as low as possible to minimize coarse precipitates during annealing.

To provide the high mechanical strength, good ductility and high thermal conductivity, the material needs to be specially heat treated and/or cold worked; the two options considered for Divertor application in ITER are as follows:

– Solution Annelead and Aged (SAA): solution annealing at 980–1,000°C for 30-60 min, water quench and aging at 460–500°C for 2–4 h.

– Solution Annealed, cold worked and Aged (SAcwA): solution annealing at 980–1,000°C for 30-60 min, water quench, followed by cold working 40–70% and aging at 450–470°C for 2–4 h.

The SAcwA treatment leads to comparatively higher tensile strength properties and is the one selected for DEMO. It is, however, more sensible to thermal treatments and the properties of the alloy in the final, as-manufactured component to depend in the end on the fabrication route. Due to this, the database for reference grades with appropriate thermomechanical treatments (TMTs) is still relatively poor.

7.3.2.2.1. Basic physical properties

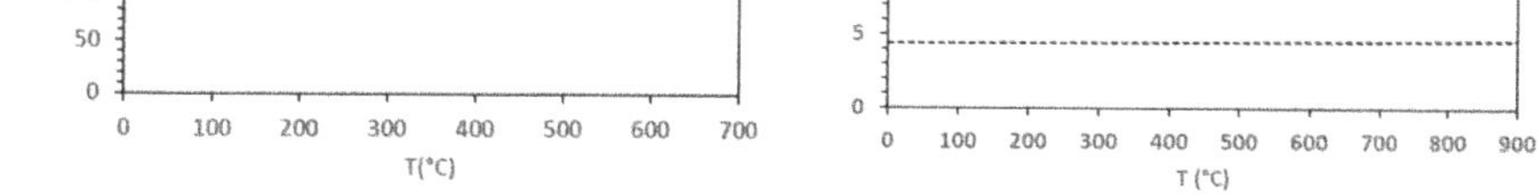

Figure 7.13. *Thermal conductivity (left) and coefficient of thermal expansion (right) of W and CuCrZr*

The thermal conductivity and CTE for CuCrZr are given in Figure 7.13 and compared with those of W. The high thermal conductivity of CuCrZr insures sufficiently low temperatures (250–300°C) at the wall/coolant interface to keep a 40% margin on the critical heat flux (You et al. 2021). The difference in CTE at the interface with the W monoblock, where temperatures are in the range of 350–450°C, would instead cause unacceptably high thermal stresses during cyclic operation. Appropriate geometries and joining procedures must be devised to accommodate the different thermal expansion during operation as described in section 7.5.

7.3.2.2.2. Basic mechanical properties

Tensile curves at room temperature for different grades of CuCrZr are given on the left side of Figure 7.14. The influence of the HT is clearly visible, but there is also a considerable scatter for products undergoing the same HT. SAcwA grade shows globally higher yield strength than the SAA grade but practically no strain-hardening capability. More importantly, these properties are modified during the manufacturing process of the component, as shown on the right side of Figure 7.14.

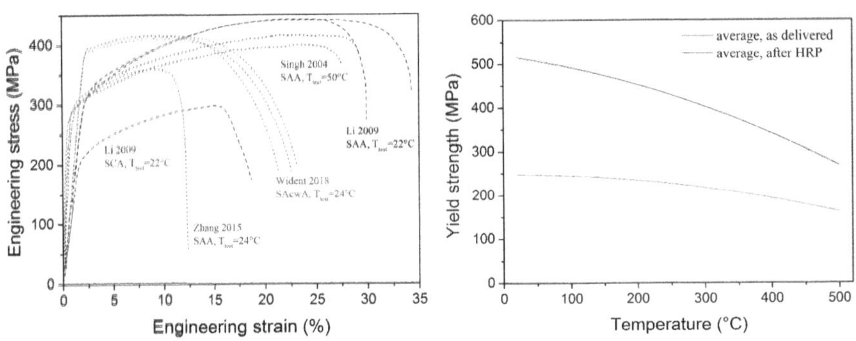

Figure 7.14. *Tensile curves for different CuCrZr grades (left) and impact of manufacturing process (hot radial pressing [HRP]) on the tensile properties of as-received products (right) (EUROfusion 2020b). For a color version of this figure, see www.iste.co.uk/balbaud/nuclearenergytoday.zip*

On the other hand, the influence of HT and specimen orientation on the fracture toughness of CuCrZr is negligible. Although there is a big scatter of experimental data, values of J-integral at crack initiation (J_Q) are globally high and above 100 kJ√m at all temperatures.

Creep of CuCrZr is already relevant, starting from 100°C, so that the effects of creep–fatigue interaction must be taken into account in the design of the Divertor (Figure 7.15). Moreover, while the fatigue properties seem to be mostly unaffected by the thermal treatments, creep properties can instead be strongly impacted.

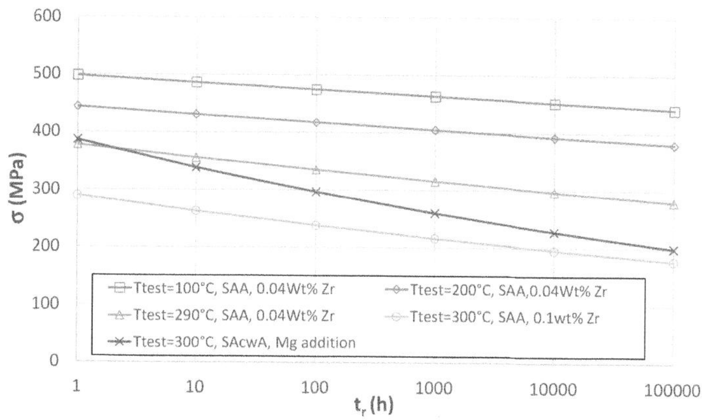

Figure 7.15. *Applied stress versus average creep rupture time of different CuCrZr alloys. For a color version of this figure, see www.iste.co.uk/balbaud/nuclearenergytoday.zip*

7.3.2.2.3. Effects of irradiation

Neutron irradiation affects both physical and mechanical properties of CuCrZr and in particular transport properties such as thermal and electrical conductivity. A decrease in thermal conductivity will have synergetic effects on the thermal stress in the high-temperature region, where strength is also affected. This must be taken into account in the design of the targets.

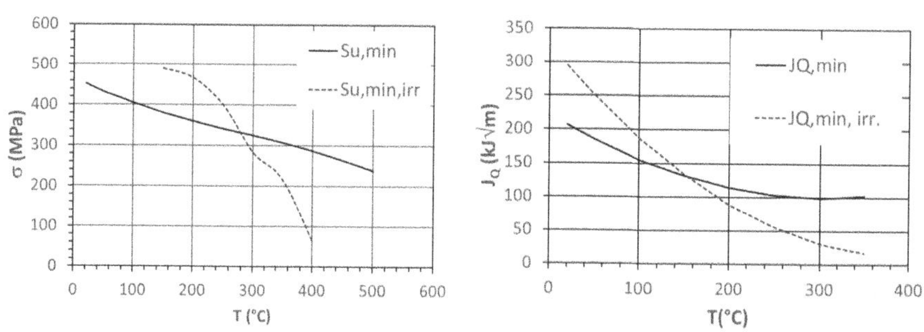

Figure 7.16. *Ultimate tensile stress (left) and fracture toughness (right) properties of SAcwA CuCrZr before and after irradiation (ITER 2013). For a color version of this figure, see www.iste.co.uk/balbaud/nuclearenergytoday.zip*

Neutron irradiation effects on mechanical properties of copper alloys depend strongly on the temperature. At temperatures extending from room temperature to 250–300°C, radiation hardening and the associated loss of work hardening capability are the dominant effect. Plastic flow localization can occur and must be prevented by more stringent design limits. Irradiation softening is instead the main concern for irradiation temperatures above 300°C. Design limits are in this case dictated by the irradiated properties. The same qualitative trend can be found for fracture toughness properties (ITER 2013).

7.4. Fabrication and joining

Due to the harsh operating environment as well as the very specific geometries and materials used in fusion reactors, manufacturing of fusion components requires the development of dedicated manufacturing techniques. For the BB, the techniques presented here are mainly those being qualified for the manufacturing of the future test blanket module (TBM), a first fully functional mock-up of a Breeder Blanket to be installed and operated in ITER. For a more in-depth overview and discussion of all the techniques considered for DEMO, the reader is referred to works in literature (Forest et al. 2020a; Neuberger et al. 2021). Divertor targets, on the other hand, have been regularly produced and tested in the past under representative loading conditions in a number of experimental facilities (Gavila et al. 2015).

7.4.1. *Manufacturing techniques for the BB*

For the fabrication of the starting sub-components, a combination of cutting and electric discharge machining (EDM) is unavoidable. EUROFER97 being ferromagnetic, a specific demagnetization step is needed after machining to avoid a magnetic blowout of arc welding and the associated risk of lack of fusion during welding. Cold forming (bending of plates up to 180°) has also shown to be possible, followed by appropriate HTs to relieve stresses accumulated during the process.

For joining of sub-components, several welding processes have been developed, based on diffusion welding, fusion welding and a combination of thereof (Zmitko et al. 2015; Forest et al. 2020a, 2020b). Diffusion welding is a solid-phase assembly process without fusion or significant deformation, where the final joint is achieved by grain growth at the interface of the pieces to join. The pieces are enclosed in a canister and placed inside a furnace where the high pressure and temperatures required for the process (~1,100°C, ~100 MPa) are achieved (hot isostatic pressure [HIP]). High quality and cleanliness of the surfaces to bond is also essential to avoid joining defects. A subsequent two-step HT of annealing at 980°C and quenching at 750°C allows the restoration of the original microstructure and the mechanical properties of the base metal. In order to achieve the complex shape of internal cooling channels, two different techniques have been developed. One is the two steps HIP process in which at first a low-pressure HIP cycle is applied to achieve leak tightness of internal channels and then a high-pressure HIP cycle to achieve final bonding. The bond line is in this case located at the mid-height of the channels (Figure 7.17). The second technique is a "YAG laser + HIP" fabrication process that consists of welding, by using a YAG laser, a thin strip (mostly in form of a mono-bloc strip) on the top of grooves machined in a base plate (see Figure 7.4) and then adding a cover plate by HIP. The final HT step allows the restoration of the properties of the base metal in the fused zone (Figure 7.18). DWTs are instead used to implement a double barrier against coolant leaks. They consist of two concentric tubes with an interface applied on the inner tube. The interface material (Fe or Cr) must be ductile enough to warrant that possible fatigue cracks are deviated or blunt before reaching the second wall. After deposition, the inner and outer tubes are joined together during a single HIP operation. The obtained DWT is then bended in its final configuration (Forest et al. 2020b).

Figure 7.17. *Left: Cooling plate fabricated by two-step HIP process: two grooved plates and a canister; right: cross-section of CP cooling channels with dimensions (Zmitko et al. 2017). For a color version of this figure, see www.iste.co.uk/balbaud/nuclearenergytoday.zip*

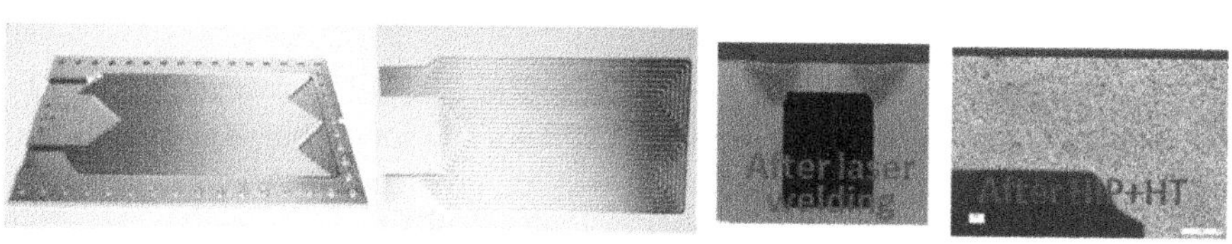

Figure 7.18. *Cooling Plate fabricated by hybrid laser+HIP process: grooved CP mock-up (left) before welding of the mono-bloc strip (center-left). Microstructure after laser welding (center-right) and after HIP+HT (right) (Zmitko et al. 2017). For a color version of this figure, see www.iste.co.uk/balbaud/nuclearenergytoday.zip*

Fusion welding processes – with or without filler metal – will mainly be used to assemble the various sub-components and achieve the final module (Figure 7.19). The choice of the final process depends on several factors, such as thicknesses to weld, geometry, accessibility of the weld, etc.

Figure 7.19. *Assembly of the final module by robotized TIG welding (Zmitko et al. 2017). For a color version of this figure, see www.iste.co.uk/balbaud/nuclearenergytoday.zip*

Regardless of the welding process used, the final as-produced weld is characterized by a "hard" fused zone (FZ) with coarse grains and a "soft" finer grained heat-affected zone (HAZ) as shown in Figure 7.20 (Rieth et al. 2021). The as-produced welds exhibit a higher DBTT (rupture mostly in the FZ) and lower tensile properties (rupture in the HAZ) than the base metal, although this effect is less pronounced with beam processes (Laser and EB) with low heat deposition. A post weld heat treatment (PWHT) is therefore necessary to restore the mechanical properties of the weld.

The standard two-step PWHT is indeed effective in recovering the full mechanical properties of the base metal, even on TIG joints, but it is feasibility on big and complex shape components that may be problematic. A single-step HT at T > 740°C, although slightly less effective is recommended in this case. The effects of repeated PWHT during manufacturing have also been assessed and found acceptable.

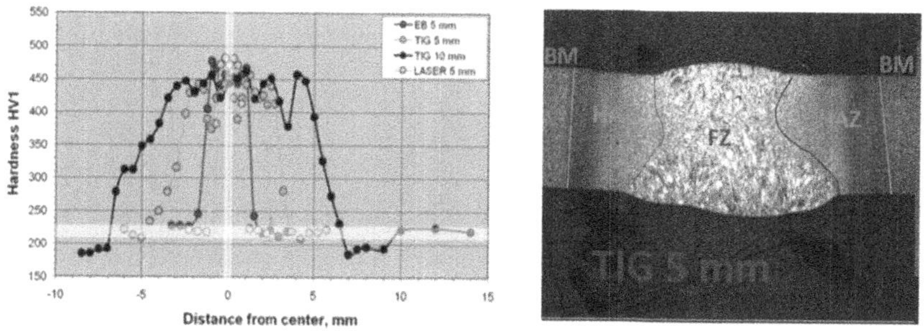

Figure 7.20. *(Left) Hardness profile of as-produced welds. Hardness level of the base material is shown in gray. (Right) Morphology of the different zones: FZ – fused zone, HAZ – heat-affected zone, BM – base metal (Rieth et al. 2021). For a color version of this figure, see www.iste.co.uk/balbaud/nuclearenergytoday.zip*

7.4.2. *Manufacturing techniques for the Divertor targets*

The main problem in the development of W/CuCrZr joints is the large difference in the coefficients of thermal expansion (and elastic modulus) between the two materials. With a conventional flat tile geometry, this difference creates very large stresses at the interface during joining and operation. The reduction of these stresses by the partial castellation of the W tiles is problematic due to the brittleness of tungsten at room temperature. Therefore, the geometry selected for the ITER and DEMO Divertor targets is the so-called "monoblock" configuration (Figure 7.21) in which the W armor is subdivided in small rectangular blocks attached to a CuCrZr tube. Moreover, a thin layer of Cu (1 mm) is applied between the CuCrZr tube and the W monoblock to act as soft compliance material in virtue of its high ductility. The Cu interlayer is first applied to the W monoblock, which is then joined to the CuCrZr heat sink.

Several different techniques, such as casting, brazing or HIPing were evaluated for the ITER Divertor (Giniyatulin et al. 2002; Saito et al. 2002). The Cu casting technique presents the best compromise between performance and costs. The good joining during casting is based on the high wettability and high creep relaxation ability of pure Cu. The essential parameters for a good bonding are the quality of the W surface, the purity of the Cu material and the temperature during the process. Oxygen free (OF) Cu is specified, with a purity of ≥99.99%. Although higher temperatures would lead to improved wettability, the maximum temperature is set to 1,200°C to avoid recrystallization of W.

Figure 7.21. *Monoblock configuration of the Divertor targets (EUROfusion 2020a). For a color version of this figure, see www.iste.co.uk/balbaud/nuclearenergytoday.zip*

The next step is joining of the W/Cu elements onto the CuCrZr cooling tubes, for which several techniques were evaluated for ITER (Hirai et al. 2017). The reference choice for DEMO is the hot radial pressing (HRP) technique (Visca et al. 2003), with HIP being considered as a possible back-up solution. The bonding temperature for these two processes (550–580°C) is in fact lower than for alternative brazing processes (>900°C), which limits the effects of the HT on the mechanical properties of the as-received CuCrZr alloy. During HRP, a (radial) diffusion bonding between the CuCrZr tube and the Cu layer of the monoblock is obtained by applying only pressure inside the tube. The main advantages of this technique w.r.t. HIP is that the required bonding pressure is lower (60 MPa instead of 100 MPa) and the canister itself is not pressurized. This allows for a much simpler design and operation of the bonding facility.

7.5. Main material limitations and solutions under evaluation

Although the specific degradation mechanisms are different, the limitations of the baseline materials selected for DEMO are related to the effects of 14 MeV neutron irradiation and the prolonged (cyclic) operation at high temperatures. Both limit the temperature window of operation and/or the lifetime of the components. For this reason, in parallel to the qualification of the "baseline" materials presented in previous chapters, an extensive R&D program is on-going for the development of advanced (or risk mitigation) materials (Pintsuk et al. 2022).

7.5.1. *Limits for structural material EUROFER97 and development of advanced/risk mitigation steels*

The choice of EUROFER97 for the first phase of DEMO is justified by the consensus that effects of irradiation up to 20 dpa can be simulated with sufficient accuracy in existing MTRs, because the level of the He production up to this fluence in a 14 MeV fusion spectrum (~few hundred appm) is deemed to not significantly affect material properties. For the second phase of DEMO (up to 50 dpa), the most serious limitation relates to the low temperature embrittlement caused both by dpa-induced damage and He transmutations, which would imply a DBTT in the operating temperature range. Based on available results, robust operation of the Blanket using EUROFER97 should only be guaranteed if the first wall structure is irradiated above 350°C, where effects of dpa-induced damage on DBTT start to disappear, but limited to a maximum He-concentration of 500 appm. This is a challenging target in case of water cooling (max. T = 315°C). On the other hand, EUROFER97 strength declines sharply above 550°C and the design creep lifetime also drops below 10^4 h for stress levels of ~100 MPa, similar to those expected in the FW. In case of He cooling, the thermodynamic efficiency of the power conversion cycle would be insufficient to reach the net electricity generation target for the DEMO reactor (300–500 MW), compromising the programmatic objectives.

Improved RAFM steels (labeled EUROFER-LT and EUROFER-HT, respectively, for low temperature and high temperature) are therefore being investigated to mitigate these risks. Two strategies are being followed in combination: (i) performing modified/additional thermal treatments or TMTs on standard EUROFER and (ii) modifying the chemical composition. More than 50 new alloys have been cast as of 2020, showing reduction of the DBTT w.r.t. standard EUROFER in the non-irradiated conditions (Figure 7.22). Preliminary screening irradiations confirmed the effectiveness of improved TMTs: in particular, the "lab-cast" EUROFER, normalized at 1,250°C/1 h and then rolled to a final rolling temperature of 850°C in six rolling steps with a reduction of 20–30% for each rolling pass, showed a shift of the transition temperature of only +50 K after irradiation at 300°C, 3 dpa, w.r.t an expected shift of +110 K on a standard EUROFER. An alternative approach is equally being followed by investigating the effects of double austenitization, quench and tempering treatments. Due to their ultra-fine grain size, the recrystallized materials are expected to have a higher radiation resistance, but their properties are lost in case of welding and cannot be recovered by PWHTs. For EUROFER-HT, the most promising route seems to be the modification of chemical composition: RAFM steels with W contents up to 3 wt% and with isotopically tailored B additions (avoiding additional He-transmutation/generation under neutron irradiation) showed a lifetime of 13,123 h in creep test at 650°C/100 MPa (Figure 7.23). Impact properties at low temperature are, however, degraded by the introduction of B and optimized HTs are still being investigated. Characterization of properties after irradiation is also needed for a final decision on the most promising alloys whose production could then be upscaled.

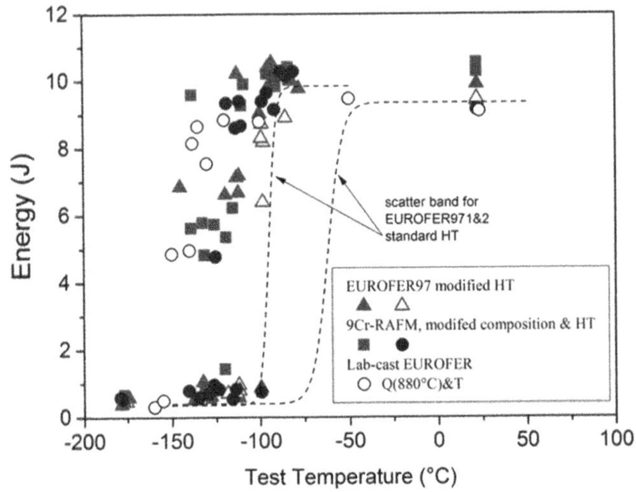

Figure 7.22. *Charpy impact test data (KLST) for optimized RAFM steels in comparison to EUROFER97 1 and 2 and standard heat treatment (HT): modified HT (triangles) – double austenized and tempered commercial EUROFER in L-T (full) and T-L (open) orientation; modified composition and HT – double austenized and tempered EUROFER with high Ta (square) and low Ta (circle) content and "lab-cast" EUROFER (open circles). For a color version of this figure, see www.iste.co.uk/balbaud/nuclearenergytoday.zip*

Additionally, in order to overcome the 50 dpa limit and open the design space for extrapolation to FPPs, oxide dispersion steels (ODS) are being investigated as advanced options, at least for the most irradiated part of the blanket (see Chapter 9). There are indeed indications that nano-structured steels break the 500 appm He limit as the dispersed oxides provide precipitate sites to trap the helium gas and He-gas bubbles, and thus prevent movement to grain boundaries and enhanced embrittlement (Yamamoto et al. 2007). Together with improved (creep) strength at high temperature, these steels could be used in areas of high temperature and high fluence. They are, however – and in particular their RAFM variants – only available at small, laboratory-scale quantities and their production involves complex TMT and powder metallurgy processes. Production of RAFM ODS grades with Cr contents ranging from 9 to 14%Cr and 0.3%Y_2O_3 has been achieved following the "conventional" mechanical alloying (MA) route in collaboration with industry. Strength and ductility values are in line with the expected performance and upscaling of the production will be pursued. An alternative production route that could avoid the time-consuming MA step (Surface Treatment of gas Atomized powder, followed by Reactive Synthesis [STARS], Figure 7.24) is being investigated (Pazos et al. 2018). By starting with atomized powder containing Y and Ti and introducing the required oxygen content in a subsequent oxidation step, an ultra-thin metastable Cr-rich oxide layer is formed on the surface of powder particles, and during subsequent HTs, this metastable oxide layer is dissociated. Accordingly, oxygen can react with Ti and Y dissolved in the ferritic matrix, forming the desired nano-oxides. The process can intrinsically scale from several to hundreds of kilograms and, consequently, production costs of ODS steels may be dramatically reduced. Development at lab scale has reached a level where powder batches whose chemical compositions are within specifications are routinely produced. The method is robust and leads to reproducible results, although improvement of the final material properties is still required.

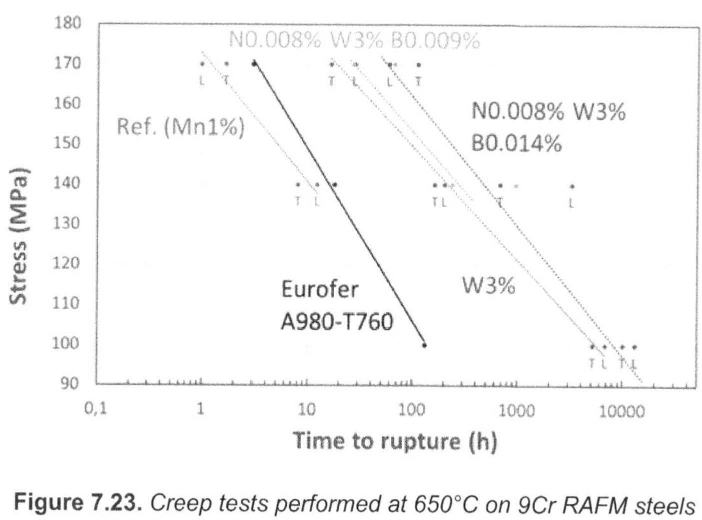

Figure 7.23. *Creep tests performed at 650°C on 9Cr RAFM steels with various W and B contents. For a color version of this figure, see www.iste.co.uk/balbaud/nuclearenergytoday.zip*

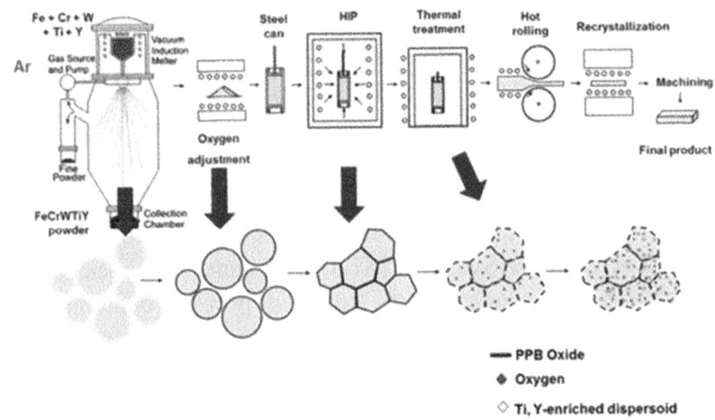

Figure 7.24. *Schematic representation of the STARS route. For a color version of this figure, see www.iste.co.uk/balbaud/nuclearenergytoday.zip*

7.5.2. *Advanced armor and heat sink materials for the Divertor targets*

The main drawbacks of W as armor material are related to the intrinsic low temperature brittleness, neutron induced embrittlement and recrystallization resistance. There are in fact indications that at the level of dpa expected in DEMO, the W monoblock would become fully brittle. Above 1,200°C, recrystallization and related material softening leads to increased susceptibility to thermal fatigue and limited lifetime. Finally, in accidental scenarios including water and/or air ingress inside the VV, radioactive tungsten oxides could form and be mobilized, which would represent a safety concern.

High ductility thin cold rolled foils of W presenting DBTT below room temperature due to nano-structuring have already been produced (Reiser et al. 2017), also in the form of laminate structures. The drawback is a low recrystallization temperature below 1,000°C due to the high stored energy. Potassium doping could be an effective solution as shown by very promising irradiation results (Terentyev et al. 2021), but its application in combination with (cold) rolling processes still requires further analyses. A second route is the development of tungsten-fiber reinforced tungsten composites with highly ductile and high temperature/ recrystallization resistant potassium-doped tungsten fibers and a pure tungsten dense or porous matrix (Neu et al. 2017). The fiber–matrix interaction mechanism introduces a pseudo-ductile behavior that avoids catastrophic brittle failure. Braided W yarns (Figure 7.25) and woven 2D composite with long fibers have already been produced. The upscaling of the production process and the stability under neutron irradiation are the main issues for these materials. Carbide and oxide particle reinforcement in conjunction with powder injection molding and sintering have also been investigated for increasing strength and recrystallization resistance but because of the decrease in ductility at low temperature, development of these materials has been discontinued. Finally, self-passivating tungsten alloys containing Cr as oxide forming element, as well as an element that stabilizes the oxidation process (e.g. Ti, Y or Zr), are being developed as FW coatings for the BB. Preliminary results show a decrease in the oxidation rates by up to a factor of 10^6 in a temperature range up to 1,000°C.

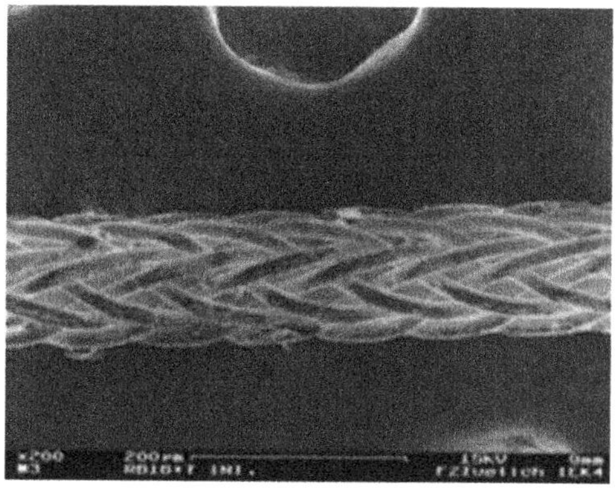

Figure 7.25. *W braided fiber yarn (16 + 7 filaments) (Pintsuk et al. 2022)*

As baseline heat sink structural material for DEMO Divertor, the limitations of CuCrZr is the very limited operational temperature range under neutron irradiation. As highlighted in previous chapters, at a temperature below 180–200°C, CuCrZr experiences hardening, loss of uniform elongation and embrittlement, while above 250–300°C, irradiation causes softening and loss of fracture toughness. While the loss of strength is acceptable at the level of irradiation expected in ITER, it would prevent operation above 250°C in DEMO, challenging the possibility of effective bonding with W. Advanced materials are therefore being investigated for improving the high temperature mechanical properties such as strength, creep and creep–fatigue to improve neutron resistance at low temperature down to 150°C corresponding to the expected coolant temperature for PCFs in the Divertor and increase the upper operational temperature limit to at least 450°C.

Tungsten reinforced Cu alloys are in particular being investigated as the most promising solution. The most developed and already industrially available materials are tungsten particle reinforced CuCr and CuCrZr materials, which show significantly improved mechanical performance (strength, fracture toughness) and stability in the non-irradiated state up to temperatures of 500–550°C. The second option is W-fiber reinforced Cu and CuCrZr composites (Muller et al. 2017). This material outperforms the particle reinforced alloys and has already been produced in the shape of plates and tubes. Preliminary results confirm that material retains superior characteristics, even after irradiation. However, further upscaling of the industrial production route is required.

In addition to these "conventional" approaches, additive manufacturing technologies using laser beam melting are also being investigated (Müller et al. 2020), exploring the modeling and CAD (computer-aided design) based development of complex tungsten matrix structures, which will subsequently be infiltrated by pure Cu or a Cu-based alloy substituting the heat sink material and providing a better adjusted thermal expansion coefficient to the plasma facing material, while still featuring high thermal conductivity (Figure 7.26).

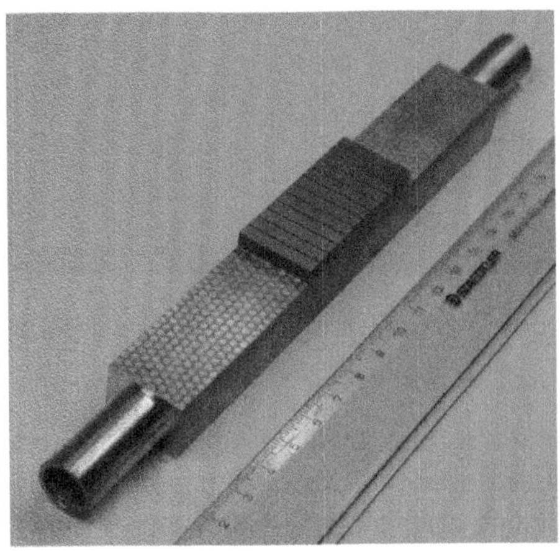

Figure 7.26. *Tungsten flat tile component with additively manufactured heat sink (Cu-infiltrated W-honeycomb structure) (Pintsuk et al. 2022)*

7.6. References

Aiello, G., Aktaa, J., Cismondi, F., Rampal, G., Salavy, J.-F., Tavassoli, F. (2011). Assessment of design limits and criteria requirements for Eurofer structures in TBM components. *J. Nucl. Mater.*, 414(1), 53–68.

Appello, M. and Fenici, P. (1988). Solution heat treatment of a Cu-Cr-Zr alloy. *Mater. Sci. Eng. A*, 102, 69–75.

Barabash, V., Peacock, A., Fabritsiev, S., Kalinin, G., Zinkle, S., Rowcliffe, A., Rensman, J.-W., Tavassoli, A.A., Marmy, P., Karditsas, P.J. et al. (2007). Materials challenges for ITER – Current status and future activities. *J. Nucl. Mater.*, 367–370, 21–32.

Carmona Gázquez, M., Hernández, M.T., Muktepavela, F., Platacis, E., Shishko, A. (2015). Magnetic field effect on the corrosion processes at the Eurofer–Pb–17Li flow interface. *J. Nucl. Mater.*, 465, 633–639.

EUROfusion (2018). European research roadmap to the realisation of fusion energy [Online]. Available at: https://www.euro-fusion.org/fileadmin/user_upload/EUROfusion/Documents/2018_Research_roadmap_long_version_01.pdf [Accessed 16 June 2022].

EUROfusion (2020a). Demo design criteria for in-vessel components – DDC-IC. EFDA_D_2P4N86.

EUROfusion (2020b). CuCrZr material properties handbook (MPH). EFDA_D_2NV3Q6.

EUROfusion (2020c). Tungsten material properties handbook (MPH). EFDA_D_2P3SPL.

EUROfusion (2022). EUROFER97 material properties handbook (MPH). EFDA_D_2NZHBS.

Federici, G., Biel, W., Gilbert, M.R., Kemp, R., Taylor, N., Wenninger R. (2017). European DEMO design strategy and consequences for materials. *Nucl. Fusion*, 57, 092002.

Federici, G., Bachmann, C., Barucca, L., Baylard, C., Biel, W., Boccaccini, L.V., Bustreo, C., Ciattaglia, S., Cismondi, F., Corato, V. et al. (2019a). Overview of the DEMO staged design approach in Europe. *Nucl. Fusion*, 59, 066013.

Federici, G., Boccaccini, L., Cismondi, F., Gasparotto, M., Poitevin, Y., Ricapito, I. (2019b). An overview of the EU Breeding Blanket design strategy as an integral part of the DEMO design effort. *Fusion Eng. Des.*, 141, 30–42.

Forest, L., Boccaccini, L.V., Cogneau, L., Li Puma, A., Neuberger, H., Pascal, S., Rey, J., Thomas, N., Tosi, J., Zmitko, M. (2020a). Test blanket modules (ITER) and Breeding Blanket (DEMO): History of major fabrication technologies development of HCLL and HCPB and status. *Fusion Eng. Des.*, 154, 111493.

Forest, L., Aktaa, J., Boccaccini, L.V., Emmerich, T., Ghidersa, B., Fondant, G., Froio, A., Li Puma, A., Namburi, H., Neuberger, H. et al. (2020b). Status of the EU DEMO Breeding Blanket manufacturing R&D activities. *Fusion Eng. Des.*, 152, 111420.

Gaganidze, E. and Aktaa, J. (2013). Assessment of neutron irradiation effects on RAFM steels. *Fus. Eng. Des.*, 88(3), 118–128.

Gaganidze, E., Chauhan, A., Schneider, H.-C., Terentyev, D., Rossaert, B., Aktaa, J. (2021). Effect of irradiation temperature on the fracture-mechanical behaviour of tungsten irradiated to 1 dpa. *J. Nucl. Mater.*, 556, 153200.

Gavila, P., Riccardi, B., Pintsuk, G., Ritz, G., Kuznetsov, V., Durocher, A. (2015). High heat flux testing of EU tungsten monoblock mock-ups for the ITER divertor. *Fusion Eng. Des.*, 98–99, 1305–1309.

Gilbert, M.R., Arakawa, K., Bergstrom, Z., Caturla, M.J., Dudarev, S.L., Gao, F., Goryaeva, A.M., Hu, S.Y., Hu, X., Kurtz, R.J. et al. (2021). Perspectives on multiscale modelling and experiments to accelerate materials development for fusion. *Journal of Nuclear Materials*, 554, 153113.

Giniyatulin, R.N., Komarov, V.L., Kuzmin, E.G., Makhankov, A.N., Mazul, I.V., Yablokov, N.A., Zhuk, A.N. (2002). Optimization of armour geometry and bonding techniques for tungsten-armoured high heat flux components. *Fusion Eng. Des.*, 61–62, 185–190.

Hirai, T., Barabash, V., Escourbiac, F., Durocher, A., Ferrand, L., Komarov, V., Merola, M. (2017). ITER divertor materials and manufacturing challenges. *Fusion Eng. Des.*, 125, 250–255.

Ibarra, A., Arbeiter, F., Bernardi, D., Krolas, W., Cappelli, M., Fischer, U., Heidinger, R., Martin-Fuertes, F., Micciché, G., Muñoz, A. et al. (2019). The European approach to the fusion-like neutron source: The IFMIF-DONES project. *Nucl. Fusion*, 59(6), 065002.

ITER (2004). Materials assessment report: 1.4. Selection of copper alloys. ITER G 74 MA 10 W0.

ITER (2013). ITER structural design criteria for in-vessel components (SDC-IC): Appendix A. Materials design limit data. G 74 MA 8 01-05-28 W 0.2.

Lebarbé, T., Pétesch, C., Muñoz-Garcia, J. (2020). Standardization of Eurofer material, a first step toward industrialization. *Fus. Eng. and Des.*, 159, 111793.

Li, M., Sokolov, M.A., Zinkle, S.J. (2009). Tensile and fracture toughness properties of neutron-irradiated CuCrZr. *J. Nucl. Mater.*, 393, 36–46.

Maday, M.-F. (2002). Comparison of the low cycle fatigue behaviour of F82H mod. and Eurofer 97 in water coolant. *Fusion Eng. Des.*, 61–62, 665–670.

Moreau, R., Bréchet, Y., Maniguet L. (2011). Eurofer corrosion by the flow of the eutectic alloy Pb–Li in the presence of a strong magnetic field. *Fusion Eng. Des.*, 86(1), 106–120.

Neu, R., Riesch, J., Müller, A.V., Balden, M., Coenen, J.W., Gietl, H., Höschen, T., Li, M., Wurster, S., You, J.-H. (1997). Tungsten fibre-reinforced composites for advanced plasma facing components. *Nuclear Materials and Energy*, 12, 1308–1313.

Neuberger, H., Hernandez, F., Ruck, S., Arbeiter, F., Bonk, S., Rieth, M., Stratil, L., Müller, O., Volker, K.-U. (2021). Advances in additive manufacturing of fusion materials. *Fusion Eng. Des.*, 167, 112309.

Özkan, F. and Aktaa, J. (2015). Creep fatigue assessment for EUROFER components. *Fusion Eng. Des.*, 100, 536–540.

Pazos, D., Cintins, A., De Castro, V., Fernández, P., Hoffmann, J., García Vargas, W., Leguey, T., Juris Purans, T., Anspoks, A., Kuzmin, A. et al. (2018). ODS ferritic steels obtained from gas atomized powders through the STARS processing route: Reactive synthesis as an alternative to mechanical alloying. *Nucl. Mater. Energy*, 17, 1–8.

Pintsuk, G., Aiello, G., Dudarev, S.-L., Gorley, M., Henry, J., Richou, M., Rieth, M., Terentyev, D., Vila, R. (2022). Materials for in-vessel components. *Fusion Eng. Des.*, 174, 112994.

Plansee (2022). Tungsten: Properties & uses [Online]. Available at: https://www.plansee.com/en/materials/tungsten.html.

Raquet, O. and Sèjourné, L. (1999). Comportement à la corrosion généralisée d'aciers ferritiques-martensitiques en milieu primaire REP. Report, CEA, Saclay.

Reiser, J., Garrison, L., Greuner, H., Hoffmann, J., Weingärtner, T., Jäntsch, U., Klimenkov, M., Franke, P., Bonk, S., Bonnekoh, C. et al. (2017). Ductilisation of tungsten (W): Tungsten laminated composites. *Int. J. Refract. Met. Hard Mater.*, 69, 66–109.

Richou, M., Tsitrone, E., Missirial, M., Magaud, P. (2020). Matériaux et composants face au plasma pour la fusion thermonucléaire. *Techniques de l'Ingénieur*, BN 3 761 v1.

Rieth, M., Dürrschnabel, M., Bonk, S., Jäntsch, U., Bergfeldt, T., Hoffmann, J., Antusch, S., Simondon, E., Klimenkov, M., Bonnekoh, C. et al. (2021). Technological processes for steel applications in nuclear fusion. *Appl. Sci.*, 11(24), 11653.

Saito, S., Fukaya, K., Ishiyama, S., Sato, K. (2002). Mechanical properties of HIP bonded W and Cu-alloys joint for plasma facing components. *J. Nucl. Mater.*, 307–311(Part 2), 1542–1546.

Sannier, J., Broc, M., Flament, T., Terlain, A. (1991). Corrosion of austenitic and martensitic stainless steels in flowing Pb17Li alloy. *Fus. Eng. Des.*, 14(3–4), 299–307.

Sawan, M.E. (2014). Transmutation of tungsten in fusion and fission nuclear environments. *Fusion Sci. Technol.*, 66(1), 272–277.

Schmunk, R.E. and Korth, G.E. (1981). Tensile and low-cycle fatigue measurements on cross-rolled tungsten. *J. Nucl. Mater.*, 103–104, 943–948.

Séran, J.-L. and Le Flem, M. (2017). Irradiation-resistant austenitic steels as core materials for Generation IV nuclear reactors. In *Structural Materials for Generation IV Nuclear Reactors*, Yvon, P. (ed.). Woodhead Publishing, Sawston, Cambridge.

Singh, B.N., Edwards, D.J., Tähtinen, S. (2004). Effect of heat treatments on precipitate microstructure and mechanical properties of CuCrZr alloy. Report, Risø National Laboratory 1436(EN), Roskilde.

Stork, D., Agostini, P., Boutard, J., Buckthorpe, D., Diegele, E., Dudarev, S.L., English, C., Federici, G., Gilbert, M.R., Gonzalez, S. et al. (2014a). Materials R&D for a timely DEMO: Key findings and recommendations of the EU Roadmap Materials Assessment Group. *Fusion Eng. Des.*, 89 (7–8), 1586-1594.

Stork, D., Agostini, P., Boutard, J.L., Buckthorpe, D., Diegele, E., Dudarev, S.L., English, C., Federici, G., Gilbert, M.R., Gonzalez, S. et al. (2014b). Developing structural, high-heat flux and plasma facing materials for a near-term DEMO fusion power plant: The EU assessment. *J. Nucl. Mater.*, 455 (1–3), 277–291.

Tavassoli, A.F. (2002). Present limits and improvements of structural materials for fusion reactors – A review. *J. Nucl. Mater.*, 302(2–3), 73–88.

Taylor, N.P. and Pampin R. (2006). Activation properties of tungsten as a first wall protection in fusion power plants. *Fusion Eng. Des.*, 81(8–14), 1333–1338.

Terentyev, D., Rieth, M., Pintsuk, G., Riesch, J., von Müller, A., Antusch, S., Mergia, K., Gaganidze, E., Schneider, H.-C., Wirtz, M. et al. (2021). Recent progress in the assessment of irradiation effects for in-vessel fusion materials: Tungsten and copper alloys. *Nucl. Fusion*, 62, 026045.

Terranova, N. and Di Pace, L. (2021). DEMO WCLL primary heat transfer system loops activated corrosion products assessment. *Fusion Eng. Des.*, 170, 112456.

Utili, M., Bassini, S., Cataldo, S., Di Fonzo, F., Kordac, M., Hernandez, T., Kunzova, K., Lorenz, J., Martelli, D., Padino, B. et al. (2021). Development of anti-permeation and corrosion barrier coatings for the WCLL Breeding Blanket of the European DEMO. *Fusion Eng. Des.*, 170, 112453.

Van der Schaaf, B., Gelles, D.S, Jitsukawa, S., Kimura, A., Klueh, R.L., Möslang, A., Odette, G.R. (2000). Progress and critical issues of reduced activation ferritic/martensitic steel development. *J. Nucl. Mater.*, 283–287(Part 1), 52–59.

Van der Schaaf, B., Tavassoli, F., Fazio, C., Rigal, E., Diegele, E., Lindau, R., LeMarois, G. (2003). The development of EUROFER reduced activation steel. *Fus. Eng. Des.*, 69(1–4), 197–203.

Van Dyck, S. and Bosch, R.-W. (2005). Environmentally assisted cracking of Eurofer 97 in water. *Fusion Eng. Des.*, 75–79, 973–977.

Visca E., Testani, C., Libera, S., Sacchetti, M. (2003). Manufacturing of small scale W monoblock mockups by hot radial pressing. *Fusion Eng. and Des.*, 66–68, 295–299.

Wident, P. (2018). Pre-irradiation material testing: Tensile and fracture-toughness properties. Report, EUROfusion.

Wirtz, M., Linke, J., Loewenhoff, Th., Pintsuk, G., Uytdenhouwen, I. (2016). Thermal shock tests to qualify different tungsten grades as plasma facing material, *Phys. Scr.*, T167, 014015.

Yamamoto, T., Odette, G.R., Miao, P., Hoelzer, D.T., Bentley, J., Hashimoto, N., Tanigawa, H., Kurtz, R.J. (2007). The transport and fate of helium in nanostructured ferritic alloys at fusion relevant He/dpa ratios and dpa rates. *J. Nucl. Mater.*, 367–370(A), 399–410.

You, J.-H., Li, M., Zhang K. (2021). Structural lifetime assessment for the DEMO divertor targets: Design-by-analysis approach and outstanding issues. *Fusion Eng. Des.*, 164, 112203.

Zhang, X.W., Zhou, X., Wang, Q.J., Liang, B. (2015). Tensile behavior of high temperature Cu-Cr-Zr alloy. *Paper presented at the International Conference on Power Electronics and Energy Engineering*, Hong Kong.

Zmitko, M., Galabert, J., Thomas, N., Forest, L., Bucci, P., Cogneau, L., Rey, J., Neuberger, H., Poitevin, Y. (2015). The European ITER test blanket modules: Current status of fabrication technologies development and a way forward. *Fusion Eng. Des.*, 96–97, 199–207.

Zmitko, M., Carin, Y., Thomas, N., Simon-Perret, M., LiPuma, A., Forest, L., Tosi, J., Aiello, G., Cogneau, L., Rey, J. et al. (2017). The European ITER test blanket modules: EUROFER97 material and TBM's fabrication technologies development and qualification. *Fus. Eng. Des.*, 124, 767–773.

8

Advanced Materials: Ceramics and Composite Materials

James BRAUN
CEA/DAM, Le Ripault, Monts, France

8.1. Introduction

Ceramics have aroused interest for nuclear applications because of their high melting temperature as well as their relative chemical inertness in harsh environments. They are already the main constituent of fuel, for example, as UO_2 or mixed oxide (MOX). Boron carbide (B_4C) is also used as a neutron absorber material in some nuclear reactors. In this chapter, a focus will be made on the use of ceramics as structural materials in current (but especially in future) nuclear concepts, where they are essential for greater safety and performance.

Among all the parameters, the damage tolerance to neutron irradiation is one of the key properties of fuel cladding or structural materials. To achieve good radiation tolerance, high density of point defect recombination centers and low point defect mobility are needed (Zinkle and Snead 2014). Some ceramics (for example, BeO or AlN) cannot be employed because they present high volumetric swelling below 600°C. Also, noncubic phase ceramics can show grain boundary cracking and strong strength decrease, such as Al_2O_3 (Keilholtz et al. 1973) or Ti_3AlC_2 and Ti_2AlC MAX phases (Clark et al. 2016). For moderate to high temperature applications (500–1,000°C), few candidate materials remain. We can mention cubic materials (silicon carbide [SiC] and MgO), hexagonal close-packed Si_3N_4 and Ti_3SiC_2 MAX phase (Zinkle 2017).

Materials and Processes for Nuclear Energy Today and in the Future,
coordinated by Fanny BALBAUD-CÉLÉRIER and Céline CABET.

SiC is of prime interest because it is one of the most radiation-tolerant materials. It has already been widely studied as it composes coating in TRistructural ISOtropic (TRISO) fuel particles, as shown in Figure 8.1 (Powers and Wirth 2010). SiC has hundreds of crystalline structures (polytypes). However, the hexagonal (α-SiC) and especially the cubic (β-SiC) are the most interesting for their limited and dose-independent swelling within the 150°C (amorphization)–1,000°C (clustering, Frank loops and void swelling) temperature range under neutron irradiation (Snead et al. 2007a). High purity materials are required because impurities, such as boron or alumina, used in the manufacturing process, decrease the irradiation tolerance.

The ceramics main drawback is their brittle behavior (low toughness), which obeys statistical mechanical failure on the contrary of steels (deterministic). Therefore, their mechanical behavior highly depends on the size of the defects they enclose and strongly decreases with the component size, precluding their use for large scale cladding and structural materials. To solve this issue and give the material sufficient damage tolerance, ceramic matrix composites (CMCs) have been developed for nuclear and aerospace applications. CMCs have two main constituents, the fibers and a matrix. A third constituent (interphase) can also be used between fibers and matrix (see Figure 8.1). Fibers in CMCs are thereby designed to accommodate matrix cracking through interface debonding and crack deflection; they enable advantageous properties of CMCs, including toughness and non-catastrophic fracture behavior. As far as fibers are concerned, two types of materials can be used for nuclear applications: carbon and SiC. C fibers are much cheaper than SiC fibers but, due to their anisotropy, present an important anisotropic swelling under irradiation.

The first SiC fibers were developed during the 1970s but their manufacturing process leads to high oxygen and carbon contents (11.7% O, 1.31 C/Si ratio). When exposed to low-dose neutron irradiations (Okamura et al. 1986), these fibers decomposed into SiC and carbon oxides, inducing fibers shrinkage and thus fibers/matrix debonding in composites (Snead et al. 1992). However, despite an initial drop in the mechanical properties, no change up to 25 dpa was observed, arousing interest for the use of these fibers. Modifications in the manufacturing process have led to a strong oxygen content decrease (0.5%), with same levels of C/Si ratios. Nonetheless, these fibers, called second-generation SiC fibers, also exhibited issues under irradiation such as shrinkage and loss of mechanical properties, mainly due to the excess in carbon, present as amorphous phases (Snead et al. 1998b). The development and availability on the market of nearly stoichiometric fibers, referred to as third-generation SiC fibers at the end of the 1990s, greatly increased the interest for nuclear applications, but also in the space and aeronautical industries. Their irradiation resistance was quickly evaluated and showed promising results up to 8 dpa in the 300–800°C temperature range (Snead et al. 2000; Katoh et al. 2003), which seemed to make them suitable for both current and future nuclear systems.

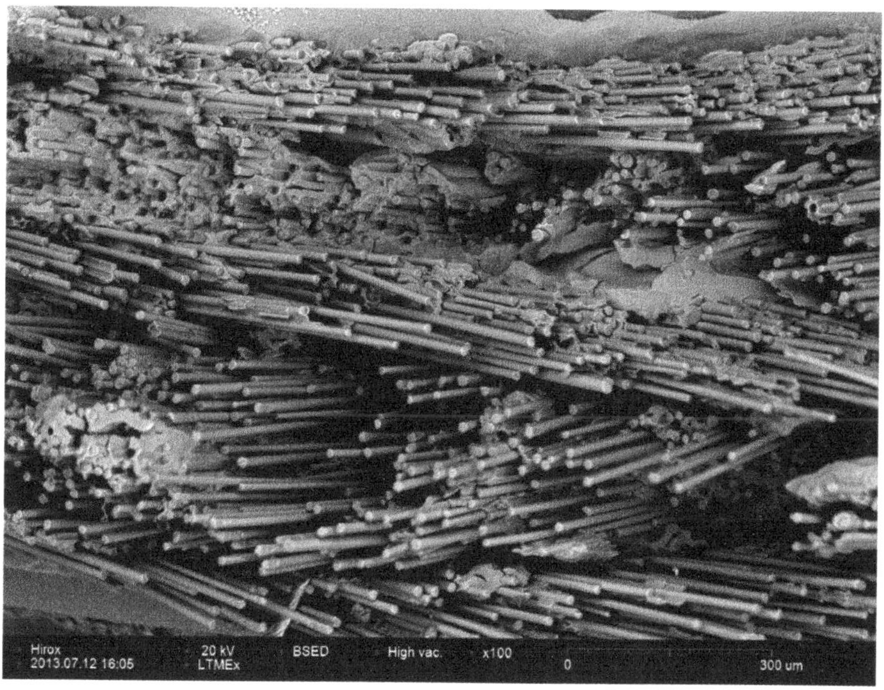

Figure 8.1. *Micrograph of a tensile tested fiber-reinforced composite*

8.2. Current interest in nuclear systems

Originally, SiC was considered as part of the TRISO fuel particles as an evolution of the BISO (bi-isotropic) fuel particles, which were limited to low temperature and burnup. The TRISO fuel (see Figure 8.2) is made of a fissile material kernel, which can be made up of uranium, plutonium, thorium or other transuranic elements, oxides, carbides, nitrides or oxycarbides. It is coated with a porous carbon buffer layer and successive thin layers of pyrocarbon (PyC), SiC and again PyC. The role of the SiC is to act as a pressure vessel and to guarantee the leak-tightness to fission products, especially gaseous and metallic ones (Powers and Wirth 2010). This fuel is the reference for high temperature gas-cooled reactors (HTGR) and very high temperature reactors (VHTR), such as the former experimental and prototype reactors: Peach Bottom and Fort St. Vrain (USA),

Dragon (UK), AVR and THTR-300 (Germany), current HTR-10 and HTR-PM (China) and HTTR (Japan); and some under-design concepts such as GTHTR300, HTMR100 or EM^2. Studies have also shown that adding a layer of ZrC (Bullock and Kaae 1983) or using ZrC instead of SiC (Minato et al. 1995) decreases the free oxygen generated during fission and could decrease particles failure (Porter et al. 2013).

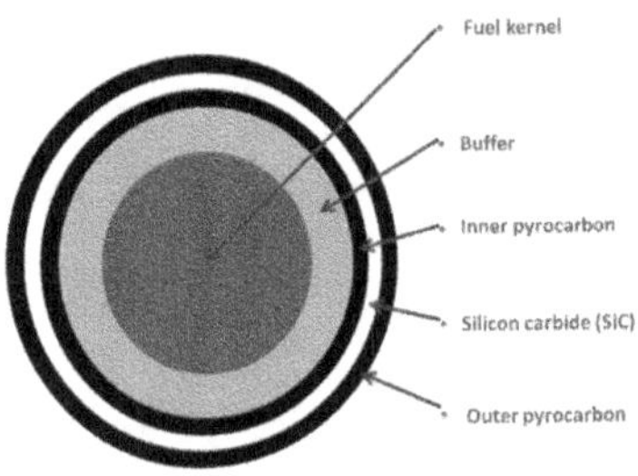

Figure 8.2. *Artistic view of a TRISO particle. For a color version of this figure, see www.iste.co.uk/balbaud/nuclearenergytoday.zip*

CMCs were primarily considered for fusion applications as core materials, such as blankets, will need to withstand harsh conditions that conventional metallic alloys could not bear long. Then, they have also been considered for fission applications in relation to their high thermomechanical strength at high temperature, high thermal conductivity, low swelling under irradiation and high melting point. Following the Generation IV International Forum, CMCs have been of prime interest for different applications because the technical feasibility of some reactor concepts is directly related to the development of high properties CMCs.

In VHTR, C/C and SiC/SiC composites were considered to manufacture the control rods sheaths (containing the boron carbide neutron absorber) and joints (Eto et al. 2011; Snead et al. 2011). However, C/C composites exhibit too strong anisotropic dimensional changes under irradiation. Therefore, they are limited to low dose applications (<1 dpa) (Bonal et al. 2009). Control rods will have to bear strong axial gradients (up to 400°C and 1,100°C in the upper and lower parts, respectively). Therefore, a 10-m long control rod will be made of several sections of ≈ 50 cm, connected by flexible joints, to guarantee its good insertion in all conditions.

In SFR, the use of hybrid SiC/SiC–metal hexagonal tubes fuel-assembly could solve the drawbacks of austenitic or ferrite-martensitic steels and improve the safety of the reactor at high temperature (Ravenet 2012). No decrease in the mechanical properties of SiC/SiC composites after immersion in liquid sodium, either pure or with high oxygen content, was observed at 550°C (Braun et al. 2021).

However, the most promising use of SiC/SiC CMC is for cladding applications in gas-cooled fast reactors (GFR) or LWR. GFR reactors have to operate at very high nominal operating temperatures (coolant temperature: 650–950°C) for several years and the cladding material has to withstand more than 100 dpa (fast spectrum) in its lifetime. SiC/SiC is currently considered as the reference material. Different concepts of core architecture were envisaged: the plate concept (Ravenet 2007) and the pin concept (similar to LWR fuel pin) (Lee and Naganuma 2009). Nevertheless, pins can be more easily manufactured and were then selected as the reference concept. Given the SiC/SiC composites thermal conductivity under irradiation (see section 8.4) and the reactor power density (100 MW/m^3), the pins' thickness has to be below 1 mm (Lee and Naganuma 2009). The current concept involves a thin layer of metal, comprised between two layers of SiC/SiC, insuring mechanical strength and leak-lightness to gases and fission products (Zabiego et al. 2013). Recently, the EM^2 system was conceptualized to operate for 30 years without refueling, shuffling or using burnable poisons (Choi and Schleicher 2017). That would only be possible using SiC/SiC composites as fuel cladding elements.

The research of new candidate materials for LWR fuel cladding has been driven since the Fukushima Daiichi accident by the objective to avoid the fast oxidation of the zirconium cladding under steam in LOCA (loss-of-coolant accident) situations. Among the candidate materials, zirconium-coated and FeCrAl claddings are the most advanced short term solutions to replace the conventional cladding (Terrani 2018). However, due to their thermomechanical properties and outstanding oxidation resistance (Opila and Hann 1997), SiC/SiC materials is the reference on the long term. The replacement of the boiling water reactors (BWRs) zirconium cladding by SiC/SiC materials is also considered and might be the first step to exploit SiC/SiC in a commercial LWR (Yueh and Terrani 2014). Although high temperature tests show a strong improvement in comparison with zirconium cladding (Steinbrueck et al. 2022) in accident situations, the hydrothermal corrosion of SiC (Terrani et al. 2015), especially under irradiation (Kondo et al. 2015), and the leak-tightness issue due to the cladding microcracking still have to be solved to duly operate in nominal conditions.

As far as the fuel is concerned, the fully ceramic microencapsulted (FCM), consisting of TRISO particles embedded in a SiC matrix, aroused interest because of its irradiation stability, incorporation of an additional barrier to fission product release and its environments stability (Terrani et al. 2012). This fuel could replace the conventional UO_2 pellets in zirconium claddings for PWR, avoiding the detrimental oxidation in a LOCA situation and could also increase the safety in HTGR in case of an air ingress accident (Terrani and Silva 2015). Moreover, it is nowadays considered as the fuel of the Micro Modular Reactor from USNC (USA), where additive manufacturing will be used to manufacture the SiC matrix.

8.3. Manufacturing and joining

As an improved version of second generation fibers with stoichiometric composition, only third-generation fibers are suitable for nuclear applications: Hi-Nicalon Type S (HNS) and Tyranno SA3 (TSA3) manufactured by NGS Advanced Fibers Co. and Ube industries in Japan, respectively, which presents a better thermal stability and thermal conductivity than the HNS (Sha et al. 2004). However, the composites manufactured with TSA3 fibers exhibit limited ductility in comparison with HNS-based composites, due to strong fibers/matrix bonding, limiting the application of the TSA3 fibers (Fellah et al. 2020). Recently, Ube Industries issued an upgraded fiber, the TSA4 fiber, that produces composites with similar pseudo-ductility to HNS-based composites (Braun and Sauder 2022). The Sylramic fiber is also considered as stoichiometric but its manufacturing process, sintering with additives like boron, prevents its use as core structural materials due to activation of ^{10}B (Newsome et al. 2007).

To produce cylindrical shapes, the fibers, gathered in tows of about 1 mm in diameter, are deposited on cylindrical mandrels. Two techniques can be employed for this purpose: filament winding and braiding. Filament winding leads to better circular shapes than braiding. This is the reason why it is often used for producing the inner layer. 2D braiding also leads to low porosity materials. 3D braiding is not considered, even if the composites have improved mechanical properties because the resulting porosity is too important for applications that require high thermal conductivity.

To confer the expected mechanical properties, an interphase has to be deposited between the fibers and the matrix, even though it was observed that it is not necessary to obtain pseudo-ductility (Fellah et al. 2020; Braun and Sauder 2022). Pyrolytic carbon (PyC) and boron nitride (BN) present lamellar structures that deviate microcracks generated in the matrix and provide debonding at the fiber/matrix interface. However, BN cannot be employed because ^{10}B would activate under neutron irradiation. Although PyC is subjected to anisotropic swelling under irradiation, no degradation of the mechanical behavior was observed for a PyC thickness below 150 nm (Newsome et al. 2007; Katoh et al. 2011), which is enough to guarantee a good fibers/matrix coupling (Braun et al. 2019). Multilayered interphases $(PyC/SiC)_n$ are also expected to exhibit a better tolerance to oxidation (Naslain et al. 2010).

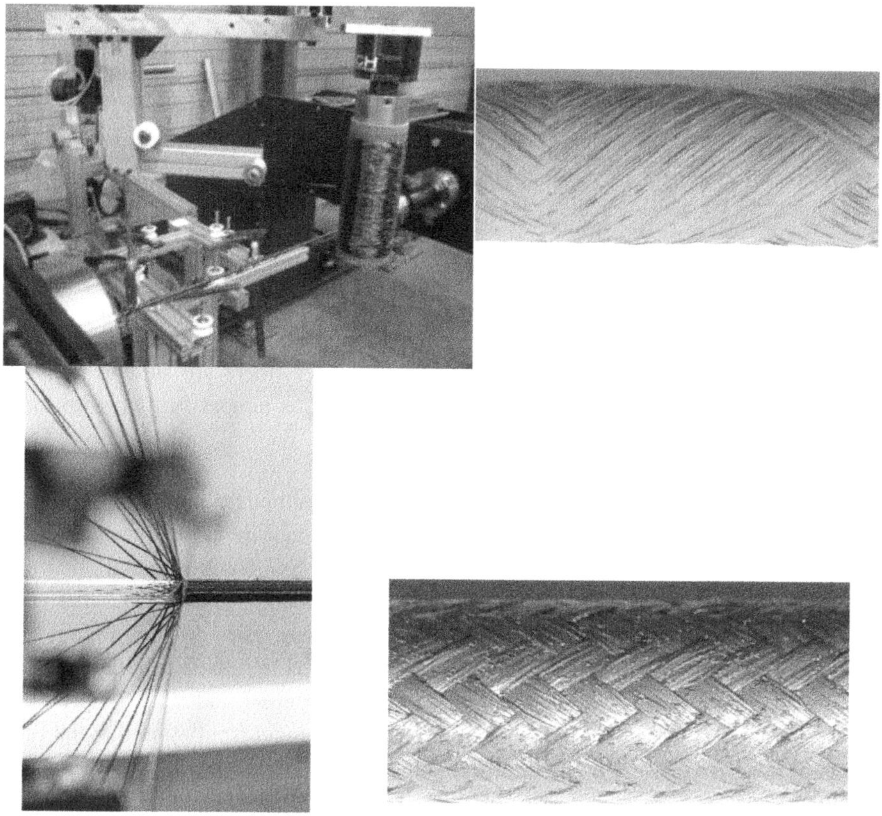

Figure 8.3. *Examples of a winding machine and a winded tube (top), and of braiding and a 2D braided tube (bottom) – credit: P.Stroppa/CEA. For a color version of this figure, see www.iste.co.uk/balbaud/nuclearenergytoday.zip*

Once the preform has been shaped, several techniques can be employed to produce the SiC matrix. The chemical vapor infiltration (CVI) is the current reference and is also the one that has to be used to deposit a thin layer of PyC. It consists in the reaction of chrolosilanes, essentially methtyltrichlorosilane ($SiCH_3Cl_3$), heated in the presence of H_2, to form SiC. As the CVI process occurs at relatively low temperatures ($T < 1{,}200°C$), the allotropic form of SiC deposited is β, isotropic, and the fibers show no thermal deterioration (Sha et al. 2007; Sha et al. 2009). The produced SiC is of high purity (Sauder 2014). However, because it occurs in the gaseous phase, the interstices between the fibers and the tows are gradually filled and it is impossible to reach a fully dense material, which decreases the final thermal conductivity, compared to monolithic β-SiC. The NITE (nano-infiltration transient eutectic phase) process is based on liquid-phase sintering (Park et al. 2007). The common SiC sintering cannot be used to manufacture composites because it occurs at too high temperatures (Stobierski and Gubernat 2003), leading to α allotropic forms, and strongly damaging the fibers (Sha et al. 2009). The solution is to make a slurry out of β-SiC nanopowders, decreasing the sintering temperature (Kohyama et al. 2002), and adding sintering additives (Al_2O_3, Y_2O_3 and SiO_2) to infiltrate the SiC texture. The pyrolysis occurs at high temperatures (about 1,800°C), limiting the process applicability to TSA3 fibers, and low compression stresses (a few dozens of MPa maximum) but leading to quasi-fully dense materials. The drawback is that the matrix contains a huge amount of oxides, which raises questions about their swelling under irradiation at high doses (Koyanagi et al. 2011). Moreover, because only TSA3 fibers are suitable for this process, the produced TAS3-reinforced composites are almost brittle, with very limited ductility (Shimoda and Hinoki 2021).

In the case of cladding materials, to guarantee the hermeticity to fission products, several solutions have been proposed. The “triplex” concept is made up of three layers: one layer of monolithic β-SiC, one layer of SiC/SiC composite and one layer of an external Chemical Vapour Deposited (CVD) coating (Feinroth 2009). The drawback of this concept is that the whole gas-tightness relies on the internal SiC tube (Ben-Belgacem et al. 2014). It is quite complicated to make such materials on several meters, so shorter sections have to be brazed together, inducing the presence of a brazing agent that could have different properties under irradiation that the SiC. Also, for low temperature applications (LWR), a strong decrease in the SiC thermal conductivity is observed under irradiation, leading to thermal gradients and, as a consequence, strong stress gradients (Cozzo and Rahman 2018). To solve the leak-tightness and the hydrothermal corrosion (see section 8.4) issues, the deposition of a thin layer of Cr on the composite surface was proposed (Ang et al. 2018). For a GFR application, CEA patented the “sandwich” concept, in which a thin layer of metal (≈100 μm), called liner, is introduced between two layers of SiC/SiC composites (Zabiego et al. 2013). Tantalum and niobium have been considered as candidate liner for GFR because of their high melting temperature, adequate ductility and limited reactivity with SiC (Braun et al. 2022). Helium permeation tests highlighted that up to the tube ultimate failure, the leak-tightness is guaranteed, while without liner, leaks are indeed detected once the first crack occurs in the matrix (Sauder 2014).

Joining of CMC is a critical issue, especially to close the cladding with end plugs (Terrani 2018). The junction has to be as radiation tolerant as the materials that have been brazed together. It can be conducted by direct deposition of SiC by CVI but the whole material has to bear the process atmosphere (especially HCl and H_2) and the premature closure of the interstices. The most common technique is brazing, but the low melting point of brazing agents in comparison to SiC, does not ensure good properties at high temperature. Diffusion bonding, that is, the chemical reaction of the ceramic with another ceramic or metal, conducted with Ti-based materials seems to be tolerant to irradiation (Katoh et al. 2014). More recently, spark plasma sintering has aroused interest for its rapid heating and short dwell times; however, it is more adapted to simple joints. Low-pressure LPS (liquid phase sintering) seems most suitable to manufacture complex shapes. These materials have retained their strength, even after exposure to 20 dpa at 530°C (Koyanagi et al. 2020).

Recently, additive manufacturing has raised interest to manufacture specific pieces, and even silicon carbide fibers (Polozov et al. 2020). The feedstock materials can be pre-ceramics polymer based (usually polycarbosilanes), powder based or reactive chemical based. All these methods show advantages and drawbacks and must be adapted to the application (Koyanagi et al. 2021). Most processes are subjected to strong volume shrinkage, generate impurities (SiO_2, C or Si) and/or require very high temperatures (up to 1,900°C) (Tu and Jiang 2018), which can damage the SiC fibers in the case of CMC, or adjacent structures. The most promising methods currently considered for nuclear applications are the binder jet-CVI hybrid SiC 3D printing (Betzler et al. 2020), the laser chemical vapor deposition (Seshadri et al. 2021) and the selective laser sintering of SiC (Karakoc et al. 2021). The produced materials are expected to be qualified within the next years to guarantee their use in the nuclear industry.

8.4. Material properties

SiC is the only stable compound in the Si–C binary system and can only be stoichiometric. It does not become liquid but decomposes at around 2,500°C (Olesinski and Abbaschian 1984). The thermal conductivity of SiC is high (≈100 W/m/K at 1,000°C), even though it decreases constantly over –70°C, and is a function of the grain size, porosities and impurities content (Snead et al. 2007b). It keeps high mechanical properties and very limited creep appears below 1,200°C (Carter et al. 1984).

One of the most interesting properties of SiC, justifying its use as core material, is its tolerance to neutron irradiation. Indeed, stoichiometric materials, that is, single crystals or CVD SiC, exhibit strong radiation tolerance. However, the presence of impurities when employing more cost-saving manufacturing processes (sintering, liquid phase sintering and Precursor infiltration and Pyrolysis (PIP) techniques) alters its radiation stability (Hollenberg et al. 1995). Amorphization is observed below 423 K (Snead et al. 1998a). Between 423 and 1,273 K, swelling increases logarithmically with dose and reaches saturation at a few dpa. Moreover, the level of swelling saturation decreases with temperature (Snead et al. 2007a). Such a behavior was attributed to enhanced recombination of Frenkel pair defects produced in cascades due to lower interstitial clustering at higher temperatures. At the highest temperatures, a non-saturable void swelling regime occurs (Snead et al. 2007a).

The oxidation behavior of SiC is decomposed into two regimes: passive and active oxidation. The transition between these two regimes is broad and is a function of temperature and oxygen potential (Harder et al. 2013; Jacobson et al. 2013). The passive oxidation occurs in high oxygen potential environments and consists of the formation of a protective silica (SiO_2) layer on the surface of SiC, for instance, following reaction [8.1]. The oxygen diffusion rate in SiO_2 is very low, limiting the growth of this layer and hence ensuring a protective layer. At very high temperatures or very low oxygen partial pressures, the active oxidation regime is favored instead: SiC oxidizes in SiO and CO (see reaction [8.2]). The reactants are both gaseous species, which hinders the presence of a protective layer and consumes the material.

$$SiC(s) + 2O_{2(g)} \rightarrow SiO_{2(s)} + CO_{2(g)} \quad [8.1]$$

$$SiC(s) + O_{2(g)} \rightarrow SiO_{(g)} + CO_{(g)} \quad [8.2]$$

The hydrothermal corrosion of SiC is a major concern for LWR applications. At high temperature, SiC rapidly oxidizes in water to form a SiO_2 layer following reactions, which depend on the potential and pH of the environment (see reactions [8.3] to [8.6]). Unfortunately, the solubility limit of silica is high in LWR water (Fournier and Rowe 1977) and SiO_2 dissolves as silicic acid following reaction [8.7] (Icenhower and Dove 2000). The water chemistry, that is, oxygen and hydrogen contents (Kim et al. 2015; Terrani et al. 2015), has a strong influence on the hydrothermal corrosion rates. In PWR nominal conditions, a recession rate of a few μm per year of SiC is observed. While it is not damageable for the cladding material itself, the deposition of corrosion products on cold parts can be an issue. Indeed, mobile elements of pumps or valve seats are tightly adjusted and changes of dimension can lead to seizing. In addition, radiolysis induced by irradiation increases the oxygen potential, which leads to an order of magnitude higher corrosion rates (Spinks and Woods 1990). Increasing the hydrogen content (by lowering the oxygen potential) may therefore lower the recession rate by an order of magnitude. To avoid the SiC hydrothermal corrosion, metallic (Cr) or ceramic

(CrN, TiN) coatings have been studied and exhibit promising results (Ishibashi et al. 2017; Ang et al. 2018).

$$SiC_{(s)} + 2H_2O_{(aq)} \rightarrow SiO_{2(s)} + CH_{4(g)} \quad [8.3]$$

$$SiC_{(s)} + 2H_2O_{(aq)} \rightarrow SiO_{2(s)} + 2H_{2(g)} + C_{(s)} \quad [8.4]$$

$$SiC_{(s)} + 3H_2O_{(aq)} \rightarrow SiO_{2(s)} + 3H_{2(g)} + CO_{(g)} \quad [8.5]$$

$$SiC_{(s)} + 4H_2O_{(aq)} \rightarrow SiO_{2(s)} + 4H_{2(g)} + CO_{2(g)} \quad [8.6]$$

$$SiO_{2(s)} + 2H_2O_{(aq)} \rightleftharpoons Si(OH)_{4(aq)} \quad [8.7]$$

The SiC thermal properties strongly rely on its microstructure and decreases with decreasing grain size and in the presence of impurities (grain boundaries). Within SiC/SiC composites and especially CVI-based ones, the numerous interfaces between the fibers and the matrix, and the presence of porosities also decrease the thermal conductivity. Moreover, when loaded over their elastic domain, the matrix multicracking generates additional interfaces, reducing again the thermal conductivity.

8.5. Main challenges and solutions under evaluation

Several issues are still at stake before massively deploying SiC/SiC composites in commercial reactors. Up to now, only two fibers, the TSA3 and HNS, were considered suitable for nuclear applications. On the one hand, the low ductility of TSA3-based composites has limited their further study for nuclear applications, even though their properties (thermal conductivity, high temperature mechanical behavior) are interesting. On the other hand, HNS-based composites meet many requirements for nuclear applications. However, recent results have shown they suffer from different swelling between the fibers and the matrix, and maintain very low thermal conductivity (<1 W/m/K) when irradiated at moderate to high dose (40–100 dpa) – especially at low temperature (300°C), relevant of LWR. Therefore, there is a necessity to develop new fibers that could meet the different criteria. The recent TSA4 fibers have shown strongly improved mechanical properties of both the fibers and the composites (Braun and Sauder 2022). The other properties still need to be characterized to see if this improvement was not conducted at the cost of the TSA3 fibers' interesting properties. New fibers are also developed using CVD processes, such as the ultra SCS fiber. As these fibers contain a glassy carbon core, they do not support irradiation, even at low doses. More recently, "Free Form Fibers" company has developed a CVD fiber without a core that may be promising for under-radiation applications. As for the TSA4 fibers, these fibers and the fiber-reinforced composites have to be evaluated in real conditions.

Concerns have also been raised about the PyC interphase tolerance to neutron irradiation, especially at moderate to high doses (40–100 dpa) at "low" temperatures of 300°C (Koyanagi et al. 2018; Nozawa et al. 2019). Interdiffusion of silicon and degradation of the PyC under irradiation leads to strong bonding at the fibers/matrix interface, promoting brittle behavior. PyC is also sensitive to oxidation at high temperature, especially if microcracking has already occurred. New interphase materials are currently investigated, such as MAX phases (Ti_3SiC_2) or oxides (Al_2O_3, ZrO_2, Y_2O_3), but their irradiation behavior still needs to be investigated.

Indeed, all those issues show that there is a lack of codes and standards for the qualification of SiC-based materials, especially regarding the influence of the irradiation on a specific material. It is then necessary to develop reliable tools to evaluate the effect of irradiation on new materials because of the difficulty to obtain experimental data. The development of standards can also improve the efficiency of the research with standard tests, which will help in certifying the materials for licensing and safety regulation.

While the benefits of SiC/SiC composites in accidental conditions of various reactors raise no doubt, their behavior in nominal operating conditions needs to be solved. The major issue is to solve the hydrothermal corrosion of SiC in LWR conditions. For this purpose, modifications of the CVI parameters may improve the SiC lifetime but it seems that the use of coatings, such as pure Cr or nitrides (CrN, TiN), would be necessary.

8.6. References

Ang, C., Kemery, C., Katoh, Y. (2018). Electroplating chromium on CVD SiC and SiCf-SiC advanced cladding via PyC compatibility coating. *J. Nucl. Mater.*, 503, 245–249.

Ben-Belgacem, M., Richet, V., Terrani, K.A., Katoh, Y., Snead, L.L. (2014). Thermo-mechanical analysis of LWR SiC/SiC composite cladding. *J. Nucl. Mater.*, 447, 125–142.

Betzler, B.R., Ade, B.J., Wysocki, A.J., Jain, P.K., Chesser, P.C., Greenwood, M.S., Terrani, K.A. (2020). Transformational challenge reactor preconceptual core design studies. *Nucl. Eng. Des.*, 367, 110781.

Bonal, J.-P., Kohyama, A., van der Laan, J., Snead, L.L. (2009). Graphite, ceramics, and ceramic composites for high-temperature nuclear power systems. *MRS Bull.*, 34, 28–34.

Braun, J. and Sauder, C. (2022). Mechanical behavior of SiC/SiC composites reinforced with new Tyranno SA4 fibers: Effect of interphase thickness and comparison with Tyranno SA3 and Hi-Nicalon S reinforced composites. *J. Nucl. Mater.*, 558, 153367.

Braun, J., Sauder, C., Lamon, J., Balbaud-Célérier, F. (2019). Influence of an original manufacturing process on the properties and microstructure of SiC/SiC tubular composites. *Compos. Part A Appl. Sci. Manuf.*, 123, 170–179.

Braun, J., Sauder, C., Rouillard, F., Balbaud-Célérier, F. (2021). Mechanical behavior of SiC/SiC composites after exposure in high temperature liquid sodium for sodium fast reactors applications. *J. Nucl. Mater.*, 546, 152743.

Braun, J., Sauder, C., Guéneau, C., Hodaj, F., Balbaud-Célérier, F. (2022). High-temperature interdiffusion of tantalum and niobium with SiC for processing hybrid metal/CMC components. *Coatings*, 12, 887.

Bullock, R.E. and Kaae, J.L. (1983). Performance of coated UO_2 particles gettered with ZrC. *J. Nucl. Mater.*, 115, 69–83.

Carter, Jr., C.H., Davis, R.F., Bentley, J. (1984). Kinetics and mechanisms of high-temperature creep in silicon carbide: II, Chemically vapor deposited. *J. Am. Ceram. Soc.*, 67, 732–740.

Choi, H. and Schleicher, R.W. (2017). The energy multiplier module (EM 2): Status of conceptual design. *Nucl. Technol.*, 200, 106–124.

Clark, D.W., Zinkle, S.J., Patel, M.K., Parish, C.M. (2016). High temperature ion irradiation effects in MAX phase ceramics. *Acta Mater.*, 105, 130–146.

Cozzo, C. and Rahman, S. (2018). SiC cladding thermal conductivity requirements for normal operation and LOCA conditions. *Prog. Nucl. Energy*, 106, 278–283.

Eto, M., Konishi, T., Shibata, T., Sumita, J. (2011). Research and developments on application of carbon-carbon composite to HTGR/VHTR. In *Japan. IOP Conf. Ser.: Mater. Sci. Eng.*, 18, 162003.

Feinroth, H. (2009). Multi-layered ceramic tube for fuel containment barrier and other applications in nuclear and fossil power plants. Patent, US2009/0032178A1.

Fellah, C., Braun, J., Sauder, C., Sirotti, F., Berger, M.-H. (2020). Influence of the carbon interface on the mechanical behavior of SiC/SiC composites. *Compos. Part A Appl. Sci. Manuf.*, 133, 105867.

Fournier, R.O. and Rowe, J.J. (1977). The solubility of amorphous silica in water at high temperatures and high pressures. *Am. Mineral.*, 62, 1052–1056.

Harder, B., Jacobson, N., Myers, D. (2013). Oxidation transitions for SiC Part II. Passive-to-active transitions. *J. Am. Ceram. Soc.*, 96, 606–612.

Hollenberg, G.W., Henager, C.H., Youngblood, G.E., Trimble, D.J., Simonson, S.A., Newsome, G.A., Lewis, E. (1995). The effect of irradiation on the stability and properties of monolithic silicon carbide and SiCf/SiC composites up to 25 dpa. *J. Nucl. Mater.*, 219, 70–86.

Icenhower, J.P. and Dove, P.M. (2000). The dissolution kinetics of amorphous silica into sodium chloride solutions: Effects of temperature and ionic strength. *Geochim. Cosmochim. Acta*, 64, 4193–4203.

Ishibashi, R., Tanabe, S., Kondo, T., Yamashita, S., Nagase, F. (2017). Improving the corrosion resistance of silicon carbide for fuel in BWR environments by using a metal coating. In *Proc. of the 2017 Water React. Fuel Perform. Meet.*, Jeju Island.

Jacobson, N., Harder, B., Myers, D. (2013). Oxidation transitions for SiC Part I. Active-to-passive transitions. *J. Am. Ceram. Soc.*, 96, 838–844.

Karakoc, O., Mao, K., Xi, J., Koyanagi, T., Liu, J., Szlufarska, I., Katoh, Y. (2021). Single-step additive manufacturing of silicon carbide through laser-induced phase separation. doi: 10.21203/rs.3.rs-1017609/v1.

Katoh, Y., Kohyama, A., Snead, L.L., Hinoki, T., Hasegawa, A. (2003). Neutron tolerance of advanced SiC-fiber/CVI-SiC composites, FT/P1-03. In *Proc. Internat. Fusion Energy Conf. 2002*. IEAE Publ., Vienna.

Katoh, Y., Nozawa, T., Snead, L.L., Ozawa, K., Tanigawa, H. (2011). Stability of SiC and its composites at high neutron fluence. *J. Nucl. Mater.*, 417, 400–405.

Katoh, Y., Snead, L.L., Cheng, T., Shih, C., Lewis, W.D., Koyanagi, T., Hinoki, T., Henager, C.H., Ferraris, M. (2014). Radiation-tolerant joining technologies for silicon carbide ceramics and composites. *J. Nucl. Mater.*, 448, 497–511.

Keilholtz, G.W., Moore, R.E., Robertson, H.E. (1973). Fast-neutron damage to polycrystalline alumina at temperatures from 60 to 1230°C. *Nucl. Technol.*, 17(3), 234–246.

Kim, D., Lee, H.-G., Park, J.Y., Park, J.-Y., Kim, W.-J. (2015). Effect of dissolved hydrogen on the corrosion behavior of chemically vapor deposited SiC in a simulated pressurized water reactor environment. *Corr. Sci.*, 98, 304–309.

Kohyama, A., Dong, S., Katoh, Y. (2002). Development of SiC/SiC composites by Nano-Infiltration and Transient Eutectoid (NITE) process. In *26th Annual Conference on Advanced Ceramics, Materials, and Structures: A: Ceramic Engineering and Science Proceedings*. John Wiley & Sons, Inc., Hoboken, NJ, 311–318.

Kondo, S., Lee, M., Hinoki, T., Hyodo, Y., Kano, F. (2015). Effect of irradiation damage on hydrothermal corrosion of SiC. *J. Nucl. Mater.*, 464, 36–42.

Koyanagi, T., Kondo, S., Hinoki, T. (2011). The influence of sintering additives on the irradiation resistance of NITE SiC. *J. Nucl. Mater.*, 417, 435–439.

Koyanagi, T., Nozawa, T., Katoh, Y., Snead, L.L. (2018). Mechanical property degradation of high crystalline SiC fiber–reinforced SiC matrix composite neutron irradiated to ~100 displacements per atom. *J. Eur. Ceram.*, 38, 1087–1094.

Koyanagi, T., Katoh, Y., Hinoki, T., Henager, C., Ferraris, M., Grasso, S. (2020). Progress in development of SiC-based joints resistant to neutron irradiation. *J. Eur. Ceram.*, 40, 1023–1034.

Koyanagi, T., Terrani, K., Harrison, S., Liu, J., Katoh, Y. (2021). Additive manufacturing of silicon carbide for nuclear applications. *J. Nucl. Mater.*, 543, 152577.

Lee, J.-K. and Naganuma, M. (2009). Design factor using a SiC/SiC composites for core component of gas cooled fast reactor. 2: Thermal stress. *Mechanical Properties and Performance of Engineering Ceramics and Composites III. Ceramic Engineering and Science Proceedings*, 28(2), 199–205.

Minato, K., Ogawa, T., Fukuda, K., Nabielek, H., Sekino, H., Nozawa, Y., Takahashi, I. (1995). Fission product release from ZrC-coated fuel particles during postirradiation heating at 1600°C. *J. Nucl. Mater.*, 224, 85–92.

Naslain, R., Pailler, R., Lamon, J. (2010). Single- and multilayered interphases in SiC/SiC composites exposed to severe environmental conditions: An overview. *Int. J. Appl. Ceram. Technol.*, 7, 263–275.

Newsome, G., Snead, L.L., Hinoki, T., Katoh, Y., Peters, D. (2007). Evaluation of neutron irradiated silicon carbide and silicon carbide composites. *J. Nucl. Mater.*, 371, 76–89.

Nozawa, T., Koyanagi, T., Katoh, Y., Tanigawa, H. (2019). High-dose, intermediate-temperature neutron irradiation effects on silicon carbide composites with varied fiber/matrix interfaces. *J. Eur. Ceram. Soc.*, 39, 2634–2647.

Okamura, K., Matsuzawa, T., Sato, M., Higashiguchi, Y., Morozumi, S., Kohyama, A. (1986). Mechanical properties of neutron irradiated SiC fibers. *J. Nucl. Mater.*, 141–143, 102–107.

Olesinski, R.W. and Abbaschian, G.J. (1984). The C-Si (carbon-silicon) system. *J. Ph. Equilibria Diffus.*, 5, 486–489.

Opila, E.J. and Hann Jr., R.E. (1997). Paralinear oxidation of CVD SiC in water vapor. *J. Am. Ceram. Soc.*, 80, 197–205.

Park, J.-S., Kohyama, A., Hinoki, T., Shimoda, K., Park, Y.-H. (2007). Efforts on large scale production of NITE-SiC/SiC composites. *J. Nucl. Mater.*, 367–370(A), 719–724.

Polozov, I., Razumov, N., Masaylo, D., Silin, A., Lebedeva, Y., Popovich, A. (2020). Fabrication of silicon carbide fiber-reinforced silicon carbide matrix composites using binder jetting additive manufacturing from irregularly-shaped and spherical powders. *Materials*, 13, 1766.

Porter, I.E., Knight, T.W., Dulude, M.C., Roberts, E., Hobbs, J. (2013). Design and fabrication of an advanced TRISO fuel with ZrC coating. *Nucl. Eng. Des.*, 259, 180–186.

Powers, J.J. and Wirth, B.D. (2010). A review of TRISO fuel performance models. *J. Nucl. Mater.*, 405, 74–82.

Ravenet, A. (2007). Macrostructured plate fuel element. Patent, WO2007017503 (A1).

Ravenet, A. (2012). Body for a nuclear fuel assembly, and nuclear fuel assembly comprising such a body. Patent, US2012201343 (A1).

Sauder, C. (2014). Ceramic matrix composites: Nuclear applications. In *Ceramic Matrix Composites: Materials, Modeling and Technology*, Bansal, N.P. and Lamon, J. (eds). John Wiley & Sons Inc., Hoboken, NJ.

Seshadri, A., Philips, B., Dave, A.J., Harrison, S., Pegna, J., Shirvan, K. (2021). Hydrothermal corrosion of laser printed SiC fibers under extreme environment. *J. Nucl. Mater.*, 548, 152805.

Sha, J.J., Nozawa, T., Park, J.S., Katoh, Y., Kohyama, A. (2004). Effect of heat treatment on the tensile strength and creep resistance of advanced SiC fibers. *J. Nucl. Mater.*, 329–333, 592–596.

Sha, J.J., Hinoki, T., Kohyama, A. (2007). Microstructural characterization and fracture properties of SiC-based fibers annealed at elevated temperatures. *J Mater Sci.*, 42, 5046–5056.

Sha, J.J., Hinoki, T., Kohyama, A. (2009). Microstructure and mechanical properties of Hi-NicalonTM Type S fibers annealed and crept in various oxygen partial pressures. *Mater. Charact.*, 60, 796–802.

Shimoda, K. and Hinoki, T. (2021). Effects of fiber volume fraction on the densification and mechanical properties of unidirectional SiCf/SiC-matrix composites. *J. Eur. Ceram. Soc.*, 41, 1163–1170.

Snead, L.L., Steiner, D., Zinkle, S.J. (1992). Measurement of the effect of radiation damage to ceramic composite interfacial strength. *J. Nucl. Mater.*, 191–194, 566–570.

Snead, L.L., Zinkle, S.J., Hay, J.C., Osborne, M.C. (1998a). Amorphization of SiC under ion and neutron irradiation. *Nucl. Instrum. Methods Phys. Res. B*, 141, 123–132.

Snead, L.L., Osborne, M.C., Lowden, R.A., Strizak, J., Shinavski, R.J., More, K.L., Eatherly, W.S., Bailey, J., Williams, A.M. (1998b). Low dose irradiation performance of SiC interphase SiC/SiC composites. *J. Nucl. Mater.*, 253, 20–30.

Snead, L.L., Katoh, Y., Kohyama, A., Bailey, J.L., Vaughn, N.L., Lowden, R.A. (2000). Evaluation of neutron irradiated near-stoichiometric silicon carbide fiber composites. *J. Nucl. Mater.*, 283–287, 551–555.

Snead, L.L., Katoh, Y., Connery, S. (2007a). Swelling of SiC at intermediate and high irradiation temperatures. *J. Nucl. Mater.*, 367–370, 677–684.

Snead, L.L., Nozawa, T., Katoh, Y., Byun, T.-S., Kondo, S., Petti, D.A. (2007b). Handbook of SiC properties for fuel performance modeling. *J. Nucl. Mater.*, 371, 329–377.

Snead, L.L., Nozawa, T., Ferraris, M., Katoh, Y., Shinavski, R., Sawan, M. (2011). Silicon carbide composites as fusion power reactor structural materials. *J. Nucl. Mater.*, 417, 330–339.

Spinks, J.W.T. and Woods, R.J. (1990). *An Introduction to Radiation Chemistry*. John Wiley & Sons Inc., Hoboken, NJ.

Steinbrueck, M., Grosse, M., Stegmaier, U., Braun, J., Lorrette, C. (2022). Oxidation of silicon carbide composites for nuclear applications at very high temperatures in steam. *Coatings*, 12, 875.

Stobierski, L. and Gubernat, A. (2003). Sintering of silicon carbide I. Effect of carbon. *Ceram. Int.*, 29, 287–292.

Terrani, K.A. (2018). Accident tolerant fuel cladding development: Promise, status, and challenges. *J. Nucl. Mater.*, 501, 13–30.

Terrani, K.A. and Silva, C.M. (2015). High temperature steam oxidation of SiC coating layer of TRISO fuel particles. *J. Nucl. Mater.*, 460, 160–165.

Terrani, K.A., Kiggans, J.O., Katoh, Y., Shimoda, K., Montgomery, F.C., Armstrong, B.L., Parish, C.M., Hinoki, T., Hunn, J.D., Snead, L.L. (2012). Fabrication and characterization of fully ceramic microencapsulated fuels. *J. Nucl. Mater.*, 426, 268–276.

Terrani, K.A., Yang, Y., Kim, Y.-J., Rebak, R., Meyer, H.M., Gerczak, T.J. (2015). Hydrothermal corrosion of SiC in LWR coolant environments in the absence of irradiation. *J. Nucl. Mater.*, 465, 488–498.

Tu, T. and Jiang, G. (2018). SiC reticulated porous ceramics by 3D printing, gelcasting and liquid drying. *Ceram. Int.*, 44, 3400–3405.

Yueh, K. and Terrani, K.A. (2014). Silicon carbide composite for light water reactor fuel assembly applications. *J. Nucl. Mater.*, 448, 380–388.

Zabiego, M., Sauder, C., Lorrette, C., Guedeney, P. (2013). Improved multilayer tube made from ceramic-matrix composite material, the resulting nuclear fuel cladding and associated production methods. Patent, WO2013017621 (A1).

Zinkle, S.J. (2017). Advanced irradiation-resistant materials for Generation IV nuclear reactors. In *Structural Materials for Generation IV Nuclear Reactors*. Yvon, P. (ed.). Woodhead Publishing, Cambridge.

Zinkle, S.J. and Snead, L.L. (2014). Designing radiation resistance in materials for fusion energy. *Annu. Rev. Mater. Res.*, 44, 241–267.

9

Advanced Materials: Oxide-Dispersion Strengthened Steels

Satoshi OHTSUKA, Takashi TANNO, Yasuhide YANO and Takeji KAITO
Oarai Research Institute, Sector of Fast Reactor and Advanced Reactor Research and Development, Japan Atomic Energy Agency, Ibaraki, Japan

9.1. General introduction to the ODS

The oxide dispersion strengthening is an effective technique for improving the mechanical strength of the alloy. The dispersed oxides prevent the gliding motion of dislocations, thus remarkably enhancing the resistance to high-temperature deformation and rupture of alloys (Bacon et al. 1973; Lund and Nix 1976; Srolovitz et al. 1984). Yttrium oxide (Y_2O_3) is often used as dispersoids of oxide-dispersion strengthened (ODS) steels. Y_2O_3 is expected to be resistant to coarsening by the Ostwald ripening (Lifshitz and Slyozov 1961; Wagner 1961) because it is thermodynamically stable; the solubilities of yttrium and oxygen in iron are extremely low (Saenko et al. 2017). Extensive efforts have been made to develop ODS steels in the fields of nuclear and fusion engineering. Research has been done to improve their performance and meet the requirements such as irradiation resistance, high-temperature strength and corrosion resistance. Based on recent research, the high-density dispersion of nanosized oxides could improve the irradiation resistance of the steels because the interface between oxide and matrix could act as sink sites for point defects (Odette 2014). The oxide

Materials and Processes for Nuclear Energy Today and in the Future,
coordinated by Fanny BALBAUD-CÉLÉRIER and Céline CABET.

dispersion produces a fine crystallographic grain matrix and inhibits the recovery of dislocations, which would also contribute to the improvement of irradiation resistance (Toloczko et al. 2014).

9.2. Actual interest in nuclear systems

For developing the advanced nuclear systems (see Chapters 4–7 of this book), advanced alloys are needed, which can be used in the harsh environment of the reactor core, that is, high-dose neutron irradiation, high temperature and corrosive media. SFR-mixed oxide (MOX) fuel system has gained operational experience in experimental reactors and prototype reactors. In this system, modified type 316 steels have been used as the core materials. When using these steels, the life-limiting factor is a dimensional instability provided by irradiation-induced swelling (Garner 2012). Ferritic steels with a body-centered-cubic (BCC) matrix are known to have superior swelling resistance to austenitic steels with a face-centered-cubic matrix (Little and Stow 1979). One shortcoming of ferritic steels is their insufficient strength at temperatures exceeding 600°C. ODS ferritic steel is a promising long-life core material with high-temperature strength and irradiation resistance. It has a swelling-resistant ferritic matrix with nanosized oxides dispersed to improve high-temperature strength (Inoue et al. 2008; Decarlan et al. 2009; Byun et al. 2014; Kim et al. 2015).

Various types of ODS steels have been developed, as shown in Table 9.1 (Hamilton et al. 2000; Klueh et al. 2005; Lindau et al. 2005; Inoue et al. 2008; Decarlan et al. 2009; Kimura et al. 2011; Byun et al. 2014; Odette 2014; Ukai et al. 2016; Ukai et al. 2017; Jaumier et al. 2019; Yamashita et al. 2019; Ohtsuka et al. 2020). The alloy MA956 is a representative legacy commercial ODS steel developed for nonnuclear and high-temperature applications. This steel has a ferritic matrix that does not transform from ferrite to austenite by elevating temperature (untransformable ferritic matrix). It has excellent thermal and corrosion resistance at an extremely high temperature of 1,000°C. However, the me chanical properties of alloy MA956 are insufficient at the temperatures assumed for sodium-cooled fast reactor (SFR) application (400°C–700°C): low mechanical strength at 600°C–700°C, poor ductility, and fracture toughness due to α/α' phase separation during thermal aging and irradiation (Kobayashi and Takasugi 2010) (Figure 9.1), which could be caused by reduced interfacial energy between oxide and matrix.

	Steels	Basic composition (wt.%)	Application	Developer
20Cr-4.5Al-ODS steel	Incoloy MA956 (Hamilton et al. 2000)	Fe–20Cr–4.8Al–0.4Ti–0.5Y_2O_3	Turbine, combustion	Special Metals (USA)
18Cr-ODS steel	CEA 18Cr-ODS (Decarlan et al. 2009)	Fe–18Cr–1W–0.26Ti–0.56Y_2O_3	Fast reactor (FR) fuel cladding tube	CEA (FRN)
16Cr-4Al-ODS steel	Super-ODS (Kimura et al. 2011)	Fe–16Cr–4Al–2W–0.15Ti–0.35Y_2O_3–Zr, Hf	Fast reactor (FR) fuel cladding tube	Kyoto University (JPN)
14Cr-ODS steels	Incoloy MA957 (Hamilton et al. 2000)	Fe–14Cr–0.3Mo–0.9Ti–0.25Y_2O_3	FR fuel cladding tube	DOE (USA)
	14YWT (Odette 2014)	Fe-14Cr-3W-0.4Ti-0.3Y_2O_3	Fusion reactor materials/FR fuel cladding tube	DOE (USA)
	CEA 14Cr-ODS (Decarlan et al. 2009; Jaumier et al. 2019)	Fe–14Cr–1W–0.3Ti–0.25Y_2O_3	FR fuel cladding tube	CEA (FRN)
12Cr-ODS steels	JAEA 12Cr-ODS (Inoue et al. 2008)	Fe–12Cr–2W–0.26Ti–0.23Y_2O_3	FR fuel cladding tube	JAEA (JPN)
	12YWT (Klueh et al. 2005)	Fe–12Cr–2.5W–0.4Ti–0.25Y_2O_3	Fusion reactor materials/FR fuel cladding tube	DOE (USA)
12Cr-6Al-ODS steel	FeCrAl-ODS for ATF (Ukai et al. 2016; Yamashita et al. 2019)	Fe–12Cr–6Al–0.5Ti–0.4Zr–0.5Y_2O_3	Accident-tolerant-fuel cladding tube	NFD/Hokkaido University (JPN)
9-11Cr-ODS steel	DOE 9Cr-ODS (Byun et al. 2014)	Fe–0.3C–9Cr–2W–0.4Ti–0.2V	Fusion reactor materials/FR fuel cladding tube	DOE (USA)
	CEA 9Cr-ODS (Decarlan et al. 2009; Jaumier et al. 2019)	Fe–0.1C –9Cr–1W–0.3Ti–0.25Y_2O_3	FR fuel cladding tube	CEA (FRN)
	JAEA 9Cr-ODS (Inoue et al. 2008)	Fe–0.13C–9Cr–2W–0.2Ti–0.35Y_2O_3	FR fuel cladding tube	JAEA (JPN)
	JAEA 11Cr-ODS (Ohtsuka et al. 2020)	Fe–0.13C–11Cr–1.4W–0.4Ni–0.2Ti–0.35Y_2O_3	FR fuel cladding tube	JAEA (JPN)
8Cr-ODS	EUROFER-ODS (Lindau et al. 2005)	Fe–8Cr–1.4W–0.4Ni–0.2Ti–0.3Y_2O_3	Fusion reactor materials	KIT (DEU)

Table 9.1. *Examples of material specifications for oxide-dispersion strengthened steels (Hamilton et al. 2000; Klueh et al. 2005; Lindau et al. 2005; Inoue et al. 2008; Decarlan et al. 2009; Kimura et al. 2011; Byun et al. 2014; Odette 2014; Ukai et al. 2016; Ukai et al. 2017; Jaumier et al. 2019; Yamashita et al. 2019; Ohtsuka et al. 2020)*

The legacy ODS steel MA957 (untransformable steel) was developed for SFR. This alloy does not contain aluminum, causing strength degradation. Chromium concentration is set to 14wt%, which is lower than that in MA956 (20%) (Hamilton et al. 2000). Researchers have studied and modified 14Cr-ODS steels for application to nuclear fission and fusion systems, including MA957; their excellent high-temperature and irradiation performance have been reported (Decarlan et al. 2009; Odette 2014). The development of 9-12Cr ODS tempered martensitic steels (TMSs) is ongoing for nuclear fission and fusion systems. The tempered martensite matrix, which has the high-density lath martensite in the BCC matrix acting as a point defect sink, could be favorable regarding irradiation resistance. Notably, chromium is a strong ferrite-forming element, so its concentration should not exceed 13wt% for maintaining the tempered martensite matrix. In most 9-14Cr-ODS steels, Y-Ti-O complex oxides are densely dispersed by the co-addition of titanium and Y_2O_3. The co-addition of Y_2O_3 and IVa elements like titanium and zirconium was beneficial for the fine oxide dispersion (Ukai and Fujiwara 2002; Kimura et al. 2011).

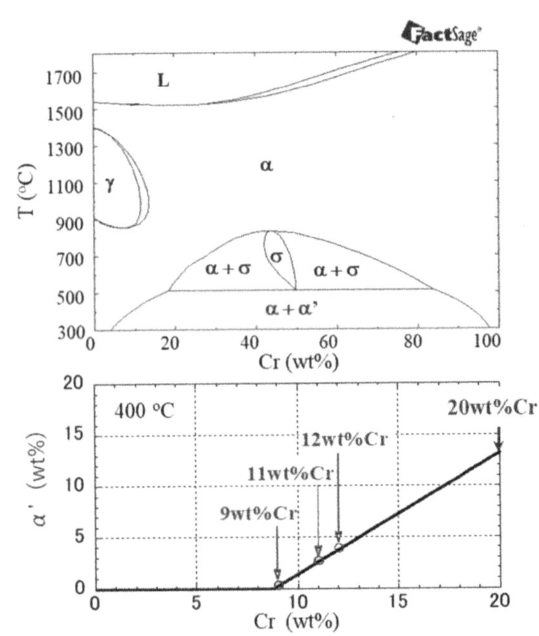

Figure 9.1. *Fe-Cr binary phase diagram calculated by FactSage and FSstel databases (Bale et al. 2016; Ukai et al. 2017). For a color version of this figure, see www.iste.co.uk/balbaud/nuclearenergytoday.zip*

The corrosion resistance against the coolants is a crucial material issue for application to fast reactors cooled by supercritical water (SCW) and lead-bismuth eutectic (LBE). As highly corrosion-resistant alloys against LBE and SCW, the FeCrAl-ODS steel designated as “Super-ODS steel” was developed by Kimura et al. (2011). In the alloy design, aluminum was added to the high-Cr steel matrix to promote a surface protective oxide layer containing aluminum (Cho et al. 2004; Hosemann et al. 2008; Takaya et al. 2010; Kimura et al. 2011), while zirconium and Y_2O_3 were co-added to produce fine and uniform dispersion of oxides containing yttrium and zirconium (Ukai and Fujiwara 2002; Kimura et al. 2011), thus improving the high-temperature strength. The R&D on FeCrAl-ODS steels modified for accident-tolerant fuel (ATF) cladding material of light-water reactor (LWR) is ongoing. In an accidental scenario of LWR, the fuel cladding tubes are required to have resistance against the high-temperature steam oxidation (Terrani et al. 2014). The problem in applying FeCrAl steel to the LWR cladding tube is the large cross-section of iron, chromium and aluminum for thermal neutron absorption. Therefore, the oxide-dispersion strengthening is crucial for maintaining structural integrity while reducing the tube wall thickness (Ukai et al. 2016; Dryepondt et al. 2018; Yamashita et al. 2019). The FeCrAl-ODS steels have weaker strength than the FeCr-ODS steels without aluminum addition. Aluminum has a higher affinity to oxygen than titanium, so the co-addition of Y_2O_3, aluminum and titanium produces Y-Al-O complex oxides rather than Y-Ti-O complex oxides. The Y-Al-O complex oxides are reported not to disperse as finely and densely as Y-Ti-O complex oxides. To overcome this issue, the addition of zirconium, having a higher affinity to oxygen than aluminum was applied to both the Super-ODS steels and the FeCrAl-ODS steels for ATF (Yamashita et al. 2019), thus producing the dense dispersion of nano-oxides containing yttrium and zirconium. Critical issues in developing the FeCrAl-ODS steels are evaluating their irradiation and thermal aging behavior. Increasing the chromium concentration and adding aluminum have both reportedly enhanced the α/α' phase separation; the optimum specification balancing the mechanical properties and corrosion resistance should be pursued by continuing careful research.

9.3. Fabrication and joining

For improving the high-temperature strength of ODS steels, various knowledge has been accumulated so far on nanostructure control, grain-boundary control and optimization of material specifications (Ukai and Fujiwara 2002; Inoue et al. 2008; Kimura et al. 2011; Odette 2014). For example, to improve the high-temperature strength and irradiation resistance of ODS steels, fine dispersion of Y-Ti-O nanosized oxides was achieved by optimizing the VIa element addition such as titanium and zirconium, and the consolidation process such as low temperature hot-extrusion (HE) (e.g. Ukai and Fujiwara 2002; Odette 2014). According to the conventional creep theory, strengthening the intra-grain by oxide dispersion inevitably reduces the creep deformation involving dislocation glide, and thus the contribution by other creep mechanisms (matrix-/grain-boundary diffusion and grain-boundary sliding) becomes relatively significant. In this case, grain-boundary control could be a key to strength improvement. When the untransformable ODS ferritic steels are consolidated by HE and rolled into a thin-walled tube by cold-rolling and heat treatment, the matrix texture will be elongated to the drawing direction. If the applied stress is lower than the threshold stress, intra-grain should be rigid. In this case, the strength has strong anisotropy, and the creep strength in the longitudinal direction is much higher than in the hoop direction (Figure 9.2) (Ukai et al. 1998). The creep strength in the hoop direction is poor, possibly due to the grain-boundary sliding, showing that grain-boundary control is a critical issue for improving the high-temperature strength of ODS steel in addition to nanostructure control. Ukai et al. (2002) showed a typical example of enhancing the creep strength in the hoop direction. The recrystallization treatment was done for the untransformable ODS ferritic steel, and adequate strength can be successfully derived both in axial and hoop directions, as shown in Figure 9.2.

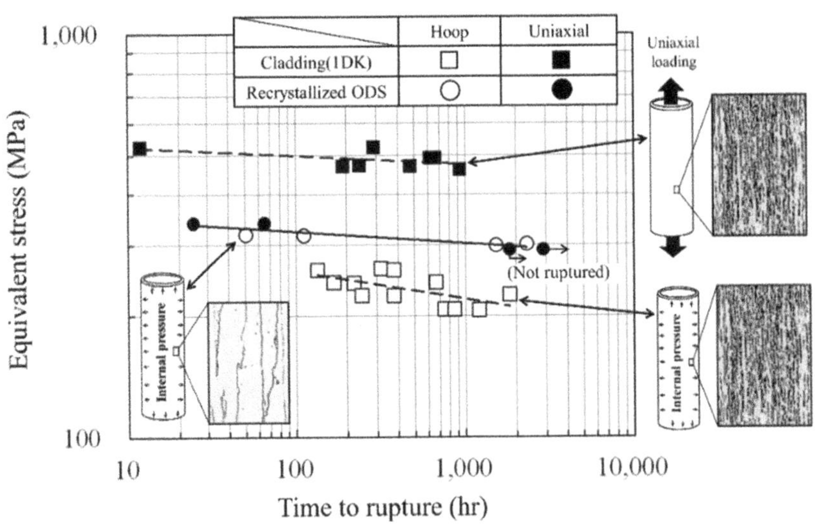

Figure 9.2. *Strength anisotropy in creep strength of ODS steels; data of circle symbols are creep strength data of recrystallized ODS steel cladding tubes having quasi-isotropic microstructure, and those of square symbols are creep strength data of unrecrystallized ODS steel cladding tube having highly anisotropic microstructures (i.e. bamboo-like structure) (Ukai et al. 1998)*

An effective method for producing a uniform and dense precipitation in a metallic matrix is to first dissolve the constituent elements in the matrix and then have them precipitated by heat treatments. However, the dissolution of Y_2O_3 in an iron-based matrix is impossible in an equilibrium condition. Then, the powder metallurgy process was applied to fabricate ODS alloys, as shown in Figure 9.3. This process has already been established at a laboratory scale. It involves the mechanical alloying, in which metallic powder and oxide powder are agitated at high-speeds with grinding balls, thus producing the nonequilibrium mixture, that is, either the solid solution of oxide forced in metallic matrix or decomposition of oxide into the small amorphized dispersed islands (Okuda and Fujiwara 1995; Suryanarayana 2001). A high density of nanosized oxides precipitates by the subsequent thermo-mechanical process (e.g. HE and HIP). The consolidated materials are machined into mother tubes, which are then drawn to the thin-walled tubes by repeating cold-rolling followed by the heat treatment for softening (Inoue et al. 2008; Decarlan et al. 2009).

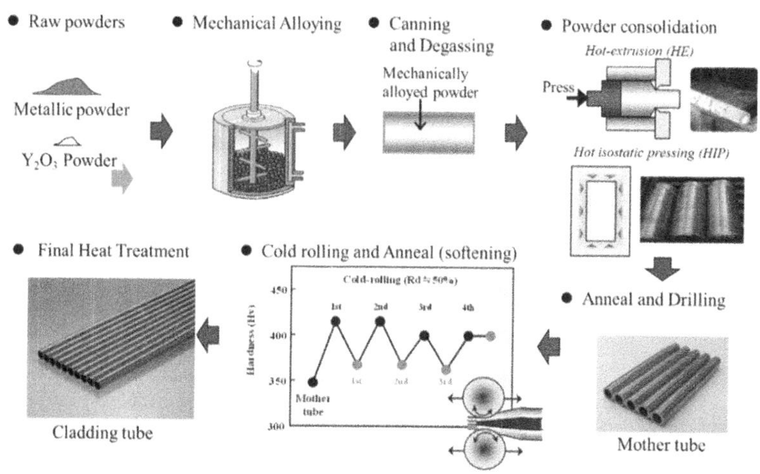

Figure 9.3. *Fabrication process of ODS steels using powder metallurgy process (Ukai et al. 2017). For a color version of this figure, see www.iste.co.uk/balbaud/nuclearenergytoday.zip*

The bonding technique is indispensable for applying the ODS steels to the engineering component. Typical welding techniques based on local fusion, such as tungsten-inert-gas welding and electron-beam welding, destroy the nanostructure of ODS steels. This results in significant degradation of mechanical properties in the welds. To avoid this degradation, the pressurized resistance welding (PRW) technique was developed for ODS steel at a laboratory scale (Seki et al. 2004). This technique is a solid-state bonding, which can be used for joining the end-plug and cladding tube of ODS steels for the fuel pin fabrication. The bonding interface by PRW was shown to keep the fine oxide dispersion and adequate mechanical strength. The neutron irradiation test proved the structural integrity of the PRW part (Kaito et al. 2013). Other solid-state bonding techniques, such as friction stir welding (FSW) (Mazumder et al. 2016) and diffusion bonding (Noh et al. 2012), are under development for application to ODS steels.

9.4. Material properties

Figure 9.4 shows the out-of-piles ultimate tensile strength (UTS) (Klueh et al. 2005; Yano et al. 2017). The oxide-dispersion strengthening significantly improved tensile strength; the UTS of ODS steels is much higher than those of non-ODS steels, particularly at temperatures exceeding 900°C. These data proved that the ODS-steel cladding tube enhances the accident tolerance of SFR fuel.

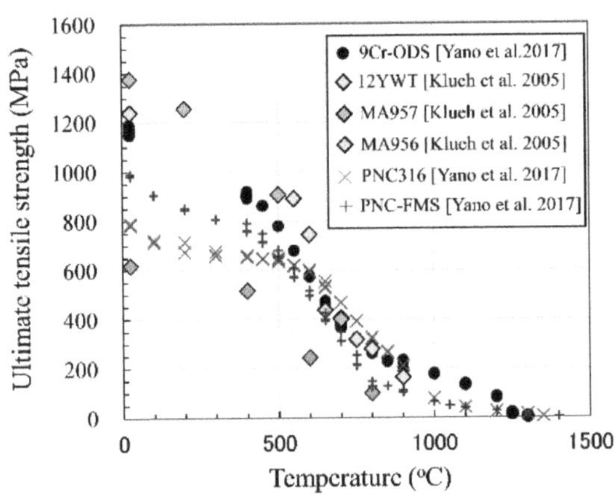

Figure 9.4. *For comparison, the ultimate tensile strength of ODS steels and non-ODS steels (Klueh et al. 2005; Yano et al. 2017). For a color version of this figure, see www.iste.co.uk/balbaud/nuclearenergytoday.zip*

Figure 9.5 (top) (Nishiyama et al. 1998; Ukai et al. 2000; Hoelzer et al. 2008; Yano et al. 2019) depicts the results of out-of-piles creep tests of ODS steels: internally pressurized creep tests of ODS steels with 9–12% Cr, which are candidates for SFR-cladding tubes, internally pressurized creep tests of FeCrAl-ODS steel (Fe-14wt%Cr-0.5Ti-6.4Al-0.47Y_2O_3) and FeCrAlZr-ODS steel (Fe-15wt%Cr-0.5Ti-6.4Al-0.37Zr-0.47Y_2O_3) and uniaxial creep tests of 14Cr-ODS steel tube (MA957) and 12Cr-ODS steel plate (12YWT). Since the stress is biaxially loaded (in both axial and hoop directions) during internal pressure creep tests, Von Mises equivalent stress is plotted in Figure 9.4 in these cases.

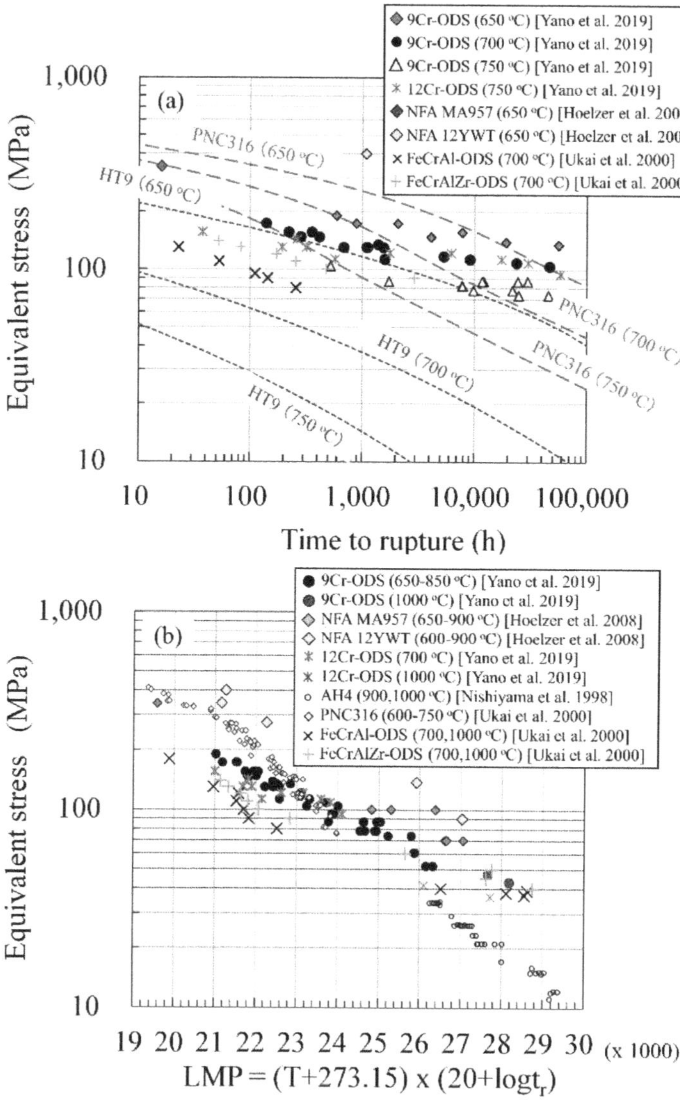

Figure 9.5. *Comparison of out-of-piles creep rupture strength of ODS steels (Nishiyama et al. 1998; Ukai et al. 2000; Hoelzer et al. 2008; Yano et al. 2019). For a color version of this figure, see www.iste.co.uk/balbaud/nuclearenergytoday.zip*

All of these ODS steels were developed for use as fuel-cladding tubes, implying that the strength in the hoop direction is important. In order to obtain satisfactory hoop strength, their matrixes were controlled by either the α to γ phase transformation (9Cr-ODS) or a recrystallization treatment (12Cr-ODS, FeCrAl-ODS, FeCrAlZr-ODS). The ODS steels are apparently advantageous in creep strength over the non-ODS steels: modified type 316 steel and ferritic steel (HT9). The stress level applied to the cladding tube depends on the type of nuclear plant and the fuel design. In the case of SFR, internal pressure produced by fission product (FP) gas gradually increases during reactor operation, reaching approximately 100 MPa in the hoop direction at the end of life. In this stress range, ODS steels have significantly longer creep lives than non-ODS steels. Figure 9.5 bottom depicts the equivalent stress versus Larson–Miller parameter. This figure includes the data at temperatures up to 1,000°C. It is seen that ODS steels have a significant advantage in creep strength even at 1,000°C, which is much higher than the heat-resistant austenitic steel AH4. Focusing on the strength of ODS steels, NFA MA957 and NFA 12YWT are superior to ODS steels with 9–12% Cr. CEA also has developed ODS steels for SFR application (9–14% Cr) (Decarlan et al. 2009; Jaumier et al. 2019; Sornin et al. 2021), whose specifications were selected by several design requirements, where much more importance appears to be attached to manufacturability and ductility.

Figure 9.6 (Ukai et al. 2000; Kaito et al. 2009) shows the in-reactor creep rupture strength of ODS steels with 9–12% Cr. No irradiation-induced degradation was recognized in the ODS steels while the creep strength apparently degrades in modified type 316 steel (PNC316) even by the irradiation to 3 dpa. This degradation is understood by the irradiation-induced recovery of high-density dislocation introduced by cold-rolling (Ukai et al. 2000). The same is true for the tensile properties of 9Cr,12Cr-ODS steels; neutron irradiation (maximum dose: 33 dpa) did not produce a significant degradation of tensile strength and uniform elongation (Yano et al. 2011). The superior behavior of ODS steels under high-temperature neutron irradiation is due to the stability of nanosized oxides. Unfortunately, the mechanical strength data of neutron-irradiated ODS steels are limited, particularly in the high-dose range. In order to validate the applicability of ODS steels up to high-dose neutron irradiation, it is essential not only to prepare and complete neutron irradiation tests but also to study the physical processes behind the irradiation effect using modeling in conjunction with ion-irradiation tests (Allen et al. 2008).

Incorporating nanosized oxide dispersion certainly improves the mechanical strength of steel, while the strength and the ductility often have a trade-off relationship. Ductility and toughness may be problematic in highly strengthened ODS steels. It has been a critical issue from the engineering point of view to seek the optimum specification balancing the strength and the ductility. The steels are known to be hardened by neutron irradiation at a lower temperature (approximately below 400°C in the case of 9Cr-ODS TMS); thus, the ductility is reduced. However, in ODS steels, an interesting irradiation response has been reported. ODS steels can maintain uniform elongation while suffering from irradiation hardening (Kimura 2005; Odette 2014).

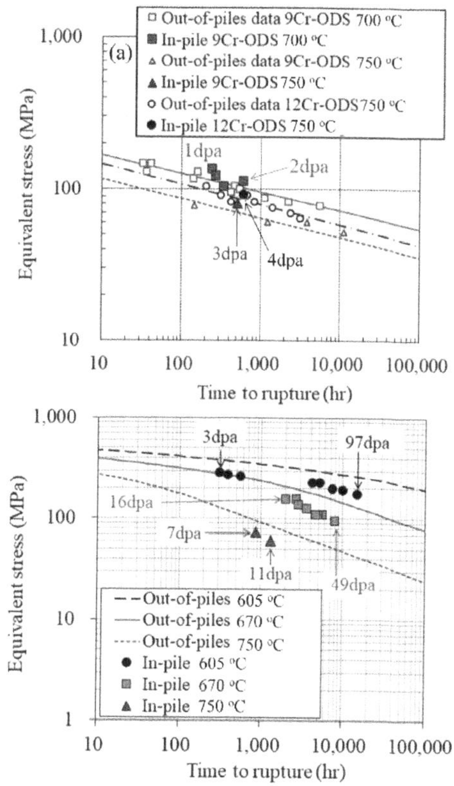

Figure 9.6. *Comparison of in-reactor and out-of-pile creep rupture strength of ODS steels (Ukai et al. 2000; Kaito et al. 2009). For a color version of this figure, see www.iste.co.uk/balbaud/nuclearenergytoday.zip*

Table 9.2 summarizes the corrosion issues against the coolants and the candidate core materials for advanced nuclear systems. Ferritic steels have adequate compatibility with liquid sodium (Furukawa et al. 2009; Ukai et al. 2017). However, when ferritic steels and austenitic steels coexist in high-temperature flowing sodium, nickel transfers from austenitic steel to ferritic steel via liquid sodium because of the relatively high nickel solubility in sodium.

Chemical environment	System	Corrosions issues	Candidate ODS steel under development
Compatibility with flowing coolant at outer surface	SFR	Mass transfer such as Ni penetration depending on the operation temperature	9-14Cr-ODS steels
	LBE-FR	Dissolution of constituent element to flowing LBE	16Cr-4Al-ODS steel
	SCW-FR	Oxidation by SCW	16Cr-4Al-ODS steel
	LWR-ATF	High-temperature steam oxidation and hydrogen release following a severe accident	12Cr-6Al-ODS steel

Table 9.2. *Main corrosion issues for oxide-dispersion strengthened steels exposed to coolants of the advanced nuclear system*

Figure 9.7 depicts the nickel-diffusion layer observed in 13Cr-ODS steel (63DSA). This sample was exposed to flowing sodium in a sodium loop test apparatus at 700°C, and dispersion of nanosized oxide was maintained in the nickel-diffusion layer, where high-temperature tensile tests showed no degradation induced by the nickel diffusion (Furakawa et al. 2009; Ukai et al. 2016). Corrosion issues are much more critical in high-temperature LBE because the solubility of iron and chromium is very high in LBE. To overcome this issue, the Super-ODS steel (16Cr-4Al-ODS steel) was developed by Kimura et al. (2011). They determined the optimized specification that can produce the protective oxide layer on the steel surface to suppress the dissolution of constituent elements (Hosemann et al. 2008; Takaya et al. 2010). Figure 9.8 presents a scanning electron microscope (SEM) image and the energy-dispersive X-ray spectrometry (EDS) line scan analysis of FeCrAl-ODS steel (15.4Cr-3.8Al-1.9W ODS steel), which was exposed to LBE containing 10^{-6} wt.% oxygen at 650°C for 5,000 h (Takaya et al. 2010). The thin protective oxide layer containing aluminum blocks the dissolution into LBE.

Notably, the protective oxide formation depends on oxygen potential in LBE and the control of oxygen potential in LBE is essential.

In the FeCrAl-ODS steels for ATF, the oxide layer containing aluminum is formed in high-temperature steam; thereby, the hydrogen generation is remarkably reduced compared with conventional cladding material (i.e. zirconium alloy) (Terrani et al. 2014). The fuel-cladding-chemical interaction (FCCI) is the corrosion phenomenon produced by the accumulation of FPs and the elevation of oxygen potential inside fuel elements (Kleykamp 1985). For enhancing the resistance to FCCI of MOX fuel assembly, a conventional means is to increase the chromium content in the steel. It was pointed out that the matrix grain size is also essential in addition to chromium concentration. According to out-of-piles FCCI simulation tests, 9Cr-ODS TMS showed a notable FCCI resistance comparable to type 316 stainless steel (17Cr) and 11Cr-ferritic steel (PNC-FMS) in spite of a lower chromium concentration. The mechanism interpreted this result that the 9Cr-ODS TMS has a fine crystallographic grain size (a few micrometers), enhancing the outward diffusion of chromium (Yutani et al. 1993; Ukai et al. 2013).

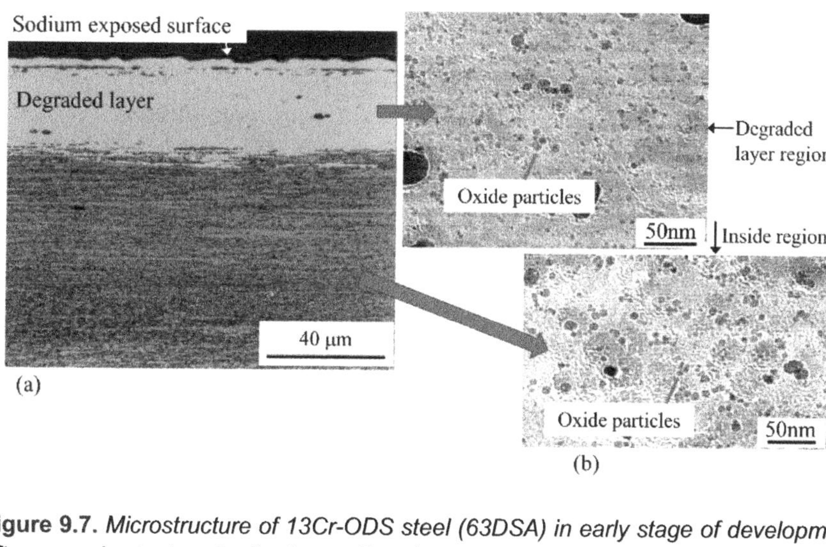

Figure 9.7. *Microstructure of 13Cr-ODS steel (63DSA) in early stage of development after corrosion test under flowing sodium (sodium flow velocity: 4.5 m/s; temperature: 700°C; time: 2,604 h). (a) cross-sectional microstructure; (b) TEM micrographs (Furakawa et al. 2009; Ukai et al. 2017). For a color version of this figure, see www.iste.co.uk/balbaud/nuclearenergytoday.zip*

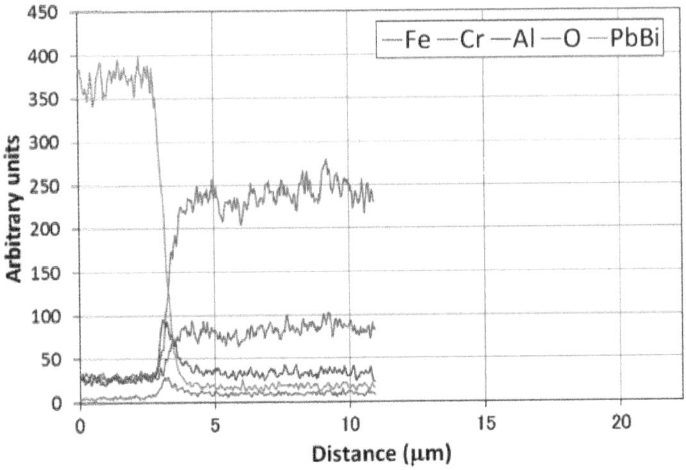

Figure 9.8. *SEM image and EDS line scan analysis of an ODS steel (15.4Cr-3.8Al-l.9W) after 5,000-h exposure to LBE containing 10^{-6} wt.% oxygen at 650°C (Takaya et al. 2010). For a color version of this figure, see www.iste.co.uk/balbaud/nuclearenergytoday.zip*

9.5. Main material challenges and solutions under evaluation

As described above, researchers have shown the predominance of irradiation resistance and high-temperature strength in ODS steels. Furthermore, extensive efforts have been made to enhance the technical knowledge to give the ODS steels a considerable corrosion resistance that could meet the design requirements of advanced nuclear systems. As for the fabrication technology, some ODS alloys are already commercialized for nonnuclear applications. But there is no experience of commercialization in the nuclear industry. The major challenges to be overcome for developing ODS steels would be as follows:

– preparation of mass-production technology of the high-quality ODS steel without microstructural heterogeneity, which can pass the strict inspection standards of nuclear components;

– demonstration of in-reactor performance (neutron irradiation, corrosion) up to high-dose level assumed in commercialized reactors.

To date, high-performance ODS steels have already been fabricated on a laboratory scale in several institutes. Scaling-up the existing production process (powder metallurgy) is essential for commercialization. However, the conventional powder metallurgy process is complex and composed of several steps (mechanical alloying, powder consolidation, machining, cold-rolling, and heat treatment for thin-walled tubing, etc.), where fabrication conditions and procedures should be carefully chosen to produce high-quality materials having fewer defects such as inclusions. Innovative processes are investigated for acceleration toward the establishment of mass production. Additive manufacturing (AM) (Wilms et al. 2018) and gas atomization reaction synthesis (GARS) (Rieken et al. 2012) are much more simple techniques, having the potential to reduce the fabrication cost and assure the quality of the component. The progress in these technologies could provide a breakthrough in ODS steel development. The in-reactor performance will have to be demonstrated to complete any new material qualification for a nuclear application. Subsequently, material strength data will have to be produced for design codes and standards. Here, a point of interest could be the stability of nanosized oxides under high-dose neutron irradiation. According to a traditional protocol, a neutron irradiation test followed by post-irradiation examination is essential. However, neutron irradiation facilities are limited globally, and testing is time consuming and costly. Furthermore, simulation techniques such as ion-irradiation and computational science would be critical for accelerating the development of new materials, including ODS steels.

9.6. References

Allen, T.R., Gan, J., Cole, J.I., Miller, M.K., Busby, J.T., Shutthanandan, S., Thevuthasan, S. (2008). Radiation response of a 9 chromium oxide dispersion strengthened steel to heavy ion irradiation. *J. Nucl. Mater.*, 375, 26–37.

Bacon, D.J., Kocks, U.F., Scattergood, R.O. (1973). The effect of dislocation self-interaction on the Orowan stress. *Phil. Mag.*, 28(6), 1241–1263.

Bale, C., Belisle, E., Chartrand, P., Decterov, S.A. (2016). FactSage thermochemical software and databases 2010–2016. *CALPHAD*, 54, 35–53.

Byun, T.S., Yoon, J.H., Hoelzer, D.T., Lee, Y.B., Kang, S.H., Maloy, S.A. (2014). Process development for 9Cr nanostructured ferritic alloy (NFA) with high fracture toughness. *J. Nucl. Mater.*, 449, 290–299.

Cho, H.S., Kimura, A., Ukai, S., Fujiwara, M. (2004). Corrosion properties of oxide dispersion strengthened steels in super-critical water environment. *J. Nucl. Mater.*, 329–333, 387–391.

De Carlan, Y., Bechade, J.-L., Dubuisson, P., Seran, J.-L., Billot, P., Bougault, A., Cozzika, T., Doriot, S., Hamon, D., Henry, J. et al. (2009). CEA developments of new ferritic ODS alloys for nuclear applications. *J. Nucl. Mater.*, 386–388, 430–432.

Dryepondt, S., Unocic, K.A., Hoelzer, D.T., Massey, C.P., Pint, B.A. (2018). Development of low-Cr ODS FeCrAl alloys for accident-tolerant fuel cladding. *J. Nucl. Mater.*, 501, 59–71.

Furukawa, T., Kato, S., Yoshida, E. (2009). Compatibility of FBR materials with sodium. *J. Nucl. Mater.*, 392, 249–254.

Garner, F.A., (2012). Radiation damage in austenitic steels. In *Comprehensive Nuclear Materials*, Konings R.J.M. (ed.). Elsevier, Oxford.

Hamilton, M.L., Gelles, D.S., Lobsinger, R.J., Johnson, G.D., Brown, W.F., Paxton, M.M., Puigh, R.J., Eiholzer, C.R., Martinez, C., Blotter, M.A. (2000). Fabrication technological development of the oxide dispersion strengthened alloy MA957 for fast reactor applications. Report, PNNL-13168. Pacific Northwest National Laboratory, Richland.

Hoelzer, D.T., Shingledecker, J.P., Klueh, R.L., Miller, M.K., Bentley, J. (2008). The microstructural stability of a ruptured thermal creep specimen of MA957. Fusion Materials Semiannual Progress Report for the Period Ending June 30, 2008, DOE-ER-0313/44, 53–61, Oak Ridge National Laboratory, Oak Ridge, TN.

Hosemann, P., Thau, H.T., Johnson, A.L., Maloy, S.A., Li, N. (2008). Corrosion of ODS steels in lead–bismuth eutectic. *J. Nucl. Mater.*, 373, 246–253.

Inoue, M., Kaito, T., Ohtsuka, S. (2008). Research and development of oxide dispersion strengthened ferritic steels for sodium cooled fast breeder reactor fuels. In *Materials Issues for Generation IV Systems*, Ghetta, V., Gorse, D., Maziere, D., Pontikis, V. (eds). Springer, Dordrecht.

Jaumier, T., Vincent, S., Vincent, L., Desmorat, R. (2019). Creep and damage anisotropies of 9%Cr and 14%Cr ODS steel cladding. *J. Nucl. Mater.*, 518, 274–286.

Kaito, T., Ohtsuka, S., Inoue, M., Asayama, T., Uwaba, T., Mizuta, S., Ukai, S., Furukawa, T., Ito, C., Kagota, E. et al. (2009). In-pile creep rupture properties of ODS ferritic steel claddings. *J. Nucl. Mater.*, 386–388, 294–298.

Kaito, T., Ohtsuka, S., Yano, Y., Tanno, T., Yamashita, S., Ogawa, R., Tanaka, K. (2013). Irradiation performance of oxide dispersion strengthened (ODS) ferritic steel claddings for fast reactor fuels. In *Proc. International Conference on Fast Reactors and Related Fuel Cycles Technologies and Sustainable Scenarios (FR13)*, Paris.

Kim, T.K., Noh, S., Kang, S.H., Jin, H.J., Kim, G.E. (2015). Development of advanced radiation resistant ODS steel for fast reactor system applications. *World J. Eng. Technol.*, 3, 125–128.

Kimura, A. (2005). Current status of reduced-activation ferritic/martensitic steels R&D for fusion energy. *Mater. Trans.*, 46(3), 394–404.

Kimura, A., Kasada, R., Iwata, N., Kishimoto, H., Zhang, C.H., Isselin, J., Dou, P., Lee, J.H., Muthukumar, N., Okuda, T. et al. (2011). Development of Al added high-Cr ODS steels for fuel cladding of next generation nuclear systems. *J. Nucl. Mater.*, 417, 176–179.

Kleykamp, H. (1985). The chemical state of the fission products in oxide fuels. *J. Nucl. Mater.*, 131, 221–246.

Klueh, R.L., Shingledecker, J.P., Swindeman, R.W., Hoelzer, D.T. (2005). Oxide dispersion-strengthened steels: A comparison of some commercial and experimental alloys. *J. Nucl. Mater.*, 341, 103.

Kobayashi, S. and Takasugi, T. (2010). Mapping of 475°C embrittlement in ferritic Fe–Cr–Al alloys. *Scripta Mater.*, 63, 1104–1107.

Lifshitz, I.M. and Slyozov, V.V. (1961). The kinetics of precipitation from supersaturated solid solutions. *J. Phys. Chem. Solids*, 19, 35–50.

Lindau, R., Moslang, A., Rieth, M., Klimiankou, M., Materna-Morris, E., Alamo A., Tavassoli, A.-A.F., Cayron, C., Lancha, A.-M., Fernandez, P. et al. (2005). Present development status of EUROFER and ODS-EUROFER for application in blanket concepts. *Fus. Eng. Des.*, 75–79, 989–996.

Little, E.A. and Stow, D.A. (1979). Void-swelling in irons and ferritic steels: II. An experimental survey of materials irradiated in a fast reactor. *J. Nucl. Mater.*, 87, 25–39.

Lund, R.W. and Nix, W.D. (1976). High temperature creep of Ni-20Cr-2ThO$_2$ single crystals. *Acta Metall.*, 24, 469–481.

Mazumder, B., Yu, X., Edmondson, P.D., Parish, M., Miller, M.K., Meyer, H.M., Feng, Z. (2016). Effect of friction stir welding and post-weld heat treatment on a nanostructured ferritic alloy. *J. Nucl. Mater.*, 469, 200–208.

Nishiyama, Y., Sawaragi, Y., Matsuda, T., Kihara, S., Kajigaya, I. (1998). Development of a new heat resistant austenitic stainless steel "NAR-AH-4". *The Sumitomo Research*, 60, 27–33.

Noh, S., Kim, B., Kasada, R., Kimura, A. (2012). Diffusion bonding between ODS ferritic steel and F82H steel for fusion applications. *J. Nucl. Mater.*, 426, 208–213.

Odette, G.R. (2014). Recent progress in developing and qualifying nanostructured ferritic alloys for advanced fission and fusion applications. *JOM*, 66, 2427–2441.

Ohtsuka, S., Tanno, T., Oka, H., Yano, Y., Tachi, Y., Kaito, T., Hashidate, R., Kato, S., Furukawa, T., Ito, C. et al. (2020). Development of ODS tempered martensitic steel for high burn up fuel cladding tube of SFR. In *Proc. GIF 2018 Symposium*, Paris.

Okuda, T. and Fujiwara, M. (1995). Dispersion behaviour of oxide particles in mechanically alloyed ODS steel. *J. Mater. Sci. Lett.*, 14, 1600–1603.

Rieken, J.R., Anderson, I.E., Kramer, M.J., Odette, G.R., Stergar, E., Haney, E. (2012). Reactive gas atomization processing for Fe-based ODS alloys. *J. Nucl. Mater.*, 428, 65–75.

Saenko, I., Fabrichnaya, O., Udovsky, A. (2017). New thermodynamic assessment of the Fe-Y system. *J. Phase Equilib. Diffus.*, 38, 684–699.

Seki, M., Hirako, K., Kono, S., Kihara, Y., Kaito, T., Ukai, S. (2004). Pressurized resistance welding technology development in 9Cr-ODS martensitic steels. *J. Nucl. Mater.*, 329–333, 1534–1538.

Sornin, D., Ehrnsten, U., Mozzani, N., Rantala, J., Walter, M., Hobt, A., Aktaa, J., Oñorbe, E., Hernandez-Mayoral, M., Ulbricht, A. et al. (2021). Creep properties of 9Cr and 14Cr ODS tubes tested by inner gas pressure. *Metall. Mater. Trans.*, 52, 3541–3552.

Srolovitz, D.J., Luton, M.J., Petkovic-Luton, R., Barnett, D.M., Nix, W.D. (1984). Diffusionally modified dislocation-particle elastic interactions. *Acta Metall.*, 32(7), 1079–1088.

Suryanarayana, C. (2001). Mechanical alloying and milling. *Progress in Materials Science*, 46, 1–184.

Takaya, S., Furukawa, T., Inoue, M., Fujisawa, T., Okuda, T., Abe, F., Ohnuki, S., Kimura, A. (2010). Corrosion resistance of Al-alloying high Cr–ODS steels in stagnant lead–bismuth. *J. Nucl. Mater.*, 398, 132–138.

Terrani, K.A., Zinkle, S.J., Snead, L.L. (2014). Advanced oxidation-resistant iron-based alloys for LWR fuel cladding. *J. Nucl. Mater.*, 448, 420–435.

Toloczko, M.B., Garner, F.A., Voyevodin, V.N., Bryk, V.V., Borodin, O.V., Mel'nychenko, V.V., Kalchenko, A.S. (2014). Ion-induced swelling of ODS ferritic alloy MA957 tubing to 500 dpa. *J. Nucl. Mater.*, 453(1–3), 323–333.

Ukai, S. and Fujiwara, M. (2002). Perspective of ODS alloys application in nuclear environments. *J. Nucl. Mater.*, 307–311, 749–757.

Ukai, S., Nishida, T., Okuda, T., Yoshitake, T. (1998). R&D of oxide dispersion strengthened ferritic martensitic steels for FBR. *J. Nucl. Mater.*, 258–263, 1745–1749.

Ukai, S., Mizuta, S., Kaito, T., Okada, H. (2000). In-reactor creep rupture properties of 20% CW modified 316. *J. Nucl. Mater.*, 278, 320–327.

Ukai, S., Okuda, T., Fujiwara, M., Kobayashi, T., Mizuta, S., Nakashima, H. (2002). Characterization of high temperature creep properties in recrystallized 12Cr–ODS ferritic steel claddings. *J. Nucl. Sci. Technol.*, 39(8), 872–879.

Ukai, S., Yamazaki, Y., Oono, N., Hayashi, S. (2013). Corrosion behavior of 9CrODS steel by simulated fission product cesium and tellurium. *J. Nucl. Mater.*, 440, 39–45.

Ukai, S., Oono, N., Kaito, T., Torimaru, T., Kimura, A., Hayashi, S. (2016). Development of FeCrAl-ODS steels for accident tolerant fuel (ATF) cladding in lightwater reactors. In *Proc. Ninth Pacific Rim International Conference on Advanced Materials and Processing (PRICM9)*, Furuhara, T., Nishida, M., Miura, S. (eds). The Japan Institute of Metals and Materials, Sendai.

Ukai, S., Ohtsuka, S., Kaito, T., De Carlan, Y., Ribis, J., Malaplate, J. (2017). Oxide dispersion-strengthened/ferrite-martensite steels as core materials for Generation IV nuclear reactors. In *Structural Materials for Generation IV Nuclear Reactors*, Yvon, P. (ed.). Woodhead Publishing, Duxford.

Ukai, S., Kato, S., Furukawa, T., Ohtsuka, S. (2020). High-temperature creep deformation in FeCrAl-oxide dispersion strengthened alloy cladding. *Mater. Sci. Eng. A*, 794, 139863.

Wagner, C. (1961). Theorie der alterung von niederschlägen durch umlösen (Ostwald-reifung). *Elekrochem.*, 65, 581–591.

Wilms, M.B., Streubel, R., Fromelc, F., Weisheita, A., Tenkamp, J., Walther, F., Barcikowski, S., Schleifenbaum, J.H., Gokceb, B. (2018). Laser additive manufacturing of oxide dispersion strengthened steels using laser-generated nanoparticle-metal composite powders. *Proc. CIRP*, 74, 196–200.

Yamashita, S., Ioka, I., Nemoto, Y., Kawanishi, T., Kurata, M., Kaji, Y., Fukahori, T., Nozawa, T., Sato, D., Murakami, N. et al. (2019). Overview of accident-tolerant fuel R&D program in Japan. *Proc. Top Fuel 2019*, 206–216.

Yano, Y., Ogawa, R., Yamashita, S., Ohtsuka, S., Kaito, T., Akasaka, N., Inoue, M., Yoshitake, T., Tanaka, K. (2011). Effects of neutron irradiation on tensile properties of oxide dispersion strengthened (ODS) steel claddings. *J. Nucl. Mater.*, 419, 305–309.

Yano, Y., Tanno, T., Oka, H., Ohtsuka, S., Inoue, T., Kato, S., Furukawa, T., Uwaba, T., Kaito, T., Ukai, S. et al. (2017). Ultra-high temperature tensile properties of ODS steel claddings under severe accident conditions. *J. Nucl. Mater.*, 487, 229–237.

Yano, Y., Sekio, Y., Tanno, T., Kato, S., Inoue, T., Oka, H., Ohtsuka, S., Furukawa, T., Uwaba, T., Kaito, T. et al. (2019). Ultra-high temperature creep rupture and transient burst strength of ODS steel claddings. *J. Nucl. Mater.*, 516, 347–353.

Yutani, T., Nomura, S., Koyama, S., Kuwajima Y., Ukai, S. (1993). High temperature chemical reactions of Fe-Cr-Ni and Fe-Cr cladding alloys by Te, I, and CsOH/CsI. *J. Nucl. Mater.*, 201, 35–45.

10

Materials Discovery

Andrea JOKISAARI, Cheng SUN and Jian GAN
Idaho National Laboratory, Idaho Falls, Idaho, USA

10.1. Materials discovery – numerical design

10.1.1. *What is materials discovery and is it needed in the nuclear industry?*

Materials discovery is the process of combining atomic elements and fabrication processing methods to create a new material with novel properties and performance characteristics. Materials discovery is closely linked with materials design, in which atomic elements and processing methods are selected to create a material to meet specific performance metrics. Historically, materials discovery has been a trial-and-error experimental process (Hautier et al. 2012; Liu et al. 2017; Cole 2020), making the development of materials iterative and slow. However, materials discovery has been recently discussed within the contexts of materials informatics (Hautier et al. 2012; Liu et al. 2017; Cole 2020), which aims to accelerate the process by leveraging high-throughput computations and large experimental data sets with (or without) the use of artificial intelligence (AI) or machine learning (ML). In many of these approaches, first-principles calculations are used to screen novel compounds for specific bulk crystalline properties (Oganov et al. 2019), looking for specific properties, such as band gap (Brunin et al. 2019), superconductivity (Eremets et al. 2008), or ultra-hardness (Cheng et al. 2014). A related approach to materials design is integrated computational materials engineering (ICME), in which a multiscale modeling approach is used to design a material, including its manufacturing pathway, to achieve specific performance requirements (Allison et al. 2006; Panchal et al. 2013). A natural integration of materials discovery (e.g. for

Materials and Processes for Nuclear Energy Today and in the Future,
coordinated by Fanny BALBAUD-CÉLÉRIER and Céline CABET.

materials with novel properties) and ICME for materials design is occurring, for example, with the development of novel magnesium alloys for the automotive industry (Allison et al. 2006).

Materials within a nuclear reactor can be divided into those that experience neutron irradiation (within the core and primary coolant loop), and those that do not (everything else, such as the secondary loop). Within the core, materials can be classed as fuel, fuel cladding and structural materials. Structural materials are present for thermomechanical, rather than neutronic, purposes. Depending on the type of reactor and where a component is located within the reactor, structural materials can be subject to corrosive environments, high temperatures, high stresses and significant neutron fluxes (Aguiar et al. 2020).

Materials must be qualified by standards bodies to be deployed in reactor service (Aguiar et al. 2020), and qualification is the final step in utilizing newly discovered and developed materials. In essence, qualification is the demonstration that the final manufactured component exhibits or exceeds the properties that were used in the design of the part. This process involves both the establishment of design properties and limits for materials, as well as the definition and validation of a process that can be used to ensure that the actual as-built material properties meet or exceed the design properties. Therefore, qualifying components require that the given fabrication techniques deliver a known set of design properties and that a means exists for certification.

Materials discovery and design is crucial in the field of nuclear energy because of the need for materials that can perform within the demanding service environments of nuclear reactors, which also have less environmental impact during fabrication and less reliance on critical materials, while still having acceptable cost. The development of materials for light water reactors (LWRs) was a decades-long process for a relatively limited set of service conditions. However, Generation IV reactor designs encompass a much wider and varied range of service conditions (Albert Sr 2021). Only six structural materials are qualified for deployment in advanced reactor conditions, while hundreds of structural materials are qualified for LWRs. Such a limited set of materials for advanced reactors severely restricts the ability of materials engineers to select functional and cost-effective materials for each reactor design, and more qualified materials are strongly needed.

10.1.2. *Historical perspective*

Materials development for nuclear reactors has historically occurred in costly and time-consuming experimental campaigns involving 10-year development cycles, with each material requiring multiple (e.g. three) cycles (Aguiar et al. 2020). These development cycles include design, fabrication, irradiation testing and characterization. Most current materials in service for nuclear reactors were developed by 1960 (Figure 10.1) (Zinkle et al. 2016), with minor compositional changes to improve performance, for example, the long-running development process of Zircaloy-2, Zircaloy-4, Zirlo and M5 for zirconium alloy fuel cladding (Zinkle et al. 2016). These massive experimental campaigns gathered data to develop empirical models of materials behavior under specific conditions. However, these empirical models lack predictive power outside the ranges of the gathered data, meaning they cannot be used to qualify materials for different service environments.

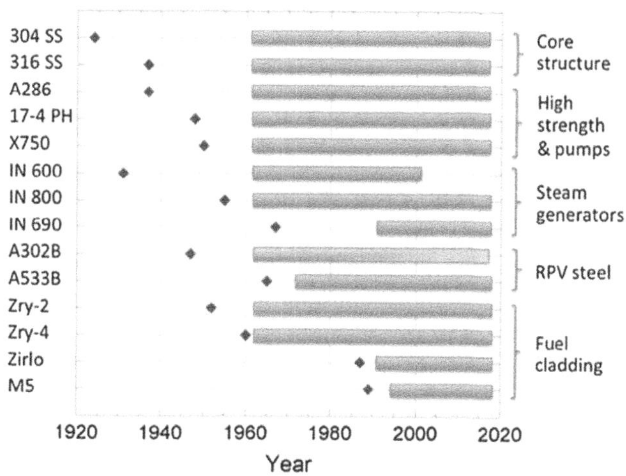

Figure 10.1. *The time at which each structural material alloy for LWRs was initially commercialized (diamonds) and the approximate timeframe that these alloys have been in service in LWRs. Reproduced from Zinkle et al. (2016). For a color version of this figure, see www.iste.co.uk/balbaud/nuclearenergytoday.zip*

The study of the embrittlement of reactor pressure vessel (RPV) steel for LWRs is a prototypical example of the extensive experimental campaign on nuclear materials (Odette and Lucas 1998) and its progression from empirical models (Odette et al. 2019) to physics-based models (Mamivand et al. 2019; Shu et al. 2019). RPV steels were developed in the 1940s through 1960s as low alloy steels. Major benefits to the nuclear industry would result if new steel alloys were adopted (Zinkle et al. 2016), but the slow pace of material development and qualification and the incremental approach to improvements in the nuclear industry have hampered the adoption of new alloys.

Similarly, zirconium alloys for LWR/CANDU fuel claddings were developed from the basic zirconium properties of low neutron cross-section and relatively good resistance to water corrosion. The development of Zircaloy alloys arose from the lucky observation that stainless steel impurities introduced during zirconium ingot melting improved corrosion resistance (Zinkle et al. 2016). Further incremental improvements over the ensuing six decades focused on minor modifications of alloy composition and changes in fabrication method to improve the behavior of fuel cladding in normal operating conditions (Sabol 2005).

10.1.3. *Modern approach*

In the current world environment, there are significant pressures to accelerate the deployment of advanced reactors, requiring an acceleration of the development of novel materials. Modern materials science has uncovered many principles governing materials properties across the atomistic, mesoscale and macroscopic length scales. The current paradigm of materials focuses on the so-called structure–property–processing–performance tetrahedron that conceptually links the formation of microstructure and composition via processing methods to materials properties and performance (McDowell and Olson 2008).

Computational materials science has proliferated enormously with the rise of computing power, allowing materials theory to move from qualitative descriptions of simplified systems to quantitative descriptions of real-world materials. In the 1980s, the Steel Research Group at Northwestern University was a pioneer of the use of computational modeling and experimental characterization for materials design. Programmatic efforts to harness computational tools for materials arose in the 1990s (McDowell and Olson 2008). These efforts linked microstructures, mechanics, hierarchical materials properties and statistics with simulations and experiments. In the United States, the DARPA Accelerated Insertion of Materials (AIM) program focused on Ni-based superalloys for gas turbine engines and composite airframe materials, while the Department of Energy-sponsored USCAR program focused on the fatigue strength of cast automotive components (McDowell and Olson 2008).

Further development of this idea led to the conceptualization of ICME, which addresses specific problems for individual products for certain applications with multiscale materials models. When fully realized, ICME enables concurrent analysis of manufacturing, design and material selection by integrating information across length and time scales for relevant materials phenomena and leveraging structure–property–processing–performance relationships (Allison et al. 2006b). A similar idea, launched in 2011, is the Materials Genome Initiative (MGI), a coordinated federal-level effort within the United States (National Science and Technology Council 2021). The MGI aims "to accelerate the discovery, design, development, and deployment of new materials, at a fraction of the cost, by harnessing the power of data and computational tools in concert with experiment" (National Science and Technology Council 2021). The MGI lays out a vision for achieving this goal through the creation of a Materials Innovation Infrastructure, a framework that relies on tightly integrated iteration of computational and experimental tools embedded in a data infrastructure. This framework allows for rapid prediction of materials properties and performance, and, ultimately, the ability to design the materials concurrently with the design of the product that will incorporate the new material. The concepts of both ICME and MGI have spread globally beyond their initial scope within the United States. For example, the Mission Innovation project, started in 2015, is a global initiative of 24 countries and the European Commission that is working to accelerate the use of clean energy technology. It includes a focus on the development of high-performance clean energy materials that are environmentally friendly and low-cost, leveraging rapid characterization and AI techniques.

The use of computational tools in the realm of nuclear materials discovery, design and qualification is still relatively nascent, but other industries have applied the ICME approach successfully. The ICME approach has been successfully deployed to lightweight materials used in vehicles (automotive and aerospace), with a particular focus on aluminum, magnesium, titanium, and advanced high strength steel alloys. Multiple automotive companies have demonstrated the use of ICME to improve aluminum powertrain castings (Allison et al. 2006b; Wang et al. 2011), while other research using physics-based modeling in wrought aluminum alloys in solute strengthening and warm formability has helped address two technological hurdles for increased deployment of aluminum (Krajewski et al. 2010; Leyson et al. 2010). Similar work has helped contribute to the development of magnesium systems (Yasi et al. 2010; Ganeshan et al. 2011) and titanium systems (Liang et al. 2020). A new titanium aerospace alloy has recently been developed using the Ti-Al-Fe system as a lower cost substitute to the canonical Ti-6Al-4V alloy (Liang et al. 2020). In this case, an alloy composition was selected using CALPHAD-based thermodynamic modeling to develop a two-phase microstructure but avoid the formation of brittle TiFe intermetallics. In addition, casting simulations using alloy-specific material properties were used to optimize the casting design to avoid casting defects. The alloy design was validated by evaluating prototype casting porosity and microstructure.

Advanced high strength steels are another major target for lightweighting of vehicles. A recent approach to developing a TRIP steel included high-fidelity multiscale modeling and advanced characterization to understand its mechanical response. The composition-dependent elastic constants of each phase were calculated via density functional theory (Fellinger et al. 2017), while crystal plasticity simulated the mechanical response of microstructure-based representative volume elements (Savic et al. 2018). In situ high energy X-ray diffraction tensile test data were used to validate the material model. This steel was deployed in an automotive design and ultimately, a 30% reduction in vehicle weight was achieved (Savic et al. 2018).

Perhaps the most notable successes of the application of ICME to discover, design, qualify and deploy a novel material included two novel steels, Ferrium C61 (Kuehmann and Olson 1998) and Ferrium S53 (Kuehmann and Olson 2011). Ferrium C61 is a high durability gear steel, and Ferrium S53 is a corrosion-resistant landing gear steel. The design of these alloys followed the concepts laid out in a flow-block diagram, which links processing steps, microstructure development, properties linked to the microstructures, and finally the performance arising from the properties. For Ferrium S53, the design process started with the identification of key design tradeoffs and the microstructural mechanisms to achieve specific properties, specifically the desire to use coherent, nanoscale M_2C carbides as strengtheners of BCC Fe. Thermodynamic and kinetic modeling was used to predict the precipitation and coarsening behavior of the M_2C carbides as a function of Mo and V content, allowing the strength to be maximized with reasonable tempering conditions. Martensite kinetic transformation was also modeled as a function of Co and Ni content; the combined M_2C precipitation and martensite start temperature was predicted. Ductile fracture also optimized by using first principles calculations of grain boundary chemistry and grain boundary pinning. Production-scale processing suitability was modeled with solidification simulations, including multicomponent diffusion to understand segregation within dendrites and post-casting homogenization; these predictions were validated with data from actual production ingots. Mechanistic models were used to predict mechanical properties and their statistical variation, allowing major reductions in the number of samples and therefore cost required to understand property design minimums (Kuehmann and Olson 2011).

The specific ICME strategy employed will vary considerably, depending on the design goals, spanning from detailed quantitative results to qualitative insights to provide guidance (Joost 2012). In all cases, the determination of a "foundational engineering problem" can help determine the required fidelity and scalability of the approach. The foundational engineering problem correctly describes the concerns to be addressed, defines the level of fidelity of the methods, and specifies the degree of reusability of the study. This approach could be used in conjunction with the flow-block diagram example discussed in the Ferrium S53 case above. Physics-based computational approaches that can be used to discover and design materials include:

– atomistic methods such as density functional theory (DFT), kinetic Monte Carlo, cluster expansion and molecular dynamics (MD), which can determine bulk phase properties, interface characteristics, diffusivities and precipitation of phases;

– spatially resolved mesoscale approaches, including dislocation dynamics, crystal plasticity, phase field and cellular automata, which observe mechanics phenomena and microstructural evolution;

– mean field mesoscale approaches include cluster dynamics, irradiation damage calculations (e.g. NRT and arc-dpa models), Mansur invariant relationships and CALPHAD-based free energy descriptions and diffusion simulations, predicting average properties;

– macroscopic approaches to model workpieces, such as casting simulations and tensile test simulations, to determine bulk behavior and manufacturability.

In the area of nuclear materials, physics-based modeling has primarily focused on understanding environmental effects that cause material degradation, rather than the development of new materials. The discovery and design of new materials for nuclear applications is still generally proceeding with conventional materials development approaches, typically experimental campaigns without the incorporation of physics-based modeling. New materials or technologies recently developed focus largely on accident-tolerant materials for LWRs that perform similarly to conventional materials under nominal conditions, and yet provide more desirable properties in off-normal conditions (Terrani 2018a). Materials under development include Cr-coated zirconium alloy cladding, FeCrAl cladding and SiC/SiC cladding (Terrani 2018b). For fuel, Cr_2O_3 doping of UO_2 fuel is being studied to enhance accident tolerance and improve high burnup performance. Chromia doping increases grain size and is predicted to improve the accident behavior of the fuel (Che et al. 2018). Carbon fiber reinforced composites and other SiC/SiC composites are being developed for extremely high-temperature regimes, such as those found in HTGR and VHTR reactor concepts, and code qualification is ongoing at a low level (Zinkle et al. 2016). Advanced steels, such as reduced activation ferritic/martensitic (RAMF) steels and oxide dispersion strengthened (ODS) steels, are also being developed with improved mechanical behavior (e.g. creep resistance) and reduced irradiation-induced swelling over existing austenitic and ferritic steels (Raman et al. 2016; Zinkle et al. 2017). Some thermodynamic modeling is being performed to increase the thermodynamic stability and density of desirable precipitates (Zinkle et al. 2017), but generally speaking, much work remains to be done to improve the manufacturability of these materials, especially the ODS steels, which obtain their desirable properties from nanoscale dispersions of second phases that are challenging to form in bulk and to perform joining operations on (Raman et al. 2016).

The very recent development of so-called "high-entropy alloys" (HEAs), also referred to as "complex concentrated alloys", "multi-principal element alloys", and other related terms, are beginning to be considered as a new material class for Generation IV fission reactor designs and fusion reactors (Pickering et al. 2021). These alloys are intended to contain roughly equimolar amounts of five or more constituent elements to access new property spaces not available from only one element as the base alloy material (Pickering et al. 2021). HEAs generally appear to have good high temperature mechanical properties and radiation stability. To-date, most alloy development and testing has been experimental in nature, though one promising computational approach has been developed to predict the formation and stability of undiscovered single-phase HEAs (King et al. 2016). It must be noted that CALPHAD-based methods are somewhat unreliable for HEAs due to the theoretical basis of the methodology, though accurate thermodynamic databases could be developed with targeted experimental campaigns characterizing that specific alloy space (Gorsee and Senkov 2018). Most of the high-fidelity modeling and simulation of HEAs has been focused on understanding and predicting the irradiation performance of this alloy class (Zhang et al. 2017).

Another very recent development in the nuclear materials area is the use of advanced manufacturing techniques, especially additive manufacturing (AM) techniques, to fabricate nuclear components (Sun et al. 2021). Advanced manufacturing involves novel fabrication methods beyond the conventional casting and bulk forming operations to provide complex component structures, functionally graded properties and fabrication of materials too challenging to form by conventional methods. Advanced manufacturing methods include novel forming methods themselves, such as laser metal AM, as well as novel on-the-fly process monitoring and parameter adjustment techniques. Advanced manufacturing techniques can also include high-performance coating technologies, such as cold spray, magnetron sputtering and chemical vapor deposition. Coatings can be applied to bulk structural materials to improve behaviors such as corrosion resistance (Wu et al. 2020) and fretting performance, though to-date, most coating studies have focused on fuel cladding (Maier et al. 2015; Alat et al. 2016; Kuprin et al. 2021). Several studies of the AM of materials relevant to nuclear materials (e.g. functionally graded 304L stainless steel and Inconel 625) have involved computational thermodynamics modeling

(Carroll et al. 2016). The use of a CALPHAD-based ICME framework can be leveraged to optimize alloys and quantify uncertainty for AM and address deviation of powder compositions for AM (Wang and Xiong 2020). The Transformational Challenge Reactor (TCR) program under the United States Department of Energy, now a part of the Advanced Materials and Manufacturing Technologies (AMMT) program, had a major focus on the use of AM to overcome materials limitations for fuel fabrication (Nelson 2019). This program spearheaded the use of a "digital twin" to link AM processing parameters to microstructure and properties of additively manufactured components. Under the AMMT program, the use of AM techniques would be pursued for the development and qualification of laser powder bed fusion-fabricated 316 stainless steel for deployment in the nuclear industry (Department of Energy 2022). The program leveraged ICME strategies to achieve these goals and to develop novel materials specifically for fabrication with AM techniques (Department of Energy 2022).

ML is another novel computational approach being taken to study nuclear materials. As described in Morgan and Jacobs (2020), ML is a broad class of computer algorithms that do not require explicit programming to solve the specific tasks they are applied to. ML is a highly active area of research for topics such as material property prediction, especially for multi-component systems that are computationally expensive for purely physics-based modeling methods, and materials discovery and design. Insights from ML can provide guidance to discover and design materials with target properties. In addition, ML can be used to develop interatomic potentials for molecular dynamics, improving the descriptions of atomic interactions by reducing constraints on the mathematical descriptions. Materials characterization can also be accelerated using ML, for example, for computer vision to classify microstructure features in micrographs (Morgan and Jacobs 2020). Very recently, ML has been applied to predict the irradiation hardening behavior of RPV steels at lower fluxes than experimentally characterized (Liu et al. 2022). A recent review of ML for nuclear materials indicates that most research to-date focuses on predicting irradiation effects, either mechanical property changes or radiation scintillator detector material (Morgan et al. 2022). ML can also be used to accelerate experiments for high-throughput processing and characterization techniques (Morgan et al. 2022), though the specific challenges presented by nuclear materials require research to address.

10.1.4. *Future needs*

The stage is set for a 21st-century approach to materials discovery in nuclear structural materials, but certain gaps remain. Nuclear materials systems are typically compositionally complex, and therefore nuclear material discovery needs to harness high-throughput atomistic calculations (especially DFT) that leverage ML techniques to predict the energetic and kinetic properties of these complex systems. ML also holds the promise of improving the accuracy of DFT and MD predictions themselves by improving their interatomic descriptions, particularly for highly correlated systems with partially filled *d* or *f* electron shells. Similar improvements in accuracy could be made for any type of model that needs to be parameterized for a specific material, such as crystal plasticity. In addition, ML could be used to develop reduced order models to bridge length and time scales.

The discovery of nuclear materials, which must be linked to its intended performance characteristics, has the added challenge of being intrinsically linked with the evolution of the material in its reactor environment. The number and type of environments is proliferating due to the desire to realize Generation IV reactor designs. However, challenges still exist to quantitatively predict materials evolution under irradiation. High-fidelity scale bridging from the atomistic, picosecond irradiation damage event to diffusional timescales is under active development. The use of accelerated damage production as a testing method for reactor service conditions is making significant strides, but important gaps must still be addressed. To fully harness accelerated testing, quantitative linking of variations in microstructure evolution and mechanical properties under high neutron flux versus low neutron flux, and especially charged particle irradiation versus neutron irradiation, must be made. This linking will likely take the form of a combined physics-based modeling and ML approach. In addition, a data framework to capture data gathered by experiments must be identified and developed so that existing and newly generated data can be leveraged by the community.

Finally, a new qualification regime is needed for advanced manufacturing approaches; advanced manufacturing offers significant promise for materials discovery and optimization. Traditionally, qualification of a process occurs. However, in the case of metal AM, any change in scanning parameters (spatial pattern, energy level, etc.) would require material requalification. Fully utilizing advanced manufacturing capabilities requires the development of models linking process conditions to materials properties via microstructure across an entire fabricated component. A microstructure-based qualification regime could leverage structure–property relations to determine whether a change in process parameters causes appreciable property changes or falls within a pre-existing envelope of material variability; this would likely entail machine vision for microstructure identification, physics-based models of microstructure evolution and ML of structure–property linkages. Uncertainty quantification in processing–properties–performance will be a major, necessary challenge to overcome, and computational methods should play a large role in addressing that issue.

10.2. Materials discovery – advanced manufacturing processes

The development of advanced nuclear systems calls for integrated material systems with multi-role application and significantly improved performance in reactor environments. AM is a process of fabricating three-dimensional objects by depositing materials layer-by-layer directly from a computational geometry model, and it eliminates the design and fabrication restrictions of conventional manufacturing methods to a large extent. As an emerging and transformative technology, AM technologies have shown the advantages of field fabrication and flexible manufacturing for nuclear components. To further increase their applications in nuclear energy, fundamental research on the environmental impact of AM materials is needed to qualify AM processes and AM materials. In this section, the current status of cutting-edge AM technologies and irradiation damage of AM materials will be summarized. The major challenges and fundamental research needed to achieve the full potential of AM technologies are discussed.

10.2.1. *AM technologies*

AM techniques have been used to fabricate a wide variety of materials, including ceramic, glass, metals, polymers and composite materials. Laser- and electron-beam-based power sources are the two major popular power sources used in AM processes. Various AM techniques, such as direct energy deposition (DED), powder-bed fusion (PBF), etc., have been successfully applied to fabricate materials and components for nuclear applications.

DED uses focused thermal energy to fuse powders or wires by melting as they are deposited on the substrate. The DED process has been used to print structural materials and coatings for nuclear core components. Figure 10.2(a) shows spherical powder particles that are used for the DED process. Figure 10.2(b) shows an example of fabricating thin-wall fuel cladding tube using the DED process at Idaho National Laboratory. The DED process has been used to fabricate Inconel-718 thin wall structures (Jinoop et al. 2019). The microstructure of DED fabricated alloys typically shows fine dendritic and cellular grains with high density of sub-grains. The grain morphology is strongly dependent on the printing orientation and printing patterns, which leads to anisotropic mechanical properties. Significant residual stress can be developed in DED fabricated components, and proper post-heat treatment is required to remove the residual stress and improve mechanical properties. HEAs are gaining interest in nuclear energy due to the enhanced irradiation tolerance and corrosion resistance in reactor environments. $Al_xCoCrFeNi$ ($x = 0.3$, 0.6 and 0.85) HEA claddings were fabricated on an austenitic stainless steel substrate using direct laser deposition process (Chao et al. 2017). A mixture of blended elemental powders was readily modified and used as feedstock. With increasing the aluminum mole fraction, the crystal structure evolves from FCC, to FCC+BCC, and then to BCC. Although DED has been widely used to fabricate nuclear structural materials, the rough finishing surfaces, porosity, high equipment cost and residual stress development are the major issues. Post-heat treatment is typically required to relieve residual stress and homogenize the microstructure.

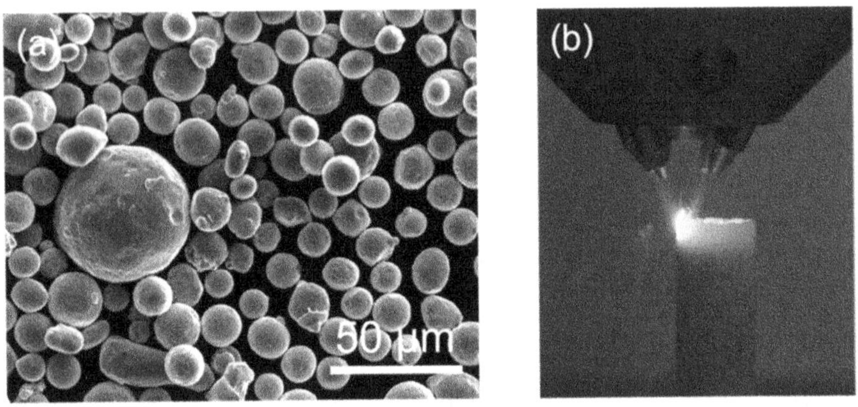

Figure 10.2. *DED process of thin-wall structures. (a) Particle morphology of 316L stainless steel powders. (Shiau et al. 2021). (b) Printing of fuel clading using DED process at Idaho National Laboratory. For a color version of this figure, see www.iste.co.uk/balbaud/nuclearenergytoday.zip*

PBF is an AM process where a heat source (laser or electron beam) selectively fuses or melts powders layer by layer to form three-dimensional objects. The powders for the PBF process require fine grains with thermoplastic properties, so that it becomes viscous, flows and solidifies quickly. The complicated heat history can result in high density of defects and flows, such as pores, impurities, dislocation networks, cracks, etc. The heterogeneous microstructure leads to the anisotropic mechanical properties of PBF components. Thin, rectangular plates of Ti-6Al-4V alloys were fabricated by PBF and the columnar microstructure exhibits both beta and alpha prime phases (Voisin et al. 2018). A pore growth mechanism was identified when the tensile axis was perpendicular to the building direction, while such a mechanism was limited when the tensile axis is parallel to the building direction. The anisotropic pore growth is attributed to the differences in the strain-to-failure of materials. The major advantages of using PBF process include the capability of rapid prototyping, elaboration of complex geometries including lattice structures and efficient recycling of un-melted powders.

AM techniques open the door to develop a new class of functionally graded materials. Unlike "traditional" AM materials, which focus on the homogeneity in structure and composition, functionally graded materials intentionally introduce heterogeneous layers to combine the advantages of each material. For example, for nuclear components made with 316 SS, surface layer coatings having stronger oxidation/corrosion resistance can improve the overall performance of the components in combined extreme environments. The DED process has been employed to fabricate bi-metallic structural materials. Three bi-metallic structures were fabricated: Cu-H13 (Fe-5.25Cr-1.35Mo-1Si-1V-0.4Mn-0.4C) joint, Cu-D22-H13 joint with D22 (Ni-2.5Si-1.4B-1Fe-0.05C) as a buffer layer and D22-H13 joint (Zhang et al. 2020b) (Figure 10.3). Directly joining Cu on H13 resulted in cracking at the interface, while with D22 as a buffer layer, Cu was successfully deposited on H13 without clear cracks. A sharp transition of elemental contents was experienced at the D22-H13 interface due to very limited interdiffusion, while a gradual transition of Cu and Ni was observed across the D22-Cu interface, indicating a successive elemental diffusion. During the tensile tests at room temperature, Cu-D22-H13 specimens fractured in the Cu zone and D22-H13 failed in the D22 region, although elongation mostly happened in the H13 section. The interfaces of both Cu-D22-H13 and D22-H13 bi-metallic structures were survived during tensile testing, indicating a strong bonding strength. Functionally graded Ti-6Al-4V/TiC composites with increasing TiC from 0% to 50% with the step of 5% were fabricated by laser metal deposition (Mahamood and Akinlabi 2015). The optimized functionally graded Ti-6Al-4V/TiC exhibits an improved wear-resistance behavior. 316L stainless steel/Inconel 718 thin-wall structure was produced by laser direct metal deposition (Shah et al. 2014). The laser energy density and powder feeding rate show a great impact on tensile strength and secondary dendritic arm spacing. The tensile strength of the functionally graded materials is inversely proportional to the laser powder and increases with powder feeding rate. Carroll et al. (2016) joined Inconel 625 and 304L stainless steel through 24 layers of gradient zone where IN625 increased by 4% each layer using the DED technique. The results revealed no sharp transition of microstructure and elemental composition along the grading route, although cracking was observed at a certain zone due to formation of a secondary phase. Thin wall SS 316L/Fe3Al tubes graded along the radial direction were achieved with good shape and metallurgical quality by optimizing process parameters (Durejko et al. 2014). Component repairing has been achieved by the DED process. AISI H13 tool steel with hemisphere-shaped defects were repaired using Co-based alloys powders as filler materials (Zhang et al. 2018). Further development of functionally graded materials by AM techniques will advance the rational design and repair of materials and components for nuclear energy systems.

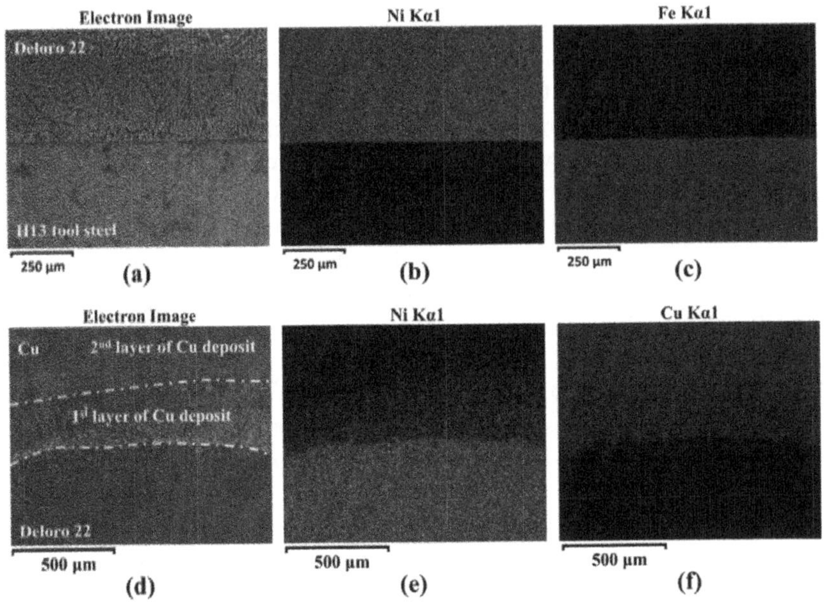

Figure 10.3. *Fabrication of copper-H13 tool steel bi-metallic structures with Deloro 22 interlayer by DED process (Zhang et al. 2020b). For a color version of this figure, see www.iste.co.uk/balbaud/nuclearenergytoday.zip*

10.2.2. *AM materials under nuclear environments*

10.2.2.1. *Irradiation damage*

Irradiation significantly degrades the performance of nuclear materials in service, and therefore the safety and reliability of nuclear reactors. Irradiation-induced damage could result in significant changes in microstructure and chemistry. Understanding the degradation mechanisms of AM materials in extreme environments is imperative for their qualification and licensing. Only a few studies have reported the irradiation response of AM materials. The current existing literature suggests that some types of defects and flaws in AM materials can have a positive impact on the irradiation tolerance of materials. Shiau et al. (2021) performed proton irradiation experiments for AM and wrought 316L stainless steels at 360°C to 1.8 and 2.5 dpa and observed that the DED fabricated 316 L SS exhibits a stronger void-swelling resistance, that is, the swelling rate was decreased from 6.0% in wrought specimen to 1.6% in AM specimen. The dislocation loop density in the AM specimen is also lower than that in its wrought counterpart. AM-induced features, such as nanopores and sub-grain boundaries, could serve as defect sinks to absorb irradiation-induced defects. The deformation behaviors of as-manufactured, annealed, and proton-irradiated AM 316L stainless steel were studied by micro-pillar compression (Shiau et al. 2021) (Figure 10.4). Manufacturing-induced defects and irradiation-introduced defects (dislocation loops and voids) show great impact on the deformation mechanisms. In the as-manufactured alloy variant, dislocation slip is the major deformation mechanism, with noticeable resistance by manufacturing-induced pore structures. In the annealed variant, hardness was reduced, although the deformation mechanism remains dislocation gliding. In the proton-irradiated variant, twinning occurs with a high density of irradiation-induced voids. Irradiation hardening could become high enough to activate the twinning process. The effect of manufacture defects on the irradiation damage of AM materials was studied by in-situ Kr ion irradiation. Shang et al. (2019) provide a direct observation of cellular dislocation wall evolution under irradiation in a TEM and conclude that such dislocation structures in the sub-grain boundaries could serve as effective defect sinks by absorbing dislocation loops. Another study by Shang et al. (2021) indicates that high-density solidification cellular structures in AM 316LN stainless steels may largely suppress the formation of irradiation-induced perfect dislocation loops and dislocation networks. Irradiation-induced solute segregation near the cellular structure is alleviated in the AM specimens, compared to that near high-angle grain boundaries.

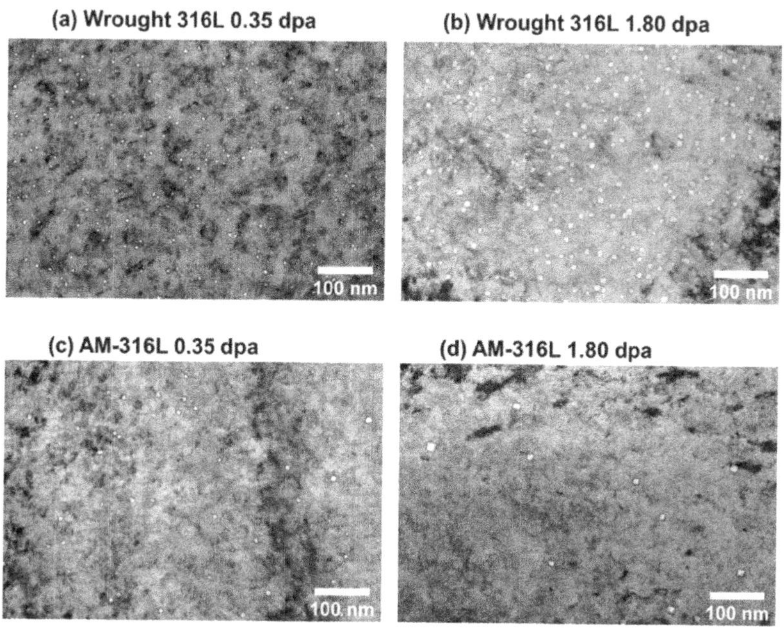

Figure 10.4. *Proton irradiation-induced void swelling in AM and wrought 316L stainless steels at 0.35 and 1.8 dpa. (a and b) Wrought 316L stainless steels. (c and d) AM 316L stainless steels. AM 316L stainless steel shows an enhanced void swelling resistance compared to its wrought counterpart (Shiau et al. 2021)*

10.2.2.2. *Stress corrosion cracking*

Stress corrosion cracking (SCC) has been reported in water-cooled nuclear reactors, including pressurized water reactors and BWRs. SCC typically results from combined reactor environments of stress, water chemistry and temperature. SCC testing was performed on PBF fabricated 316L stainless steels in a simulated BWR environment at 288°C (Lou et al. 2017a). With proper heat treatment, AM 316L exhibits SCC growth rates similar to its wrought counterpart. An anisotropic SCC response was observed in the stress-relieved (SR) AM 316L stainless steels, that is, cracking preferentially occurs along the building direction, due to the non-equilibrium heterogeneous microstructure developed during printing. Corrosion fatigue cracking behavior of PBF fabricated AM 316L stainless steels was studied in oxygenated water at 288°C (Lou et al. 2017b). SR specimens show a higher crack growth rate than annealed specimens and the influence of material microstructure becomes more significant when the loading frequency is decreased. High-temperature recrystallization treatment is recommended to improve the resistance to corrosion fatigue cracking.

One aspect of particular importance in light-water reactors is the increased susceptibility of materials to SCC under irradiation, termed irradiation-assisted stress corrosion cracking (IASCC) (McMurtrey et al. 2021). Initial results suggest that AM 316L stainless steel fabricated by DED shows improved resistance to IASCC at 360°C under proton irradiation with plastic strain up to 4%. As seen in Figure 10.5, severe cracking was present in the wrought 316L stainless steel. The AM 316L was much less susceptible to IASCC than wrought 316L, although IASCC was observed in the AM 316L when tested with the tensile axis parallel to the build direction. The lower density of radiation-induced voids and dislocation loops in AM 316L was believed to contribute to the IASCC resistance by decreasing the localized deformation in the irradiated specimens.

Post-heat treatment has great impact on the IASCC susceptibility. Irradiation-assisted stress corrosion cracking length was quantitatively measured and compared in SR and hot isostatic pressing (HIP) AM 316L (Song 2019). In SR AM316L samples, crack length per unit area is higher when pulled perpendicular to the printing direction (PD) than pulled parallel to the PD. The cracking susceptibility of SR AM 316L was comparable to that of the forged material, while HIP processing significantly reduces the cracking susceptibility of AM 316L. Figure 10.6 compares the IASCC susceptibility of AM and forged 316L after proton and Fe ion irradiation.

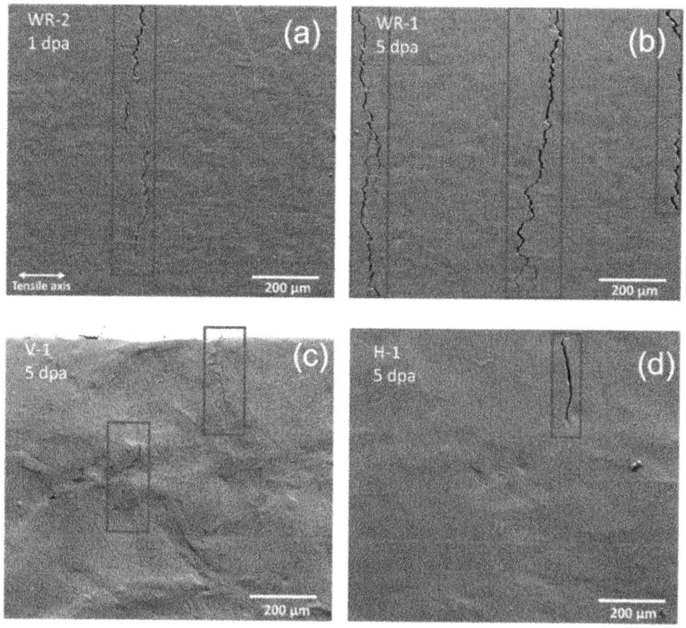

Figure 10.5. *Irradiation-assisted stress corrosion cracking in AM and wrought 316L stainless steels. (a and b) Wrought 316L stainless steel. (c and d) AM 316L stainless steels at vertical and horizonal orientations (McMurtrey et al. 2021). For a color version of this figure, see www.iste.co.uk/balbaud/nuclearenergytoday.zip*

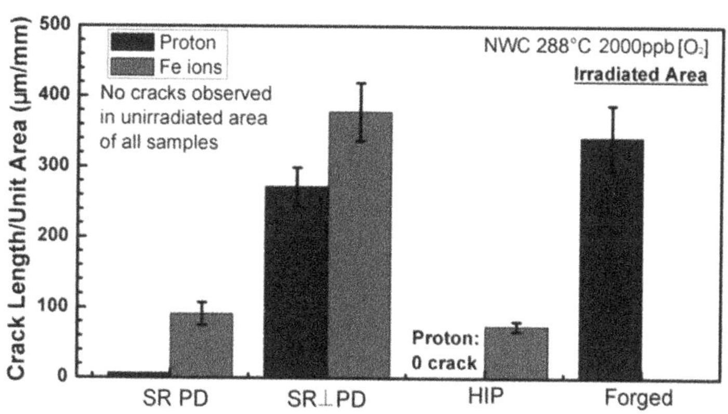

Figure 10.6. *Comparison of IASCC susceptibility of AM and forged 316L stainless steels after proton and Fe ion irradiation (Song 2019). For a color version of this figure, see www.iste.co.uk/balbaud/nuclearenergytoday.zip*

10.2.3. *Challenges of AM processes for nuclear materials*

10.2.3.1. *Residual stress*

As aforementioned, laser-based AM is a fusion-based process in which materials are locally melted and solidified with the aid of a laser heat source. A laser source can develop a large thermal gradient near the molten pools. Powders and droplets typically experience repeated rapid heating and cooling processes due to the layer-by-layer deposition fashion. Significant residual stress can be developed in AM materials, leading to the local cracking in the components.

The effect of inter-layer dwell time on the residual stress in laser-based directed energy deposited Ti6Al4V and Inconel 625 was studied (Denlinger et al. 2015). In AM Inconel 625, the residual stress decreases with increasing dwell times from 0 to 40 s. The residual stress in AM Ti-6Al-4V decreases with lower dwell times, and the residual stress reaches a minimum in Ti-6Al-4V alloy without dwell time. The residual stress was measured by the hole-drilling method. Deposition pattern also plays a significant role in affecting residual stress and deflection. A raster deposition pattern with lines oriented 90 from the beam's long axis leads to the lowest stresses and deflections in beam-shaped substrates. For plate-shaped substrate, the spiral pattern scanned from the outside to the inside produces low and uniform deflections (Nickel et al. 2001).

In general, excessive residual stress can be partially relieved by optimizing AM process parameters or the post-thermal annealing. Zhang et al. (2020a) measured the residual stress in DED-fabricated 316L stainless steel using synchrotron X-ray scattering. The as-printed SS316L exhibits compressive residual stress of –197.4 MPa. The residual stress is greatly relieved to –53.8 MPa after annealing at 1,093°C. In addition, in fabricating FGM, dissimilar materials may have significant variations in thermal expansion (Zhang et al. 2020a), which give rise to a significant amount of residual stress that, if not controlled, may result in severe distortion of the part.

10.2.3.2. *Surface distortion*

In laser-based AM fabrication, the large temperature gradients and nonhomogeneous cooling near the molten pool could induce considerable thermal stress, leading to surface distortion in the final building parts. Such distortion could result in heterogeneous properties of materials and instability of structural components. Nozar et al. (2018) investigated the effects of laser power on the dimension changes of AM products and revealed that a higher laser power causes nonuniform shrinkage and curling of final parts, while lower laser power causes incomplete sintering of the powders. He et al. (2021) simulated the surface profiles of PBF fabricated fuel cell components with Inconel 718 and 316L stainless steel, as shown in Figure 10.7. The simulation shows that laser power and scanning speed exhibit a significant impact on the surface distortion. Increasing the scan speed reduces the maximum distortion with a nonlinear correlation, while the laser power's impact is almost linear, and the distortion decreases with increasing laser power. 316L stainless steel components show much less distortion compared to Inconel 718 components. Despite the efforts, understanding and controlling the surface distortion remains challenging.

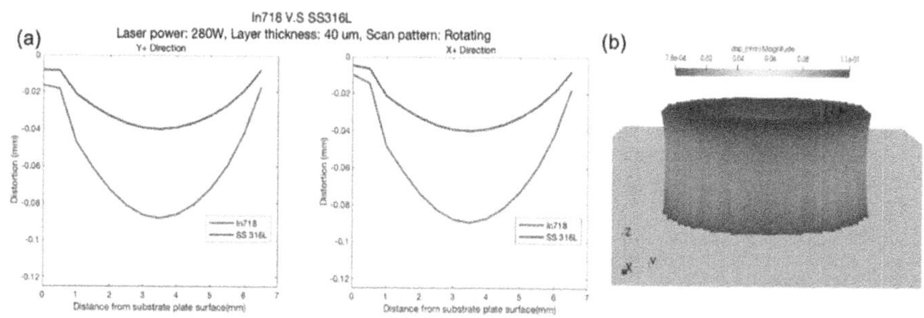

Figure 10.7. *Materials distortion in laser-based additive manufactured fuel cell components with In718 and 316L stainless steel (He et al. 2021). For a color version of this figure, see www.iste.co.uk/balbaud/nuclearenergytoday.zip*

10.2.3.3. *Heterogeneous microstructure*

Heterogeneous microstructures are typically formed in laser-based additive manufactured materials. The rapid heating and cooling rate causes the microstructure heterogeneity in terms of grain morphology and phase components. In Inconel 718 thin-walled builds, produced using direct laser deposition, heterogeneous grain structures and phases were observed (Parimi et al. 2014). At lower laser power, the microstructure shows a bi-modal distribution of grain sizes with a weak texture. At higher power, a strong <100> textured columnar structure was generated. Similarly, in DED fabricated 304 stainless steel, a mixture of elongated and equiaxed grains were formed (Li et al. 2022), as seen in Figure 10.8. A kinetic Monte Carlo modeling reveals that larger pores can easily truncate grains and thereby reduce grain size, while smaller pores are embedded in the grain interiors, resulting in irregular columnar grains. In addition, the fraction of equiaxed grains increases as the laser transverse speed increases (Li and Soshi 2019). Heterogeneous phase components could also be created in AM materials. In direct laser deposited Inconel 718, significant Laves phase precipitates were observed. When increasing the laser power, the Laves phase particles become larger. δ phase and carbides are also created at higher laser power (Parimi et al. 2014).

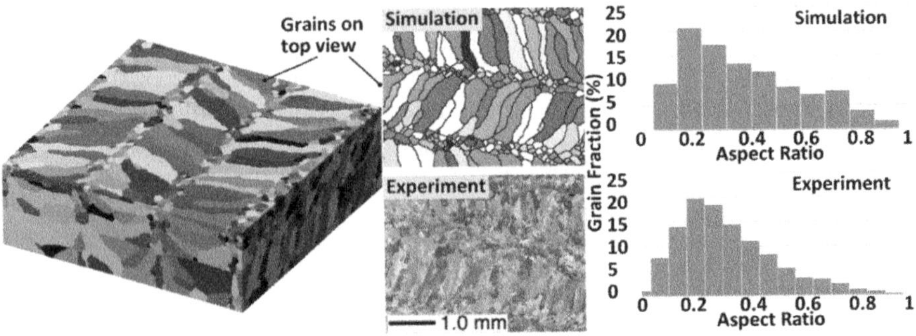

Figure 10.8. *Heterogeneous grain morphology in 304 stainless steels fabricated by direct energy deposition (Li et al. 2022). For a color version of this figure, see www.iste.co.uk/balbaud/nuclearenergytoday.zip*

10.2.4. *Summary*

AM techniques have shown great potential for the field fabrication of complex nuclear components. Several AM processes have been used to print prototype components for nuclear applications, while controlling microstructure and properties remains challenging. Qualifying and licensing AM processes and materials require more fundamental research on the correlation of microstructure, residual stress, surface quality and properties of additive manufactured materials. Despite the recent efforts on ion irradiation of additive manufactured materials, their performance in reactor environments (e.g., neutron irradiation, molten salt corrosion and irradiation creep) has not been fully studied. The aforementioned fundamental research is needed to achieve full potential of AM in nuclear energy.

10.3. Materials discovery – high-throughput characterization

High-throughput characterization (HTC) is essential to accelerate materials research and development (R&D) for broad applications, including the development of advanced materials for innovative nuclear energy systems. Historically, materials development for nuclear energy systems takes an even longer time than the already lengthy R&D period for conventional materials for non-nuclear applications. HTC for nuclear materials development typically may involve materials fabrication and sample preparation, microstructural characterization, mechanical property evaluation, irradiation and post-irradiation examination (PIE). Each of these areas will be discussed briefly in the following sections. HTC is expected to generate data at unprecedented speeds and quantities. With the advances of experimental and modeling capabilities, along with ML and AI in automated data analytics, the full potential of HTC may be realized and effectively utilized. However, some areas face significant technical challenges and will require substantial effort for improvement.

10.3.1. *Material fabrication and sample preparation*

High-throughput materials fabrication and sample preparation is critical for HTC. Materials R&D aims to optimize microstructure and microchemistry to ensure properties for enhanced performance, such as mechanical strength and ductility, dimensional stability, high-temperature strength, radiation tolerance, corrosion resistance and weldability. While modifying the material composition is often considered the first step in the search for the optimum alloy, a difference in thermomechanical processing (TMP) of a material with the same composition can lead to drastically different microstructures, which can significantly impact the performance. The recent review by Miracle et al. (2021) on emerging capabilities for HTC of structural materials is quite comprehensive and covers many relevant topics. It emphasizes the challenge of using combinatorial high-throughput testing for structural materials development because their properties are extremely sensitive to both composition and microstructure. The use of materials fabricated with continuous composition gradients or a series of small samples with discrete compositions becomes attractive due to the unique advantage of combinatorial synthesis over conventional bulk materials fabrication. However, the latter offers the advantage of building a sample library of a specific composition with different, controlled microstructures obtained by different TMP. Caution should be taken when using laser-based AM to build sample arrays with various compositions. The lack of control of the sample microstructure, plus the low feasibility of post-AM processing with TMP to refine the microstructure, are the two major concerns for this approach.

Even while the major elements of a material composition are retained, the addition of minor elements, often in trace amounts, can make a difference in fine-tuning the material properties, such as for the advanced stainless steels. It is the combination of major alloy elements, plus minor element additions, along with different routes of TMP, which determines the material properties. Ultimately, it is not only the bulk composition but the local microstructure and microchemistry that affect the material properties and performance, such as phase domains, ordered structures, grain texture and character, elemental segregation, dislocation configuration, precipitate distribution and morphology, and fine features and defects at the atomic scale. It is a daunting challenge to develop a sample library to cover all these material conditions, even just for a very limited composition space.

10.3.2. *Microstructural characterization*

Microstructural characterization from the macroscale (cm) to the atomic scale (Å) can reveal information about grain size and texture, crystal structure, dislocation configuration, second phase precipitates, internal stress state, elemental segregation at interfaces, phase decomposition, radiation-induced defects and phase changes, cavities, etc. All these microstructural features can contribute to macroscopic property changes, including hardening and embrittlement, loss of mechanical strength and ductility, softening, dimensional instability, creep and swelling, surface degradation and enhanced corrosion. The major instruments for microstructural characterization include optical microscopy for preliminary analysis over large sample areas, SEM for detailed characterization with composition, structural and grain texture and orientation distribution, and TEM for high-resolution defect analysis, structural and composition characterization down to the atomic scale, and charge state analysis in a compound via Electron Energy Loss Spectroscopy (EELS). The use of X-ray for 3D computed tomography (CT) or structural analysis with diffractometry is particularly helpful for bulk phase analysis. Atom probe tomography (APT) characterization is powerful for near-atomic resolution chemical and isotopic 3D distributions and segregation at interfaces, and it elucidates the detailed composition of small features down to a few nanometers, which is not easily accessible by TEM. For characterization of small vacancy clusters, positron annihilation spectroscopy (PAS) is the only proven technique.

Recent developments with focused ion beam (FIB) for site-specific ion cutting and polishing, probe-corrected transmission electron microscopy (TEM) with sub-angstrom feature resolution, and APT for near-atomic, 3D element distribution, have made transformational changes to advanced microstructural characterization (Gan et al. 2010; Sun et al. 2018; Bachhav et al. 2019). SEM analysis of the sectioned areas produced by FIB can reveal non-disturbed pristine microstructural features not accessible from samples produced by conventional preparation with wet mechanical polishing (Gan et al. 2014; Jadernas et al. 2018). This is critical for electron backscatter diffraction (EBSD) characterization of samples that are extremely sensitive to surface oxidation. TEM analysis with samples prepared from site-specific FIB processing at a typical size of $10 \times 10\ \mu m^2$ can provide detailed microstructure and chemical information that is unprecedented using conventional TEM sample preparation. This is extremely important when dealing with highly radioactive samples such as irradiated nuclear fuels because of the significant reduction in radioactivity by nearly million times by using a FIB lamella, compared to a conventional 3.0 mm TEM disc. This not only reduces personal radiation exposure from sample handling and analysis but also makes it possible to perform detailed chemical analysis without saturating and damaging the EDS detector. In addition, FIB TEM lamella preparation for cross sectional characterization has overcome the challenges of traditional cross-section sample preparation, making these samples now routinely achievable. The main disadvantage of using FIB lamella for TEM characterization is the ion beam damage from ion cutting and polishing. This can be a problem particularly for low dose irradiation studies because the FIB ion damage could overshadow the irradiation damage. However, this problem could be managed by using a low energy ion beam (~ 2 keV) or using a Xe plasma FIB for final ion thinning and polishing. It has been demonstrated that post-FIB processing of TEM lamella samples with precision argon ion polishing at an ion energy down to few hundred eV can effectively remove the ion damage from FIB processing (Aitkaliyeva et al. 2015).

Some characterization tasks for high-throughput microstructural characterization can be automated if the materials properties are relatively consistent and sample-to-sample instrument running conditions can be kept the same (or follow the same procedure), and a standard characterization routine can be applied. It remains a challenge to conduct high-throughput TEM analysis, even with multiple TEM lamellae mounted on a single FIB grid. This is due to the difficulty in automating defect characterization, which often requires special imaging conditions for different types of defects. Nevertheless, several groups have demonstrated high-throughput microstructural characterization. Ju et al. (2021) conducted comprehensive characterization using a macro-nano-atomic-scale high-throughput approach with both SEM and TEM analysis focused on the effect of carbides on material performance in a nickel-base single crystal superalloy. Their terabyte panoramic data over large areas with many orders of magnitude difference in length scale (cm to sub-Å), combined with ML for carbide identification and conventional electron microscopy, enabled a simultaneous multiscale analysis of carbide evolution during creep. They were able to analyze carbide type, location, composition, size, shape and relationship with the matrix, providing sample-scale quantitative statistical data and obtaining a precise insight into the effect of carbides in the superalloy in a way not previously possible. The essential part of their HTC was the high-speed data collection and high-throughput image analysis using ML. Thousands of BSE images were collected and automatically stitched into a montage, and ML identified both the primary carbides at the microscale (MC) and secondary carbides at the nanoscale ($M_{23}C_6$). Detailed analysis with TEM on selected areas was performed to link the atomic information and detailed dislocation configuration to the features identified in SEM. A total of 8,449 microstructural images over five samples were collected and analyzed.

Kane et al. (2022) recently demonstrated high-throughput non-destructive microstructural characterization using X-ray CT 3D scanning for several TRISO fuel compacts with volumetric packing fractions from 25% to 48%. A total of 4,501 radiographs were acquired over 360° of sample rotation for each compact. 3D X-ray CT analysis enabled the examination of thousands of embedded TRISO fuel particles simultaneously for size, morphology and position, prior to any destructive analysis, as shown in Figure 10.9. In their work, a total of 36,701 TRISO particles were analyzed from a total of 10 compacts. This 3D non-destructive HTC technique can be used to significantly improve the quality control for TRISO fuel compact fabrication and process optimization. It also has great potential for more general applications such as quality inspection of welds along the joint line, and screening of small parts with internal defects.

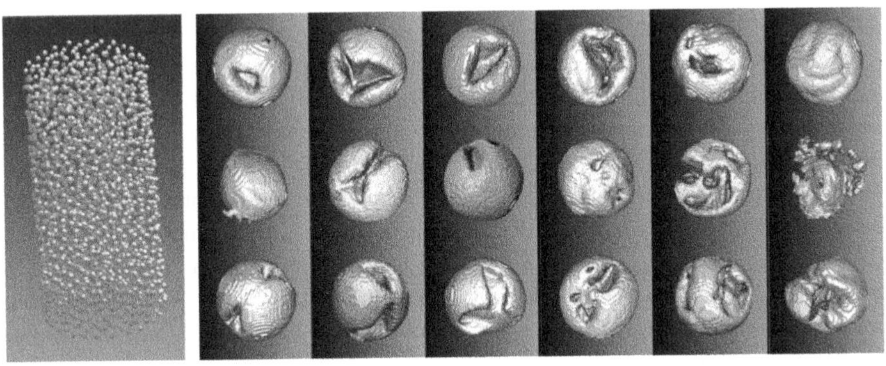

Figure 10.9. *3D view of a TRISO fuel compact (dia. = 12.3 mm, height = 24.7 mm) with 25% packing fraction (left) and 18 images of abnormal fuel particles with UCO kernels (~475 μm) outside the porosity and sphericity tolerance intervals collected from multiple compacts (right) (duplicated from Kane et al. 2022). For a color version of this figure, see www.iste.co.uk/balbaud/nuclearenergytoday.zip*

10.3.3. *High-throughput mechanical and corrosion property evaluation*

Mechanical and corrosion properties are critical for structural material applications. When evaluating mechanical properties, sample size can affect the test results, where large samples characterize averaged, bulk mechanical behavior and miniature specimens produce responses sensitive to local microstructural features. There is a strong interest in correlating the mechanical properties obtained from miniature specimens to bulk mechanical properties. For model development, it is important to have a comprehensive set of data from local mechanical property evaluation corresponding to local microstructural information about dislocations, grain boundaries, phase boundaries, precipitate and cavity morphology and distribution, etc. It is the collective contribution of local mechanical response and interactions under a specific stress condition that dictates the bulk mechanical response.

High-throughput mechanical property evaluation often requires high-throughput sample preparation. Harvey et al. (2021) demonstrated the use of a femtosecond laser in fabricating miniature tensile bars (on the orders of micrometers to millimeters) of 316 stainless steel (SS) and additively manufactured Grade 91 steel. The materials were chosen to benchmark the laser machining technique for a simple alloy 316 SS with a large database for its mechanical properties, and to compare with that of the more complex alloy 91 made by AM with heterogeneous microstructure. A laser beam with a spot size of ~ 20 μm, pulse width of 300 fs, pulse energy of 16 μJ and a repetition rate of 50 kHz was used for laser machining. The tensile bar had a gauge cross section of 0.2×0.1 mm^2 with a gauge length of 0.7 mm. The results from the mesoscale tensile bars test showed reasonable agreement with that of bulk samples for 316 SS. However, a large disparity was observed for AM alloy 91, where the total elongations for the miniature specimens was only approximately half of that of the bulk samples. It was believed that the disparity in test results of miniature and bulk samples for AM alloy 91 is largely due to the complex and heterogeneous microstructure that varies significantly between the miniature specimens. This study shows the importance of carefully choosing the miniature tensile bar size and geometry to link mesoscale tensile properties to macroscale tensile properties.

High-throughput mechanical property evaluation for Al-Al bond line strength from hot isostatic pressing (HIP) bonding of U-10Mo monolithic plate fuel has been demonstrated by Frazer et al. (2022) using plasma-FIB for micro-tensile bar fabrication. The final dimensions of the micro-tensile bars had a gauge size of $7 \times 7 \times 20$ μm^3 with a notch, resulting in a micro-tensile specimen width of approximately 3–5 μm at the bond line. Given the relatively large grain size of the Al, it is evident that the fabricated micro-tensile bars were mostly bicrystals with a single grain on either side of the bond line. Microtensile testing was performed using a Hyistron PI-88 picoindenter mounted in a plasma-FIB. The results show the ductile and brittle failure mode around bond lines for the unirradiated and the neutron irradiated samples, respectively, as shown in Figure 10.10. This work demonstrated the value of high-throughput site-specific mechanical property evaluation via microscale tensile testing, which is critical when the use of millimeter size macroscale samples is not practical.

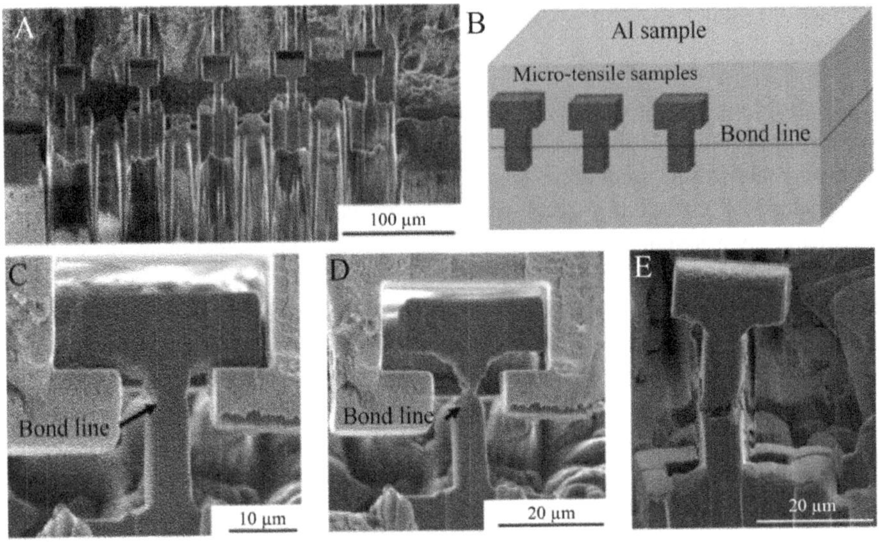

Figure 10.10. *SEM secondary electron image showing a group of finished micro-tensile bars of the Al-Al bond line ready for micro tensile testing (A). A schematic showing the location of the micro-tensile bars (B). An unirradiated micro-tensile bar with a notch at the bond line for stress concentration (C). Ductile failure mode in the unirradiated aluminum around the bond line (D). Brittle failure mode after micro-tensile testing of the neutron irradiated specimen (E) (duplicated from Frazer et al. 2022). For a color version of this figure, see www.iste.co.uk/balbaud/nuclearenergytoday.zip*

HTP testing and characterization for corrosion property evaluation is critical for molten salt reactor R&D. An innovative approach was discussed by Couet (2022) in a recent short review paper. To perform HTP molten salt corrosion testing, a set of 3D printed samples with 5 × 5 composition matrix was produced on a base plate of 10 × 10 cm^2 and a solid salt pill made up of LiCl-KCl-$EuCl_3$ (43wt.%-55wt.%-2wt.%) was placed on top of each 3D-printed sample. The test assembly was then heated to 500°C in an inert environment ($O_2 < 2$ ppm, $H_2O < 0.1$ ppm) and all the solid salt pills transformed to salt droplets after reaching to the salt melting point of 353°C, and the molten corrosion test at 500°C were performed for 96 h. After the corrosion test, the test assembly cooled down to room temperature, the solidified salt was removed, and the sample surface was cleaned for each alloy/salt couple. The corrosion products for each corrosion couple sample were analyzed using an inductively coupled plasma mass spectroscopy (ICP-MS) and a glow discharge optical emission spectroscopy (GDOES). ICP-MS is used to measure the dissolved corrosion products in the entire salt droplet extracted from a specific corrosion couple, and GDOES is used to determine the concentration profiles of alloying elements as function of sample depth. With this HTP molten salt corrosion testing and characterization, 115 CrFeMnNi compositionally complex alloys (CCAs) were tested in about three months. It demonstrated the effectiveness of HTP corrosion property characterization in support of materials development for MSR.

10.3.4. *High-throughput irradiation and characterization*

High-throughput irradiation studies are necessary to expedite the discovery of new materials for nuclear applications. Compared to time-consuming and high-cost neutron irradiation in a test reactor, ion irradiation offers cost-effective and well-controlled irradiation conditions (temperature, dose rate, dose and He-appm/dpa ratio) for irradiation studies with quick turnaround times. Heavy ion irradiation can achieve very high doses with significantly better temperature and dose rate control than is practically achievable with current neutron irradiation facilities. In addition, proton irradiation can provide sufficient uniformly irradiated sample depths for micromechanical testing. It is well known that ion irradiation has made significant contributions to the study of irradiated microstructures and materials performance under irradiation in both fundamental and applied research, and has resulted in numerous literature publications.

Moorehead et al. (2021, 2022) developed a high-throughput ion irradiation system and performed high-throughput ion irradiation of additively manufactured CCAs. AM by directed energy deposition (DED) was used to produce a compositional array of unary, binary, ternary and quaternary alloys, including several CCAs, which span the Cr-Fe-Mn-Ni composition space. The DED system used was an Optomec LENS MR-7 equipped with a 1-kW Nd:YAG laser with a spot size of 600 μm and four independently controlled powder hoppers. Each hopper was filled with single elemental powders in the diameter range of 45–150 μm that were produced by gas atomization. Laser printing of the 5 × 5 alloy array was conducted in an argon environment with the oxygen content controlled below 10 ppm onto a 316 stainless steel substrate kept at room temperature. Each printed alloy had dimensions of 10 × 10 mm^2 and was roughly 5-mm thick. After homogenization at 1,000°C for 24 h in a vacuum of ~ 10^{-6} torr, the printed alloy array was irradiated with 4 MeV Ni^{2+} ions to 50 dpa in the peak damage region, using a defocused beam at room temperature, as shown in Figure 10.11. Each alloy was characterized using X-ray fluorescence (XRF), X-ray diffraction (XRD), scanning electron microscopy (SEM), energy dispersive spectroscopy (EDS) and nanoindentation. The unirradiated half of each printed alloy was used as a reference. To evaluate the change in the mechanical properties from irradiation, nanoindentation measurements were performed using a Bruker Hysitron TI-950 Triboindenter, equipped with a Berkovich tip, using load control mode with the maximum applied load fixed at 6,000 μN. Substantial time saving was realized by high-throughput material fabrication, ion irradiation and characterization. It was emphasized that high-throughput techniques will always need to be coupled to more precise low throughput techniques for more detailed characterization. However, this implementation serves as an enabling approach to accelerate the development of novel materials for advanced nuclear technologies in general.

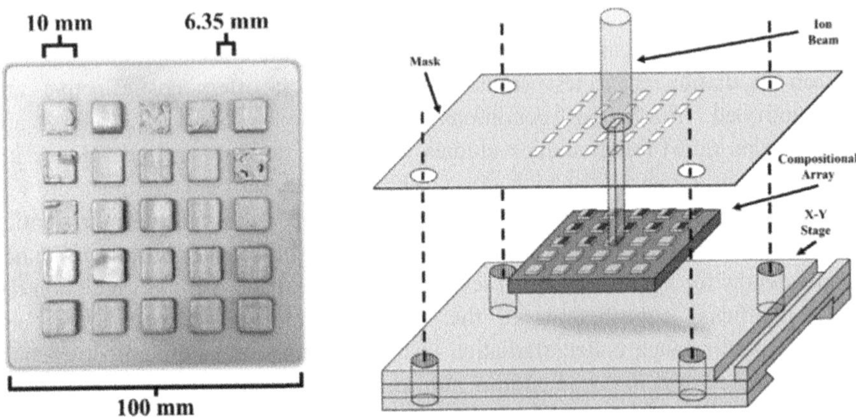

Figure 10.11. *The photo of the as-built Cr-Fe-Mn-Ni alloy array laser printed on a 316 SS plate (left) and the schematic showing the ion irradiation stage arrangement (right) to irradiate only half of each alloy print (red) (duplicated from Moorehead et al. 2022). For a color version of this figure, see www.iste.co.uk/balbaud/nuclearenergytoday.zip*

High-throughput post-irradiation examination (HTP-PIE) is still in its infancy. Preliminary efforts of PIE with automated SEM characterization covering a large sample area of irradiated nuclear fuel with high resolution images has been demonstrated successfully using Maps 3.3, a sophisticated SEM imaging software developed by FEI, and on PIE of irradiated TRISO fuel compact with X-ray 3D CT analysis by Kane and Chuirazzi (2022). The X-ray 3D CT analysis of the fuel compact was performed with a photon energy around 110 keV, and each fuel compact consisted of ~1,918 fuel kernels. The highest dose rate from each specimen at contact was 12 R/h for gamma and >50 R/h for beta. The initial data sets were quite large, requiring approximately 35 GB per low-resolution scan, and data analysis was performed on a customized workstation with a 2.80 GHz AMD EPYC 7282 16-Core CPU, 2 TB RAM, two NVIDIA RTX A6000 GPUs with dedicated GPU memory of 44.5 GB and shared each GPU memory of 1,018 GB. The X-ray radiography at low and high beam energy settings, along with the CT of the UCO fuel kernels, is shown in Figure 10.12. The non-destructive X-ray CT analysis demonstrated the effectiveness and efficiency of high-throughput PIE for cost and time savings, compared to previous PIE approaches involving destructive analysis of irradiated fuel particles and compacts by sectioning for detailed SEM and TEM characterization (Fu et al. 2020). Best practices leveraging HTP-PIE will likely involve using 3D tomography for quick sample assessments and identification of areas of interest, which can guide detailed SEM and TEM characterization with destructive PIE.

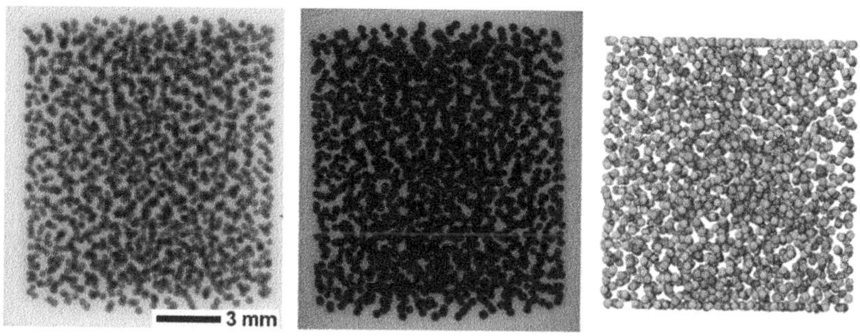

Figure 10.12. *X-ray radiographs of the irradiated AGR-3/4 TRISO Compact 12-4 acquired with high (left) and low (middle) X-ray energy beam settings. Volume renderings of the UCO fuel kernels of AGR-3/4 Compact 7-1 with no 3D perspective where the kernels are colored by sphericity: gray kernels are highly spherical, and blue kernels exhibit low sphericity (right) (courtesy of Kane and Chuirazzi (2022)). For a color version of this figure, see www.iste.co.uk/balbaud/nuclearenergytoday.zip*

10.4. Summary

Accelerated materials discovery through numerical design, advanced manufacturing and HTC have been briefly discussed in the previous sections. All of these three areas are closely related, and an integrated effort is the key to address the technical challenges to meet the requirement for a modern approach for materials R&D for the advanced nuclear energy system. Many of the topics discussed here are also applicable for materials R&D for energy applications beyond nuclear reactor structural materials. The advancement of the computational power and the new instrument capabilities make it possible to conduct R&D at high speed of modeling and simulation, high-throughput alloy design-processing-testing, data production with unprecedent details, experiment and characterization for materials properties and behavior cross broad time and dimensional scales. Advanced characterization can provide atomic resolution of the microstructural details to support the model validation and development. Computational science can help understand the materials' behavior in time and dimensional scale not accessible with current experimental capabilities. Advanced manufacturing can play a critical role on the accelerated materials R&D.

10.5. References

Aguiar, J., Jokisaari, A., Kerr, M., Allen Roach, R. (2020). Bringing nuclear materials discovery and qualification into the 21st century. *Nature Communications*, 11(1), 1–3.

Aitkaliyeva, A., Madden, J.W., Miller, B.M., Cole, J.I., Gan, J. (2015). Comparison of preparationtechniques for nuclear materials for transmission electron microscopy (TEM). *J. Nucl. Mater.*, 459, 241–246.

Alat, E., Motta, A.T., Comstock, R.J., Partezana, J.M., Wolfe, D.E. (2016). Multilayer (TiN, TiAlN) ceramic coatings for nuclear fuel cladding. *Journal of Nuclear Materials*, 478, 236–244.

Albert Sr., M. (2021). Advanced reactor materials development roadmap. Electric Power Research Institute.

Allison, J., Backman, D., Christodoulou, L. (2006a). Integrated computational materials engineering: A new paradigm for the global materials profession. *Jom*, 58(11), 25–27.

Allison, J., Li, M., Wolverton, C., Su, X. (2006b). Virtual aluminum castings: An industrial application of ICME. *Jom*, 58(11), 28–35.

Arkhurst, B.M., Park, J.-J., Lee, C.-H., Kim, J.H. (2017). Direct laser deposition of 14Cr oxide dispersion strengthened steel powders using Y_2O_3 and HfO_2 dispersoids. *Korean J. Met. Mater.*, 55(8), 550–558.

Bachhav, M., Gan, J., Keiser, D., Giglio, J., Jadernas, D., Leenaers, A., Berghe, S. (2019). A novel approach to determine the local burnup in irradiated fuels using atom probe tomography (APT). *J. Nucl. Mater.*, 528, 151853.

Bhavar, V., Kattire, P., Patil, V., Khot, S., Gujar, K., Singh, R. (2017). A review on powder bed fusion technology of metal additive manufacturing. In *Additive Manufacuring Handbook*, Badiru, A.B., Valencia, V.V., Liu, D. (eds). CRC Press, Boca Raton, FL.

Bobbio, L.D., Bocklund, B., Otis, R., Borgonia, J.P., Dillon, R.P., Sapiro, A.A. (2018). Characterization of a functionally graded material of Ti-6Al-4V to 304L stainless steel with an intermediate V section. *J. Alloy Compd.*, 742, 1031–1036.

Boman, K. (2015). What kind of potential does 3D printing hold for oil, gas? [Online]. Available at: https://www.rigzone.com/news/oil_gas/a/141296/.

Brunin, G., Ricci, F., Ha, V., Rignanese, G., Hautier, G. (2019). Transparent conducting materials discovery using high-throughput computing. *Npj Computational Materials*, 5(1), 1–13.

Calignano, F., Tommasi, T., Manfredi, D., Chiolerio, A. (2015). Additive manufacturing of a microbial fuel cell – A detailed study. *Scientific Report*, 5(1), 1–10.

Carroll, B.E., Otis, R.A., Borgonia, J.P., Suh, J.-O., Dillon, R.P., Shapiro, A.A., Hofmann, D.C., Liu, Z.K., Beese, A.M. (2016). Functionally graded material of 304L stainless steel and inconel 625 fabricated by directed energy deposition: Characterization and thermodynamic modeling. *Acta Materilia*, 108, 46–54. doi: 10.1016/j.actamat.2016.02.019.

Chao, Q., Guo, T., Jarvis, T., Wu, X., Hodgson, P., Fabijanic, D. (2017). Direct laser deposition cladding of AlxCoCrFeNi high entropy alloys on a high-temperature stainless steel. *Surf. Coat Technol.*, 332, 440–451.

Che, Y., Pastore, G., Hales, J., Shirvan, K. (2018). Modeling of Cr_2O_3-doped UO_2 as a near-term accident tolerant fuel for LWRs using the BISON code. *Nuclear Engineering and Design*, 337, 271–278.

Chen, K., Huang, R., Li, Y., Lin, S., Zhu, W., Tamura, N. (2020). Rafting-enabled recovery avoids recrystallization in 3D-printing-repaired single-crystal superalloys. *Adv. Mater.*, 32(12), 1907164.

Cheng, X.Y., Chen, X. Q., Li, D.Z., Li, Y.Y. (2014). Computational materials discovery: The case of the W–B system. *Acta Crystallographica Section C: Structural Chemistry*, 70(2), 85–103.

Cole, J. (2020). A design-to-device pipeline for data-driven materials discovery. *Accounts of Chemical Research*, 53(3), 599–610.

Couet, A. (2022). Integrated high-throughput research in extreme environments targeted toward nuclear structural materials discovery. *Journal of Nuclear Materials*, 559, 153425.

Dass, A. and Moridi, A. (2019). State of the art in direct energy deposition from additive manufacturing to materials design. *Coatings*, 9(7), 418.

Denlinger, E.R., Heigel, J.C., Michaleris, P., Palmer, T. A. (2015). Effect of inter-layer dwell time on distortion and residual stress in additive manufacturing of titanium and nickel alloys. *Journal of Materials Processing Technology*, 215, 123–131.

Department of Energy (2022). Advanced materials and manufacturing technologies roadmap. Unpublished communication.

Durejko, T., Zietala, M., Polkowski, W., Czujko, T. (2014). Thin wall tubes with Fe3Al/SS316L graded structure obtained by using laser engineered net shaping technology. *Materials & Design*, 63, 766–774. doi: 10.1016/j.matdes.2014.07.011.

Eremets, M., Trojan, I., Medvedev, S., Tse, J., Yao, Y. (2008). Superconductivity in hydrogen dominant materials: Silane. *Science*, 319(5869), 1506–1509.

Fellinger, M., Hector Jr, L., Trinkle, D. (2017). Ab initio calculations of the lattice parameter and elastic stiffness coefficients of bcc Fe with solutes. *Computational Materials Science*, 126, 503–513.

Frazer, D., Teng, F., Murray, D., Pomo, A., Winston, A., Cole, J. I., Jue, J.F., Giglio, J. (2022). Micro-tensile testing of the bond line in hot isostatic pressed aluminum. *J. Nucl. Mater.*, 561, 153532.

Fu, Z., van Rooyen, I.J., Bachhav, M., Yang, Y. (2020). Microstructure and fission products in the UCO kernel of an AGR-1 TRISO fuel particle after post irradiation safety testing. *J. Nucl. Mater.*, 528, 151884.

Gan, J., Keiser Jr., D.D., Wachs, D.M., Robinson, A.B., Miller, B.D., Allen, T. R. (2010). Transmission electron microscopy characterization of irradiated U-7Mo/Al-2Si dispersion fuel. *J. Nucl. Mater.*, 396, 234–239.

Gan, J., Miller, B.D., Keiser Jr., D.D., Robinson, A.B., Madden, J.W., Medvedev, P.G., Wachs, D.M. (2014). Microstructural characterization of irradiated U-7Mo/Al-5Si dispersion fuel to high fission density. *J. Nucl. Mater.*, 454, 434–445.

Ganeshan, S., Hector Jr, L., Liu, Z. (2011). First-principles calculations of impurity diffusion coefficients in dilute Mg alloys using the 8-frequency model. *Acta Materialia*, 59(8), 3214–3228.

Gorsse, S. and Senkov, O. (2018). About the reliability of CALPHAD predictions in multicomponent systems. *Entropy*, 20(12), 899.

Harvey, C., Torrez, A.J., Lam, S., Kim, H., Maloy, S.A., Gigax, J.G. (2021). Demonstration of a high-throughput tensile testing technique using femtosecond laser-fabricated tensile bars in AISI 316 and additively manufactured Grade 91 steel. *JOM*, 73, 4240–4247.

Hautier, G., Jain, A., Ong, S. (2012). From the computer to the laboratory: Materials discovery and design using first-principles calculations. *Journal of Materials Science*, 47(21), 7317–7340.

He, P., Sun, C., Wang, Y. (2021). Material distortion in laser-based additive manufacturing of fuel cell component: Three-dimensional numerical analysis. *Additive Manufacturing*, 46, 102188.

Hinojos, A., Mireles, J., Reichardt, A., Frigola, P., Hosemann, P., Murr, L. (2016). Joining of Inconel 718 and 316 stainless steel using electron beam. *Materials Design*, 94, 17–27.

Jadernas, D., Gan, J., Keiser, D., Madden, J., Bachhav, M., Jue, J.-F., Robinson, A. (2018). Microstructural characterization of as-fabricated and irradiated U-10Mo fuel using SEM/EBSD. *J. Nucl. Mater.*, 509, 1–8.

Jinoop, A.N., Paul, C.P., Mishra, S.K., Bindra, K.S. (2019). Laser additive manifacturing using directed energy deposition of Inconel-718 wall structure with tailed characteristics. *Vacuum*, 166, 270–278.

Joost, W. (2012). Reducing vehicle weight and improving US energy efficiency using integrated computational materials engineering. *Jom*, 64(9), 1032–1038.

Ju, Y., Li, S., Yuan, X., Cui, L., Godfrey, A., Yan, Y., Cheng, Z., Zhong, X., Zhu, J. (2021). A macro-nano-atomic-scale high-throughput approach for material research. *Science Advances*, 7, eabj8804.

Kane, J. J. and Chuirazzi, W.C. (2022). X-ray computed tomography of irradiated and unirradiated AGR-3/4 compacts. INL/RPT-22-67308. Idaho National Laboratory, Idaho Falls, ID.

Kane, J.J., Marshall, D.W., Cordes, N.L., Chuirazzi, W.C., Kombaiah, B., van Rooyen, I.J., Stempien, J.D. (2022). 3D analysis of TRISO fuel compacts via X-ray computed tomography. *J. Nucl. Mater.*, 565, 153745.

King, W.E., Anderson, A.T., Ferencz, R.M., Hodge, N.E., Kamath, C., Khairallah, S.A. (2015). Laser powder bed fusion additive manufacturing of metals: Physics, computational and materials challenges. *Appl. Phys. Rev.*, 2(4), 041304.

King, D., Middleburgh, S., McGregor, A., Cortie, M. (2016). Predicting the formation and stability of single phase high-entropy alloys. *Acta Materialia*, 104, 172–179.

Krajewski, P., Hector Jr, L., Du, N., Bower, A. (2010). Microstructure-based multiscale modeling of elevated temperature deformation in aluminum alloys. *Acta Materialia*, 58(3), 10074–10086.

Kuehmann, C. and Olson, G. (1998). GEARSTEELS: Designed by computer. *Advanced Materials & Processes*, 153, 40–43.

Kuehmann, C. and Olson, G. (2011). ICME: Success stories and cultural barriers. In *Models, Databases and Simulation Tools Needed for Realization of Integrated Computational Materials Engineering*, Arnold, S. and Wong, T. (eds). ASM International, Almere.

Kuprin, A., Vasilenko, R., Tolstolutskaya, G., Voyevodin, V., Belous, V., Ovcharenko, V., Kopanets, I. (2021). Irradiation resistance of chromium coatings for ATFC in the temperature range 300–550°C. *Journal of Nuclear Materials*, 549, 152908.

Lacey, S.D., Kirsch, D.J., Li, Y., Morgenstern, J.T., Zarket, B.C., Yao, Y. (2018). Extrusion-based 3D printing of hierarchically porous advanced battery electrodes. *Adv. Mater.*, 30(12), 1705651.

Leyson, G., Curtin, W., Hector, L., Woodward, C. (2010). Quantitative prediction of solute strengthening in aluminium alloys. *Nature Materials*, 9(9), 750–755.

Li, W. and Soshi, M. (2019). Modeling analysis of the effect of laser transverse speed on grain morphology during directed energy deposition process. *The International Journal of Advanced Manufacturing Technology*, 103, 3279–3291.

Li, L., Bellehumeur, C., Gu, P. (2001). Composite modeling and analysis of FDM prototypes for design and fabrication of functionally graded parts. *International Solid Freeform Fabrication Symposium*, Austin, TX.

Li, W., Qian, D., Nagaraja, M., Sun, C. (2022). Modeling study of porosity's effect on grain evolution during directed energy deposition. *Materials Letters*, 307, 131009.

Liang, Z., Miao, J., Sachdev, A., Williams, J., Luo, A. (2020). Titanium alloy design and casting process development using an integrated computational materials engineering (ICME) approach. In *MATEC Web of Conferences*. EDP Sciences, Les Ulis.

Liu, Y., Zhao, T., Ju, W., Shi, S. (2017). Materials discovery and design using machine learning. *Journal of Materiomics*, 3(3), 159–177.

Liu, Y., Wu, H., Mayeshiba, T., Afflerbach, B., Jacobs, R., Perry, J., George, J., Cordell, J., Xia, J., Yuan, H. et al. (2022). Machine learning predictions of irradiation embrittlement in reactor pressure vessel steels. *Npj Computational Materials*, 8(1), 1–11.

Lou, X. and Gandy, D. (2019). Advanced manufacturing for nuclear energy. *JOM*, 71(8), 2834–2836.

Lou, X., Othon, M.A., Rebak, R.B. (2017a). Corrosion fatigue crack growth of laser additively-manufactured 316L stainless steel in high temperature water. *Corrosion Science*, 127, 120–130.

Lou, X., Song, M., Wmigh, P.W., Othon, M.A., Andresen, P.L. (2017b). On the stress corrosion crack growth behaviour in high temperature water of 316L stainless steel made by laser powder bed fusion additive manufacturing. *Corrosion Science*, 128, 140–153.

Mahamood, R.M. and Akinlabi, E.T. (2015). Laser metal deposition of functionally graded Ti6Al4V/TiC. *Materials & Design*, 84, 402–410. doi: 10.1016/j.matdes.2015.06.135.

Maier, B.R., Garcia-Diaz, B.L., Hauch, B., Olson, L.C., Sindelar, R.L., Sridharan, K. (2015). Cold spray deposition of Ti2AlC coatings for improved nuclear fuel cladding. *Journal of Nuclear Materials*, 466, 712–717.

Mamivand, M., Wells, P., Ke, H., Shu, S., Odette, G., Morgan, D. (2019). CuMnNiSi precipitate evolution in irradiated reactor pressure vessel steels: Integrated cluster dynamics and experiments. *Acta Materialia*, 180, 199–217.

McDowell, D. and Olson, G. (2008). Concurrent design of hierarchical materials and structures. In *Scientific Modeling and Simulations*, Yip, S. and Diaz de la Rubia, T. (eds). Springer, Dordrecht.

McMurtrey, M.D., O'Brien, R., Sun, C., Shiau, C., Teng, F. (2019). Irradiation damage and IASCC of printed 316L for use as fuel cladding. OSTI.GOV, Boston.

McMurtrey, M., Sun, C., Rupp, R.E., Shiau, C.-H., Hanbury, R., Jerred, N., O'Brien, R. (2021). Investigation of the irradiation effects in additively manufactured 316L steel resulting in decreased irradiation assisted stress corrosion cracking susceptibility. *Journal of Nuclear Materials*, 545, 152739. doi: 10.1016/j.jnucmat.2020.152739.

Miracle, D.B., Li, M., Zhang, Z., Mishra, R., Flores, K. M. (2021). Emerging capabilities for the high-throughput characterization of structural materials. *Annu. Rev. Mater. Res.*, 51, 131–164.

Moorehead, M., Queylat, B., Zhang, H., Kriewaldt, K., Couet, A. (2021). Development of a versatile, high-temperature, high-throughput ion irradiation system. *Nucl. Inst. and Meth. in Phys. Res., A*, 1020, 165892.

Moorehead, M., Nelaturu, P., Elbakhshwan, M., Parkin, C., Zhang, C., Sridharan, K., Couet, A. (2022). High-throughput ion irradiation of additively manufactured compositionally complex alloys. *J. Nucl. Mater.*, 547, 152782.

Morgan, D. and Jacobs, R. (2020). Opportunities and challenges for machine learning in materials science. *Annual Review of Materials Research*, 50, 71–103.

Morgan, D., Pilania, G., Couet, A., Uberuaga, B., Sun, C., Li, J. (2022). Machine learning in nuclear materials research. *Current Opinion in Solid State and Materials Science*, 26(2), 100975.

National Research Council (2008). *Integrated Computationl Materials Engineering: A Transformational Discipline for Improved Competitiveness and National Security*. National Academies Press, Washington, DC.

National Science and Technology Council (2021). *Materials Genome Initiative Strategic Plan*. Executive Office of the President of the United States, Washington, DC.

Nelson, A. (2019). *Features that Further Performance Limits of Nuclear Fuel Fabrication: Opportunities for Additive Manufacturing of Nuclear Fuels*. Oak Ridge National Laboratory, Knoxville, TN.

Nickel, A.H., Barnett, D.M., Prinz, F.B. (2001). Thermal stresses and deposition patterns in layered manufacturing. *Materials Science and Engineering: A*, 317, 59–64.

Nozar, M., Zetková, I., Hronek, O. (2018). Searching for favourable powder bed fusion settings in sintering of maraging steel MS1. In *Proceedings of the 29th International DAAAM Symposium*, Zadar.

Odette, G. and Lucas, G. (1998). Recent progress in understanding reactor pressure vessel steel embrittlement. *Radiation Effects and Defects in Solids*, 144(1–4), 189–231.

Odette, G., Yamamoto, T., Williams, T., Nanstad, R., English, C. (2019). On the history and status of reactor pressure vessel steel ductile to brittle transition temperature shift prediction models. *Journal of Nuclear Materials*, 526, 151863.

Oganov, A., Pickard, C., Zhu, Q., Needs, R. (2019). Structure prediction drives materials discovery. *Nature Reviews Materials*, 4(5), 331–348.

Panchal, J., Kalidindi, S., McDowell, D. (2013). Key computational modeling issues in integrated computational materials engineering. *Computer-Aided Design*, 45(1), 4–25.

Parimi, L.L., Ravi, G.A., Clark, D., Attallah, M.M. (2014). Microstructural and texture development in direct laser fabricated IN718. *Materials Characterization*, 89, 102–111.

Pickering, E., Carruthers, A., Barron, P., Middleburgh, S., Armstrong, D., Gandy, A. (2021). High-entropy alloys for advanced nuclear applications. *Entropy*, 23(1), 98.

Raman, L., Gothandapani, K., Murty, B. (2016). Austenitic oxide dispersion strengthened steels: A review. *Defence Science Journal*, 66, 316–322.

Sabol, G. (2005). ZIRLO – An alloy development success. In *Zirconium in the Nuclear Industry: Fourteenth International Symposium*, Rudling, P. and Kammenzind, B. (eds). ASTM, West Conshohocken, PA.

Savic, V., Hector, L., Singh, H., Paramasuwom, M., Basu, U., Basudhar, A., Stander, N. (2018). Development of a lightweight third-generation advanced high-strength steel (3GAHSS) vehicle body structure. *SAE International Journal of Materials and Manufacturing*, 11(4), 303–314.

SBIR.gov. (2017). 3D Printing of nuclear fuel assembly bottom nozzles [Online]. Available at: https://www.sbir.gov/sbirsearch/detail/1332851 [Accessed 12 June 2022].

Shah, K., Haq, I.U., Khan, A., Shah, S.A., Khan, M., Pinkerton, A.J. (2014). Parametric study of development of Inconel-steel functionally graded materials by laser direct metal deposition. *Materials & Design*, 54, 531–538. doi: 10.1016/j.matdes.2013.08.079

Shang, Z., Fan, C., Xue, S., Ding, J., Li, J., Voisin, T., Zhang, X. (2019). Response of solidification cellular structures in additively manufactured 316 stainless steel to heavy ion irradiation: An in situ study. *Materials Research Letters*, 7, 290–297.

Shang, Z., Fan, C., Ding, J., Xue, S., Gabriel, A., Shao, L., Zhang, X. (2021). Heavy ion irradiation response of an additively manufactured 316LN stainless steel. *Journal of Nuclear Materials*, 546, 152745.

Shiau, C.-H., McMurtrey, M.D., O'Brien, R., Jerred, N.D., Scott, R.D., Lu, J., Sun, C. (2021). Deformation behavior and irradiation tolerance of 316 L stainless steel fabricated by direct energy deposition. *Materials & Design*, 204, 109644.

Shiau, C.-H., Sun, C., McMurtrey, M., O'Brien, R., Garner, F.A., Shao, L. (2022). Orientation-selected micro-pillar compression of additively manufactured 316L stainless steels: Comparison of as-manufactured, annealed, and proton-irradiated variants. *Journal of Nuclear Materials*, 566, 153739.

Shu, S., Wells, P., Odette, G., Morgan, D. (2019). A kinetic lattice Monte Carlo study of post-irradiation annealing of model reactor pressure vessel steels. *Journal of Nuclear Materials*, 524, 312–322.

Song, M., Wang, M., Lou, X., Rebak, R.B., Was, G.S. (2019). Radiation damage and irradiation-assisted stress corrosion cracking of additively manufactured 316L stainless steels. *Journal of Nuclear Materials*, 513, 33–44.

Sun, C., Sprouster, D.J., Kattar, K., Ecker, L.E., He, L., Gao, Y., Gan, J. (2018). Formation of tetragonal gas bubble superlattice in bulk molybdenum under helium ion implantation. *Scripta Materialia*, 149, 26–30.

Sun, C., Wang, Y., McMurtrey, M., Jerred, N., Liou, F., Li, J. (2021). Additive manufacturing for energy: A review. *Applied Energy*, 282, 116041.

Terrani, K. (2018a). Accident tolerant fuel cladding development: Promise, status, and challenges. *Journal of Nuclear Materials*, 501, 13–30.

Terrani, K. (2018b). Advanced manufacturing and materials to enable advanced nuclear energy. *Transactions of the American Nuclear Society*, 118(1), 1575.

ThermoFisher Scientific. (n.d.). Maps 3 Software: Correlative electron microscopy and cross-platform imaging automation software [Online]. Available at: https://www.thermofisher.com/us/en/home/electron-microscopy/products/software-em-3d-vis/maps-software.html [Accessed 29 May 2022].

Turner, B.N., Strong, R., Gold, S.A. (2014). A review of melt extrusion additive manufacturing processes I: Process design and modeling. *Rapid Prototyping Journal*, 20(3), 192–204.

Uddin, S.Z., Murr, L.E., Terrazas, C.A., Morton, P., Roberson, D.A., Wicker, R.B. (2018). Processing and characterization of crack-free aluminum 6061 using high-temperature heating in laser powder bed fusion additive manufacturing. *Additive Manufacturing*, 22, 405–415.

Verhoef, L.A., Budde, B.W., Chockalingam, C., Garcia, N.B., van Wijk, A. (2018). The effect of additive manufacturing on global energy demand: An assessment using a bottom-up approach. *Energy Policy*, 112, 349.

Voisin, T., Calta, N.P., Khairallah, S.A., Forien, J.-B., Balogh, L., Cunningham, R.W., Rollett, A.D., Wang, Y.M. (2018). Defects-dictated tensile properties of selective laser melted Ti-6Al-4V. *Materials & Design*, 158, 113–126.

Wang, X. and Xiong, W. (2020). Uncertainty quantification and composition optimization for alloy additive manufacturing through a CALPHAD-based ICME framework. *Npj Computational Materials*, 6(1), 1–11.

Wang, Q., Jones, P., Wang, Y., Gerard, D. (2011). Advances in computational tools for virtual casting of aluminum components. In *Proceedings of the 1st World Congress on Integrated Computational Materials Engineering (ICME)*, Allison, J., Collins, P., Spanos, G. (eds). John Wiley & Sons, New York.

Wang, R., Zhang, K., Davies, C., Wu, X. (2017a). Evolution of microstructure, mechanical and corrosion properties of AlCoCrFeNi high-entropy alloy prepared by direct laser fabrication. *J. Alloy Compound*, 694, 971–981.

Wang, Y., Chen, C., Xie, H., Gao, T., Yao, Y., Pastel, G. (2017b). 3D-printed all-fiber Li-ion battery toward wearable energy storage. *Adv. Funct. Mater.*, 27(43), 1703140.

Westinghouse Global Technology Office. (2020). Advanced manufacturing flysheet-2020 [Online]. Available at: https://westinghousenuclear.com/Portals/0/operating%20plant%20services/components/manufacturing/Advanced%20Manufacturing%20Flysheet%20-%202020.pdf [Accessed 12 June 2022].

Wu, Z., Zhao, X., Liu, Y., Cai, Y., Li, J.Y., Chen, H., Wan, Q., Yang, D., Tan, J., Liu, H.D. et al. (2020). Lead-bismuth eutectic (LBE) corrosion behavior of AlTiN coatings at 550 and 600°C. *Journal of Nuclear Materials*, 539, 152280.

Xiu, P., Massey, C.P., Kelsy Green, T.M., Taller, S., Isheim, D., Sridharan, N., Field, K.G. (2022). Microchemical evolution of irradiated additive-manufactured HT9. *J. Nucl. Mater.*, 559, 153410.

Yasi, J., Hector Jr, L., Trinkle, D. (2010). First-principles data for solid-solution strengthening of magnesium: From geometry and chemistry to properties. *Acta Materialia*, 58(17), 5704–5713.

Zhang, Y., Zhao, S., Weber, W., Nordlund, K., Granberg, F., Djurabekova, F. (2017). Atomic-level heterogeneity and defect dynamics in concentrated solid-solution alloys. *Current Opinion in Solid State and Materials Science*, 21(5), 221–237.

Zhang, X., Li, W., Chen, X., Cui, W., Liou, F. (2018). Evaluation of component repair using direct metal deposition from scanned data. *The International Journal of Advanced Manufacturing Technology*, 95(9), 3335–3348.

Zhang, X., McMurtrey, M.D., Wang, L., O'Brien, R.C., Shiau, C.-H., Wang, Y., Scott, R., Ren, Y., Sun, C. (2020a). Evolution of microstructure, residual stress, and tensile properties of additively manufactured stainless steel under heat treatments. *JOM*, 72, 4167–4177.

Zhang, X., Sun, C., Pan, T., Flood, A., Zhang, Y., Li, L., Liou, F. (2020b). Additive manufacturing of copper – H13 tool steel bi-metallic structures via Ni-based multi-interlayer. *Additive Manufacturing*, 36, 101474.

Zhong, Y., Rannar, L.-E., Liu, L., Wikman, S., Olsen, J. (2017). Additive manufacturing of 316L stainless steel by electron beam melting for nuclear fusion applications. *J. Nucl. Mater.*, 486, 234–245.

Zinkle, S., Terrani, K., Snead, L. (2016). Motivation for utilizing new high-performance advanced materials in nuclear energy systems. *Current Opinion in Solid State and Materials Science*, 20(6), 401–410.

Zinkle, S., Boutard, J., Hoelzer, D., Kimura, A., Lindau, R., Odette, G., Rieth, M., Tan, L., Tanigawa, H. (2017). Development of next generation tempered and ODS reduced activation ferritic/martensitic steels for fusion energy applications. *Nuclear Fusion*, 57(9), 092005.

List of Authors

Giacomo AIELLO
EUROfusion – Programme Management Unit
Garching
Germany

Hervé ANTONY
Orano Recyclage
Etablissement de la Hague
La Hague
France

Tai ASAYAMA
Japan Atomic Energy Agency
Oarai
Japan

Fanny BALBAUD-CÉLÉRIER
CEA
Service de recherche en Corrosion et Comportement des Matériaux
Université Paris-Saclay
Gif-sur-Yvette
France

James BRAUN
CEA/DAM
Le Ripault
Monts
France

Céline CABET
CEA
Service de Recherche en Matériaux et Procédés Avancés
Université Paris-Saclay
Gif-sur-Yvette
France

Yann DE CARLAN
CEA
Université Paris-Saclay
Gif-sur-Yvette
France

Eric CAVALETTI
Orano Recyclage
Chatillon
France

Pierre CHAMBRETTE
Orano Recyclage
Chatillon
France

Pierre FAUVET
CEA
Service de recherche en Corrosion et Comportement des Matériaux
Université Paris-Saclay
Gif-sur-Yvette
France

Jian GAN
Idaho National Laboratory
Idaho Falls
Idaho
USA

Serguei GAVRILOV
SCK CEN Belgian Nuclear Research Centre
Mol
Belgium

Nathalie GRUET
CEA
Service de recherche en Corrosion et Comportement des Matériaux
Université Paris-Saclay
Gif-sur-Yvette
France

Benoit GWINNER
CEA
Service de recherche en Corrosion et Comportement des Matériaux
Université Paris-Saclay
Gif-sur-Yvette
France

Andrea JOKISAARI
Idaho National Laboratory
Idaho Falls
Idaho
USA

Laurent JUNOD
Orano Projets
Cherbourg-en-Cotentin
France

Takeji KAITO
Oarai Research Institute
Sector of Fast Reactor and Advanced Reactor Research and Development
Japan Atomic Energy Agency
Ibaraki
Japan

Fraser KING
Integrity Corrosion Consulting Ltd
Nanaimo BC
Canada

Pierre LAGHOUTARIS
CEA
Service de recherche en Corrosion et Comportement des Matériaux
Université Paris-Saclay
Gif-sur-Yvette
France

Laure MARTINELLI
CEA
Université Paris-Saclay
Gif-sur-Yvette
France

Jean-Paul MASSOUD
EDF/DIPNN/DT
France

Eric MOLINIÉ
EDF/DIRES/R&D
France

Satoshi OHTSUKA
Oarai Research Institute
Sector of Fast Reactor and Advanced Reactor Research and Development
Japan Atomic Energy Agency
Ibaraki
Japan

Beatriz PUGA
CEA
Service de recherche en Corrosion et Comportement des Matériaux
Université Paris-Saclay
Gif-sur-Yvette
France

Valentin ROHR
Orano Recyclage
Etablissement de la Hague
La Hague
France

Cheng SUN
Idaho National Laboratory
Idaho Falls
Idaho
USA

Takashi TANNO
Oarai Research Institute
Sector of Fast Reactor and Advanced Reactor Research and Development
Japan Atomic Energy Agency
Ibaraki
Japan

Hayato YAMASHITA
Japan Atomic Energy Agency
Oarai
Japan

Yasuhide YANO
Oarai Research Institute
Sector of Fast Reactor and Advanced Reactor Research and Development
Japan Atomic Energy Agency
Ibaraki
Japan

Index

F

G, H

I, L

M, N

O, P

R, S

T, W, Z